Analysis and Purification Methods in Combinatorial Chemistry

CHEMICAL ANALYSIS

A SERIES OF MONOGRAPHS ON ANALYTICAL CHEMISTRY AND ITS APPLICATIONS

Editor
J. D. WINEFORDNER

VOLUME 163

Analysis and Purification Methods in Combinatorial Chemistry

Edited by

BING YAN

WILEY-INTERSCIENCE

A JOHN WILEY & SONS, INC., PUBLICATION

Library of Congress Cataloging-in-Publication Data:
Analysis and purification methods in combinatorial chemistry / edited by Bing Yan.
        p.   cm.—(Chemical analysis; v. 1075)
Includes bibliographical references and index.
     ISBN 0-471-26929-8 (cloth)
  1. Combinatorial chemistry.   I. Yan, Bing.   II. Series.
     RS419.A53 2004
     615′.19—dc22

                                                            2003014606

Printed in the United States of America.

10 9 8 7 6 5 4 3 2 1

CONTENTS

v

PREFACE

More than 160 volumes of *Chemical Analysis: A Series of Monographs on Analytical Chemistry and Its Applications* have been published by John Wiley & Sons, Inc. since 1940. These volumes all focused on the most important analytical issues of their times. In the past decade one of the most exciting events has been the rapid development of combinatorial chemistry. This rapidly evolving field posed enormous analytical challenges early on. The two most-cited challenges are requirements for very high-throughput analysis of a large number of compounds and the analysis of polymer-bound compounds. Very impressive achievements have been made by scientists working in this field. However, there are still formidable analytical challenges ahead. For example, the development of highly parallel analysis and purification technologies and all methods associated with analysis to ensure combinatorial libraries are "synthesizable," "purifiable," and "drugable." For these evident reasons, I almost immediately agreed to edit a volume on the analysis and purification methods in combinatorial chemistry when the series editor Professor J. D. Winefordner asked me a year ago.

In the past year it has been a great pleasure for me to work with all contributors. The timely development of this volume is due entirely to their collaborative efforts. I have been impressed with their scientific vision and quality work throughout the year. To these contributors, I owe my very special thanks. I also owe a great debt to my colleagues especially Dr. Mark Irving, and Dr. Jiang Zhao for their assistance in my editorial work. Finally I wish to thank staff at Wiley for their professional assistance throughout this project.

Part I of the book includes six chapters describing various approaches to monitor reactions on solid support and optimize reactions for library synthesis: Lucas and Larive give a comprehensive overview of the principle and application of quantitative NMR analysis in support of synthesis in both solution and solid phase. Salvino describes in detail the application of ^{19}F NMR to monitor solid-phase synthesis directly on resin support. Cournoyer, Krueger, Wade, and Yan report on the single-bead FTIR method applied in monitoring solid-phase organic synthesis. Guinó and de Miguel report on HR-MAS NMR analysis of solid-supported samples.

A parallel analysis approach combined with chemometrics analysis in materials discovery and process optimization is presented by Potyrailo, Wroczynski, Lemmon, Flanagan, and Siclovan. Enjalbal, Lamaty, Martinez, and Aubagnac report their work on monitoring reactions on soluble polymeric support using mass spectrometry.

Part II of the book is dedicated to high-throughput analytical methods used to examine the quality of libraries. Hou and Raftery review the development of high-throughput NMR techniques and their own work on parallel NMR method. Simms details the theory and application of micellar electrokinetic chromatography as a high-throughput analytical tool for combinatorial libraries. Zhang and Fitch describe Affymax's approach on quality control and encoding/decoding of combinatorial libraries via single-bead analysis methods.

In Part III, various high-throughput purification techniques are discussed. Zhao, Zhang, and Yan review the chromatographic separation and their application in combinatorial chemistry. Hochlowski discusses various purification methods and the high-throughput HPLC and SFC methods developed at Abbott. God and Gumm present the new generation of parallel analysis and purification instruments and methods.

In Part IV, analytical methods applied in postsynthesis and postpurification stages are reviewed. Morand and Cheng report studies on stability profile of compound archives. Tseng, Chang, and Chu discuss a novel quartz crystal microbalance method to determine the binding between library compounds and biological targets. Faller reviews high-throughput methods for profiling compounds' physicochemical properties. Lipinski presents a detailed study of solubility issue in drug discovery and in combinatorial library design. Villena, Wlasichuk, Schmidt Jr., and Bao describe a high-throughput LC/MS method for the determination of log D value of library compounds.

BING YAN

August 2003
South San Francisco, California

CONTRIBUTORS

Jean-Louis Aubagnac, Laboratoire des aminocides, peptides et protéines, UMR 5810, Université de Montpellier II, 34095 Montpellier Cedex 5, France, E-mail: aubagnac@univ-montp2.fr

James J. Bao, Ph.D., Theravance, Inc., 901 Gateway Blvd., S. San Francisco, CA 94080, E-mail: jbao@theravance.com

I-Nan Chang, ANT Technology Co., Ltd., 7F-4, No. 134, Sec. 1, Fushing S Road, Taipei 106, Taiwan, ROC

Xueheng Cheng, Abbott Laboratories, Global Pharmaceutical Research and Development Division, Department R4PN, Building AP9A, 100 Abbot Park Road, Abbot Park IL 60064-6115, E-mail: xueheng.cheng@abbott.com

Yen-Ho Chu, Department of Chemistry, National Chung-Cheng University, Chia-Yi, Taiwan 621, Republic of China, E-mail: cheyhc@ccunix.ccu.edu.tw

Jason Cournoyer, ChemRx Division, Discovery Partners International, Inc., 385 Oyster Point Blve., South San Francisco, CA 94080

Yolanda de Miguel, Ph.D., Organic Chemistry Lecturer, Royal Society Dorothy Hodgkin Fellow, Chemistry Department, King's College London, Strand, London WC2R 2LS, E-mail: yolanda.demiguel@kcl.ac.uk

Christine Enjalbal, Laboratoire des aminocides, peptides et protéines, UMR 5810, Université de Montpellier II, 34095 Montpellier Cedex 5, France

Bernard Faller, Ph.D., Technology Program Head, Novartis Institute for Biomedical Research, WKL-122.P.33, CH-4002 Switzerland, Bernard.faller@pharma.novartis.com

William L. Fitch, Roche BioScience, Palo Alto, CA 94304

William P. Flanagan, General Electric, Combinatorial Chemistry Laboratory, Corporate Research and Development, PO Box 8, Schenectady, NY 12301-0008

Ralf God, Ph.D., Arndtstraß 2, D-01099 Dresden (Germany), E-mail: r.god@gmx.de

Meritxell Guinó, Organic Chemistry Lecturer, Royal Society Dorothy Hodgkin Fellow, Chemistry Department, King's College London, Strand, London WC2R 2LS

Holger Gumm, SEPIAtec GmbH, Louis-Bleriot-Str. 5, D-12487 Berlin, Germany, E-mail: hgumm@sepiatec.com

Jill Hochlowski, Abbott Laboratories, Dept. 4CP, Bldg. AP9B, 100 Abbott Park Road, Abbott Park, IL 60064-3500, E-mail: Jill.hochlowski@abbott.com

Ting Hou, Department of Chemistry, West Lafayette, IN 47907-1393

Clinton A. Krueger, ChemRX Division, Discover Partners International, Inc., South San Francisco CA 94080

Cynthia K. Larive, University of Kansas, Department of Chemistry, 2010 Malott Hall, 1251 Wescoe Hall Rd, Lawrence, KS 66045, E-mail: clarive@ku.edu

Frederic Lamaty, Laboratoire des aminocides, peptides et protéines, UMR 5810, Université de Montpellier II, 34095 Montpellier Cedex 5, France, E-mail: frederic@ampir1.univ-montp2.fr

John P. Lemmon, General Electric, Combinatorial Chemistry Laboratory, Corporate Research and Development, PO Box 8, Schenectady, NY 12301-0008

Christopher A. Lipinski, Ph.D., Pfizer Global R&D, Groton Labs, Eastern Point Road, MS 8200-36, Groton, CT 06340, E-mail: christopher_a_lipinski@groton.pfizer.com

Laura H. Lucas, Dept of Chemistry, 2010 Malott Hall, University of Kansas, Lawrence KS 66045

J. Martinez, Laboratoire des aminocides, peptides et protéines, UMR 5810, Université de Montpellier II, 34095 Montpellier Cedex 5, France, E-mail: mlorca@univ-montp2.fr

Kenneth Morand, Procter & Gamble Pharmaceuticals, Health Care Research Center, 8700 Mason-Montgomery Road, Mason, OH 45040, E-mail: morand.kl@pg.com

Radislav A. Potyrailo, General Electric, Combinatorial Chemistry Laboratory, Corporate Research and Development, PO Box 8, Schenectady, NY 12301-0008, E-mail: potyrailo@crd.ge.com

Daniel Raftery, Department of Chemistry, West Lafayette, IN 47907-1393, E-mail: raftery@purdue.edu

Joseph M. Salvino, Ph.D., Rib-X Pharmaceuticals, Inc., 300 George St., New Haven, CT 06511, E-mail: jsalvino@Rib-x.com

Donald E. Schmidt, Jr., Theravance, Inc., 901 Gateway Blvd., S. San Francisco, CA 94080

Oltea P. Siclovan, General Electric, Combinatorial Chemistry Laboratory, Corporate Research and Development, PO Box 8, Schenectady, NY 12301-0008

Peter J. Simms, Ribapharm Inc., 3300 Hyland Ave., Costa Mesa, CA 92626, E-mail: pjsimms@icnpharm.com

Ming-Chung Tseng, Dept of Chemistry and Biochemistry, National Chung Cheng University, 160 San-Hsing, Min-Hsiung, Chia-Yi 621, Taiwan, ROC

Jenny D. Villena, Theravance, Inc., 901 Gateway Blvd., S. San Francisco, CA 94080

Janice V. Wade, ChemRX Division, Discover Partners International, Inc., South San Francisco CA 94080

Ken Wlasichuk, Theravance, Inc., 901 Gateway Blvd., S. San Francisco, CA 94080

Ronald J. Wroczynski, General Electric, Combinatorial Chemistry Laboratory, Corporate Research and Development, PO Box 8, Schenectady, NY 12301-0008

Bing Yan, ChemRX Division, Discover Partners International, Inc., South San Francisco CA 94080, E-mail: byan@chemrx.com

Jing Jim Zhang, Ph.D., Affymax, Inc., 4001 Miranda Avenue, Palo Alto, CA 94304, E-mail: jim_zhang@affymax.com

Lu Zhang, ChemRx Division, Discovery Partners International, Inc., 9640 Towne Centre Drive, San Diego, CA 92121

Jiang Zhao, ChemRx Division, Discovery Partners International, Inc., 385 Oyster Point Blve., South San Francisco, CA 94080, E-mail: jzhao@chemrx.com

PART

I

ANALYSIS FOR FEASIBILITY AND OPTIMIZATION OF LIBRARY SYNTHESIS

CHAPTER

1

QUANTITATIVE ANALYSIS IN ORGANIC SYNTHESIS WITH NMR SPECTROSCOPY

LAURA H. LUCAS and CYNTHIA K. LARIVE

1.1. INTRODUCTION

The development of combinatorial methods of synthesis has created a great need for robust analytical techniques amenable to both small- and large-scale syntheses as well as high-throughput analysis. A variety of spectroscopic methods such as mass spectrometry (MS), infrared spectroscopy (IR), and nuclear magnetic resonance (NMR) spectroscopy have been widely used in the combinatorial arena.[1–4] Many of these methods have been coupled online with high-performance liquid chromatography (HPLC), and such hyphenated techniques afford high-resolution structural data on library compounds.[5–7] NMR has an advantage over other spectroscopic techniques because it is a universal analytical method that potentially reveals the identity and purity of any organic compound. Since organic molecules contain ^1H and ^{13}C, nuclei that are NMR active, no analyte derivitization is necessary. NMR can also distinguish isomers or structurally similar compounds that may coelute when analyzed by HPLC. This selectivity is especially important when attempting to resolve and quantitate structures of library compounds or natural products.[8] Furthermore the non-invasive nature of NMR allows the sample to be recovered for additional analyses or used in a subsequent reaction.

Although traditionally thought of as a low-sensitivity technique, technological improvements in NMR instrumentation have significantly reduced sample mass requirements and experiment times. Sensitivity enhancements have been achieved with higher field magnets, small-volume flow probes for solution-phase analysis,[9] the introduction of cryogenically cooled NMR receiver coils and preamplifiers,[10–13] and high-resolution magic-angle spinning (HR-MAS) NMR technology for solid-phase systems, making routine analysis of μg quantities (or less) possible.[14,15] These advancements com-

Analysis and Purification Methods in Combinatorial Chemistry, Edited by Bing Yan.
ISBN 0-471-26929-8 Copyright © 2004 by John Wiley & Sons, Inc.

bined with developments in automated sample handling and data processing have improved the throughput of NMR such that entire 96-well microtitre plates can be analyzed in just a few hours.[16]

Besides the structural information provided by NMR, quantitation is possible in complex mixtures even without a pure standard of the analyte, as long as there are resolved signals for the analyte and reference compound. This is a particular advantage in combinatorial chemistry, where the goal is the preparation of large numbers of new compounds (for which no standards are available). Since the NMR signal arises from the nuclei (e.g., protons) themselves, the area underneath each signal is proportional to the number of nuclei. Therefore the signal area is directly proportional to the concentration of the analyte:

$$\frac{\text{Area}}{\text{Number of nuclei}} \propto \text{Concentration} \qquad (1.1)$$

This relationship holds true when experimental parameters are carefully optimized, which may require some additional experimental time.

Consider the example shown in Figure 1.1 for the quantitation of a maleic acid solution, containing a known concentration of the primary standard potassium hydrogen phthalate (KHP). A primary analytical standard is a material of very high purity for which the mass of the compound can be used directly in the calculation of solution concentration. The area of the KHP peak is divided by 4.0, the number of protons that contribute to the KHP aromatic resonances, so that its normalized area is 1773.1. Similarly the maleic acid peak area is normalized by dividing by 2.0 to give a normalized area of 1278.3. A simple proportion can then be established to solve for the concentration of maleic acid:

$$\frac{\text{Normalized area}_{(\text{Maleic acid})}}{[\text{Maleic acid}]} = \frac{\text{Normalized area}_{(\text{KHP})}}{[\text{KHP}]},$$

$$[\text{Maleic acid}] = \frac{[\text{KHP}] \times \text{Normalized area}_{(\text{Maleic acid})}}{\text{Normalized area}_{(\text{KHP})}},$$

$$[\text{Maleic acid}] = \frac{28.5\,\text{mM} \times 1278.3}{1773.1},$$

$$[\text{Maleic acid}] = 28.5\,\text{mM} \times 0.72094.$$

The concentration of maleic acid in this solution is therefore about 72% that of the primary standard KHP. In this example the concentration of the KHP is 28.5 mM and the calibrated concentration of maleic acid is 20.5 mM. Even though maleic acid is not a primary standard, this standardized maleic

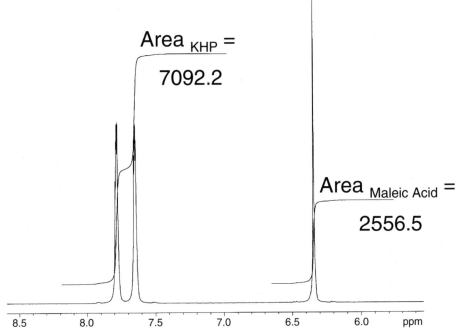

Figure 1.1. 600 MHz ^{1}H NMR spectrum of potassium hydrogen phthalate (KHP) and maleic acid dissolved in D$_2$O. The KHP is a primary standard by which the maleic acid concentration can be quantitated. The data represents 8 FIDs coadded into 28,800 points (zero-filled to 32K points) across a spectral width of 7200.1 Hz. An exponential multiplier equivalent to 0.5 Hz line broadening was then applied.

acid solution can now be used to quantitate additional samples in a similar manner (e.g., ibuprofen as discussed in more detail below). This approach allows the selection of a quantitation standard based on its NMR spectral properties and does not require that it possess the properties of a primary analytical standard.

1.2. FUNDAMENTAL AND PRACTICAL ASPECTS OF THE QUANTITATIVE NMR EXPERIMENT

1.2.1. Experimental Parameters

For a detailed description of NMR theory and practice, the reader is encouraged to see one of the many excellent books on the subject.[17–20] A brief description of the NMR experiment is presented below, with an emphasis

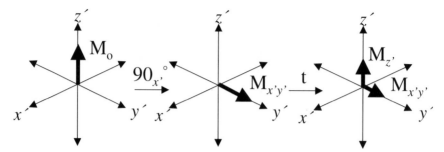

Figure 1.2. The equilibrium population difference of the nuclei in an applied magnetic field can be represented as a vector $\mathbf{M_o}$. Disturbing this macroscopic magnetization away from its equilibrium position creates the NMR signal. This is accomplished by application of a radio frequency pulse ($90°$ along the x' axis in this example) to tip the magnetization into the $x'y'$ plane of the rotating frame of reference. During a time delay t, following the pulse the detectable magnetization $\mathbf{M_{x'y'}}$, decays by T_2 relaxation in the transverse plane while the longitudinal magnetization $\mathbf{M_{z'}}$ begins to recover through T_1 relaxation.

on the important parameters for proper acquisition and interpretation of spectra used for quantitation. When the sample is placed in the magnetic field, the nuclei align with the field, by convention along the z' axis in the rotating frame of reference, as illustrated in Figure 1.2 by the vector, $\mathbf{M_o}$, representing the macroscopic magnetization. A radio frequency pulse (B_1) applied for time t_p tips the magnetization through an angle, θ:

$$\theta = \gamma B_1 t_p, \tag{1.2}$$

where γ is the gyromagnetic ratio of the nucleus of interest. A $90°$ radio frequency (rf) pulse along the x' axis tips the magnetization completely into the $x'y'$ plane to generate the observable signal, $\mathbf{M_{x'y'}}$. After the pulse the magnetization will relax back to equilibrium via two mechanisms: spin-spin (T_2, transverse) relaxation and spin-lattice (T_1, longitudinal) relaxation. This is shown in Figure 1.2 as a decrease in the magnitude of the $x'y'$ component of the vector $\mathbf{M_o}$ and an increase in the magnitude of the z' component at time t following the pulse. Acquisition of $\mathbf{M_{x'y'}}$ for all nuclei in the sample as a function of time results in the free induction decay (FID), which upon Fourier transformation is deconvolved into the frequency domain and displayed as the familiar NMR spectrum. The FID decays exponentially according to T_2, the spin-spin or transverse relaxation time. In addition to the natural T_2 relaxation times of the nuclei that comprise the FID, magnetic field inhomogeneity contributes to the rate at which the transverse magnetization is lost. The apparent transverse relaxation time, T_2^*, is the summation of the natural relaxation time and the component induced by magnetic field inhomogeniety and can be calculated from the width at half-height of the NMR signals:

$$w_{1/2} = \frac{1}{\pi T_2^*}.$$ (1.3)

The acquisition time during which the FID is detected is often set to three to five times T_2^* to avoid truncation of the FID.

1.2.2. T_1 Relaxation

Just as the NMR signal decays exponentially in the transverse plane, the magnetization also relaxes back to its equilibrium state in an exponential fashion during the time t following the rf pulse. For the longitudinal component ($M_{z'}$), this occurs as a first-order rate process:

$$M_{z'} = M_o\left[1 - \exp\left(\frac{-t}{T_1}\right)\right].$$ (1.4)

Equation (1.4) reveals that when the magnetization is fully tipped into the $x'y'$ plane by a 90° pulse, the magnetization will recover to 99.3% of its equilibrium value along z' at $t = 5T_1$. The T_1 relaxation time can be measured with an inversion-recovery experiment,[21] where a 180° pulse inverts the magnetization to the negative z' axis and allows the magnetization to recover for various times until the curve described by Eq. (1.5) is adequately characterized:

$$M_{z'} = M_o\left[1 - 2\exp\left(\frac{-t}{T_1}\right)\right].$$ (1.5)

Equation (1.5) is equivalent to (1.4), except the factor of 2 reflects the use of a 180° pulse, meaning that the magnetization (signal intensity) now should take twice as long to recover as when a 90° pulse is used. The results of the inversion-recovery experiment are illustrated in Figure 1.3, where the recovery of ibuprofen and maleic acid resonances are shown as a series of spectra and as signal intensities (inset) fit to Eq. (1.5).

It should be noted that if the sample contains multiple components, the 90° pulse will not be exactly the same for all spins. Equations (1.3) to (1.5) will not hold rigorously, leading to errors in T_1 measurements. Correction factors can be performed mathematically[22] or by computer simulation,[23] depending on the complexity of the sample. In practice, the average 90° pulse or the 90° pulse for the peaks of interest is used.

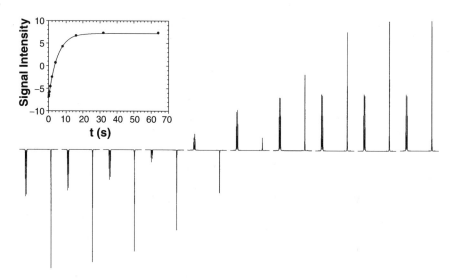

Figure 1.3. Inversion-recovery experimental results for determining T_1 relaxation times of ibuprofen and maleic acid. The spectral region from 5.6 to 7.6 ppm is displayed as a function of the variable delay time t. The t values used were (from left to right across the figure): 0.125 s, 0.25 s, 0.50 s, 1.0 s, 2.0 s, 4.0 s, 8.0 s, 16 s, 32 s, and 64 s. The acquisition and processing parameters are the same as for Figure 1.1. The inset shows the signal intensities of the maleic acid peak (5.93 ppm) fit to Eq. (1.5).

1.2.3. Repetition Time

As mentioned previously, the T_1 relaxation time affects the repetition time (acquisition time + relaxation delay), which must be carefully chosen to ensure that the magnetization returns to equilibrium between pulses. If the time between FIDs is insufficient to allow for complete relaxation, the intensity of the detected magnetization will be reduced:

$$\mathbf{M}_{x'y'} = \frac{\mathbf{M}_o\left[1-\exp\left(\frac{-t_r}{T_1}\right)\right]\sin\theta}{\left[1-\exp\left(\frac{-t_r}{T_1}\right)\right]\cos\theta},\qquad(1.6)$$

where t_r is the repetition time and θ is the tip angle.[24] If the sample contains multiple components, t_r should be set based on the longest T_1 measured for results to be fully quantitative. The ^{1}H spectrum in Figure 1.4 for an ibuprofen-maleic acid mixture measured using an average 90° pulse shows that the ibuprofen T_1 relaxation times range from 0.721 to 2.26 s, but maleic acid

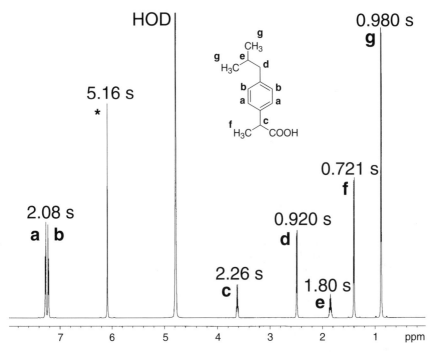

Figure 1.4. 600 MHz ¹H NMR spectrum of the ibuprofen-maleic acid mixture. The average T_1 relaxation times are displayed above each peak. The asterisk (*) indicates the maleic acid peak while the ibuprofen peaks are labeled *a* to *g* to correspond with the structure. The acquired data were zero-filled to 32 K, and other acquisition and processing parameters are as listed for Figure 1.1.

has a T_1 of 5.16 s. Therefore, to quantitate the ibuprofen concentration using the internal standard maleic acid, the repetition time should be >25 s if a 90° pulse is used. Because multiple relaxation mechanisms exist (e.g., dipole-dipole interactions, chemical shift anisotropy, and spin-rotation relaxation),[17] T_1 values for different functional groups may vary significantly, as illustrated by ibuprofen.

The consequences of a truncated repetition time are shown in Figure 1.5 for the ibuprofen aromatic protons and maleic acid singlet. Repetition times of 0.7, 1, 3, and $5T_1$ (based on the measured T_1 of maleic acid, 5.16 s) were used. The intensities of the ibuprofen aromatic proton resonances are less affected by decreased repetition times, since their T_1 values are much shorter, and appear to have recovered fully when a repetition time of $3T_1$ (15.5 s) is used. The intensity of the maleic acid singlet is more significantly affected because its T_1 is longer. The integral of the maleic acid resonance

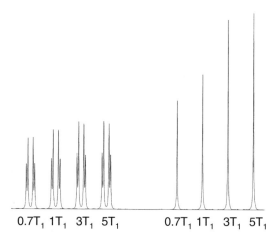

$0.7T_1$ $1T_1$ $3T_1$ $5T_1$ $0.7T_1$ $1T_1$ $3T_1$ $5T_1$

Figure 1.5. ^{1}H spectra for the ibuprofen-maleic acid mixture acquired with different repetition times. Only the spectral region including the aromatic ibuprofen protons and maleic acid singlet is shown. The repetition times were selected on the basis of the measured T_1 of maleic acid (5.16s) and are as follows: $0.7T_1$ (3.57s), $1T_1$ (5.16s), $3T_1$ (15.5s), and $5T_1$ (25.8s). All spectra were acquired and processed as described for Figure 1.1.

is the basis for the quantitation of the ibuprofen concentration, and repetition times shorter than $5T_1$ lead to reduced integral values for maleic acid. This results in positive systematic errors in the calculated ibuprofen concentration as shown in Table 1.1. The quantitation errors are greater when shorter repetition times are used, resulting in a gross overestimation of the ibuprofen concentration.

To efficiently utilize instrument time, it may be necessary to use repetition times less than $5T_1$. Theoretically the signal intensities obtained using a repetition time less than $5T_1$ should be correctable according to Eq. (1.1). For example, the data acquired with a repetition time of $1T_1$ (5.16s) should give a maleic acid signal that is 63.2% recovered. The ibuprofen aromatic proton signals are 91.6% recovered at this repetition time because of their faster T_1 recovery. The signal areas can be corrected to determine their value at 100% recovery. Using the corrected integrals, an ibuprofen concentration of 50.4mM was obtained. This result reflects an approximate 10% error relative to the fully relaxed data (acquired with a repetition time of $5T_1$) and may result from pulse imperfections as well as T_1 differences for the two ibuprofen aromatic protons (which were not completely baseline resolved and hence integrated together). These results show that the repetition time is probably the most important experimental parameter in

Table 1.1. Quantitation Accuracy of Ibuprofen against Maleic Acid as an Internal Standard Using Various Repetition Times and Tip Angles

Repetition Time	Tip Angle (°)	[Ibuprofen] (mM)	Measured Error[a]
$5T_1$ (25.8 s)	90	56.6	N/A[b]
$3T_1$ (15.5 s)	90	58.0	2.5%
$1T_1$ (5.16 s)	90	73.0	29.0%
$1T_1$ (5.16 s)	51	63.1	11.5%
$0.7T_1$ (3.57 s)	90	80.5	42.2%

[a] Measured errors in concentration were determined relative to the fully relaxed spectrum acquired with a repetition time of $5T_1$.
[b] Not applicable.

quantitation by NMR, and that proper quantitation can take time. (The fully relaxed spectrum was acquired in 27.5 minutes.)

1.2.4. Signal Averaging

Proper quantitation may require even longer experimental times when a limited amount of sample is available and signal averaging is necessary to improve the spectral signal-to-noise ratio (S/N). Coherently adding (coadding) successive single-pulse experiments increases the S/N by a factor of $\sqrt{N}$, where N is the number of FIDs added together. This is shown by the improvements in the spectral S/N in Figure 1.6 for the ibuprofen and maleic acid spectra as more FIDs are coadded. As shown in Table 1.2, the increases in S/N observed for increasing numbers of FIDs are very close to those predicted. The quantitation precision is not significantly affected by the number of FIDs co-added in this example, since both components are present at fairly high (millimolar) concentrations. However, the precision with which resonance intensities are determined directly depends on the error in the integrals. Therefore the precision of concentration determinations may be limited by the spectral S/N for more dilute solutions unless signal averaging is employed.

The signal intensity lost due to incomplete relaxation when repetition times less than $5T_1$ are used can often be regained by signal-averaging.[25] As shown by Rabenstein, when a 90° tip angle is used, S/N reaches a maximum when the repetition time is $1.25T_1$ (compared to using a $5T_1$ repetition time in an equivalent total experimental time).[24,26] The S/N is affected by both the repetition time and the tip angle. A tip angle less than 90° generates less signal (since the magnetization is not fully tipped into the $x'y'$ plane),

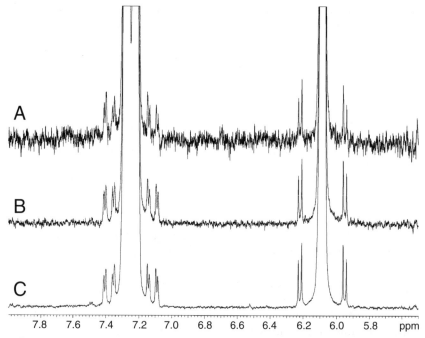

Figure 1.6. ^{1}H spectra for the ibuprofen-maleic acid mixture acquired with 1 (*A*), 8 (*B*), and 64 (*C*) FIDs coadded to improve the spectral signal-to-noise ratio (*S/N*). Again, only the ibuprofen aromatic and maleic acid peaks are displayed. The actual *S/N* values for each spectrum are shown in Table 1.2. Acquisition and processing parameters match those given for Figure 1.1.

Table 1.2. Improvements in the Signal-to-Noise Ratio (*S/N*) Gained by Co-adding Successive FIDs for Ibuprofen (IB) and Maleic Acid (MA)

Number of FIDs	*S/N* IB	*S/N* MA	Predicted *S/N* Increase	Actual *S/N* Increase IB	Actual *S/N* Increase MA	Concentration IB
1	987.3	2,041.7	N/A[a]	N/A[a]	N/A[a]	56.9 mM
8	2,737.5	6,450.3	2.83[b]	2.77[b]	3.16[b]	56.5
64	7,164.8	17,358	8.00[b]	7.26[b]	8.50[b]	56.6

[a] Not applicable.
[b] Calculated relative to the result for 1 FID.

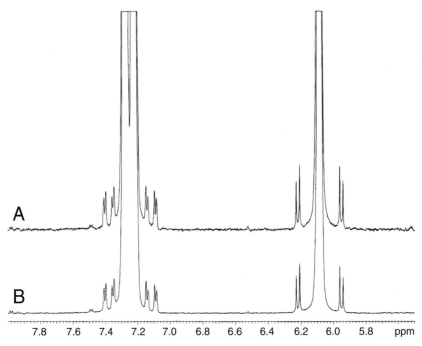

Figure 1.7. ^{1}H NMR spectra for the ibuprofen-maleic acid mixture acquired in 27.5 minutes. Spectrum (A) was acquired with 64 FIDs and a repetition time of $5T_1$ (25.8s), using a 90° tip angle. Spectrum (B) was acquired with 320 FIDs and a repetition time of $1T_1$ (5.16s), with a 50.8° tip angle as calculated by Eq. (1.7).

but less time is required for the magnetization to relax to equilibrium, affording more time for signal averaging. If the T_1 of the analyte and desired repetition time are known, the optimum tip angle (i.e., the Ernst angle) is calculated with

$$\cos\theta = \exp\left(\frac{-t}{T_1}\right). \tag{1.7}$$

Using the known T_1 of maleic acid (5.16s) and a desired repetition time of $1T_1$, the Ernst angle calculated for this solution is 50.8°. Figure 1.7 shows the improved S/N achieved when a shorter tip angle is used. Spectrum A represents 64 FIDs acquired with a 90° pulse and a repetition time of $5T_1$. The total experimental time was 27.5 minutes. When a 50.8° pulse and $1T_1$ repetition time were used, 320 FIDs could be co-added in an equivalent experimental time (Figure 1.7B). The S/N ratios achieved by acquiring 320

FIDs with a reduced tip angle increased by a factor of 1.74 for ibuprofen and 1.54 for maleic acid relative to the S/N obtained with 64 FIDs and a 90° tip angle. This is again attributed to the differences in T_1 for the two molecules and leads to an inflated ibuprofen concentration (63.1 mM compared to 56.6 mM). While the error might be less if the T_1 relaxation times are nearly the same for the standard and the analyte, in many quantitation experiments a 90° tip angle and repetition time of $\geq 5T_1$ are used, especially if extensive signal averaging is not required.[26]

1.2.5.   Defining Integrals

When considering the inversion-recovery experiment, it is illustrative to monitor the exponential recovery of signal intensity. However, relative peak heights vary based on their widths and shapes, so peak areas must be measured for quantitation.[26] Because NMR signals exhibit Lorentzian lineshapes, the peaks theoretically extend to infinity on either side of the actual chemical shift. Given that the peaks are phased properly and digital resolution is sufficient, Rabenstein has reported that 95% of the Lorentzian peak area can be described if the integral region encompasses 6.35 times the line width at half-height.[26] In most cases this will be impractical since other resonances will be present in this frequency range. But does it make a difference how wide the integral regions are? Figure 1.8 shows the integrals determined for ibuprofen and maleic acid using three different methods: truncating the integrals on either side of the main peak, truncating the integrals where the ^{13}C-satellites are baseline resolved, and extending the integral regions on either side of the peak so that the total region integrated is three times the width between the ^{13}C satellites. Table 1.3 shows that including ^{13}C satellites does make a small difference in the peak areas but extending the integral regions past the satellites does not. This result is not surprising, especially if baseline correction is applied near the peaks that may truncate the Lorentzian line shapes. As shown in Table 1.3, failing to apply baseline correction results in significant errors in quantitation (10% in this example) when wider integral regions are used. Since line widths are not constant for all peaks, it is important to define integral regions based on individual line widths rather than a fixed frequency range.[26] For example, if a constant integral region of 92 Hz is used to quantitate the example shown in Figure 1.8, the resulting ibuprofen concentration is 56.6 mM. Since the ibuprofen and maleic acid peaks have similar line widths in this example, the resulting concentration is not drastically affected. Significantly different results could be expected if products tethered to solid-phase resins were analyzed, since the heterogeneity of the

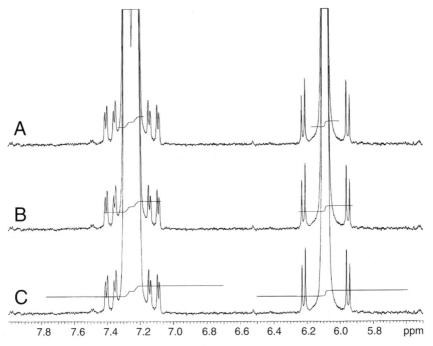

Figure 1.8. Defining integral regions for quantitating the ibuprofen concentration against maleic acid. The region from 5.5 to 8.0 ppm of the spectrum described in Figure 1.4 is shown. In (A), only the main peak is defined by the integral. This region is extended in (B) to include the ^{13}C satellites. The integrals are extended into the baseline (which has been automatically corrected using a spline function) in (C) such that the total integral covers three times the frequency range defined in (B). Table 1.3 provides the corresponding peak areas and resulting ibuprofen concentrations from this data.

Table 1.3. Effects of Integral Region Width and Baseline Correction[a] on Ibuprofen (IB) Quantitation against Maleic Acid (MA) as an Internal Standard

Integration Method	Baseline Correction	Integral IB	Integral MA	Concentration IB
Peak only	Yes	7296.5	3967.3	56.5 mM
Peak + ^{13}C Satellites	Yes	7396.5	4018.1	56.6
3 × (Peak + ^{13}C Satellites)	Yes	7405.5	4021.5	56.6
Peak Only	No	7121.3	3803.6	57.5
Peak + ^{13}C Satellites	No	7014.36	3700.5	58.3
3 × (Peak + ^{13}C Satellites)	No	6240.5	3077.5	62.3

[a] A spline function was used to automatically correct the baseline in these examples.

swollen resin leads to broader resonances than those typically encountered in solution-phase samples.

1.3. GENERAL STRATEGIES FOR QUANTITATION IN NMR

1.3.1. Desired Properties of Standards

Although pure standards are not required for quantitation by NMR, the standard should yield a simple NMR spectrum with resonances resolved from those of the analyte of interest so that accurate and precise integrals can be measured. It is helpful if the standard selected has a T_1 relaxation time similar to the resonances of the analyte to be used for quantitation. Furthermore it is important that the standard is chemically inert and has solubility properties similar to those of the analyte. The standard concentration is often selected to be near the sample concentration to ensure high precision of signal integrals used in quantitation. Generally, the standard may be used as an internal reference (dissolved with the sample) or an external reference (contained in a sealed capillary placed within the NMR tube containing the analyte).

1.3.2. Types of Standards

Internal Standards

The maleic acid used to quantitate ibuprofen as described above is an example of an internal standard. Such standards are added at a known concentration to a fixed sample volume in the NMR tube. KHP, maleic acid, and trimethylsilylpropionic acid (TSP) are common standards for aqueous samples, while cyclohexane or benzene are suitable for samples soluble in organic solvents. A more comprehensive list of standards has been provided by Kasler.[27] This is a convenient way to perform quantitative analyses as additional solutions (e.g., those required when generating a calibration curve for other methods) are not necessary, although the sample is contaminated by the internal standard.

External Standards

Many of these same standards can also be used as external standards, which are contained in situ in a sealed capillary. The capillary (inner diameter 1–1.5 mm for use in a conventional 5-mm NMR tube) contains an unknown but constant volume of a standard of known concentration. The external standard has an advantage over the internal standard as contamination of the sample is avoided. An additional advantage when using a standard that is not a primary analytical standard is that the capillary need only be standardized once for use in many subsequent quantitation experiments, provided that the compound is stable and does not decompose or adsorb to the capillary walls over time.[28] The capillary containing the external standard must be placed in the NMR tube concentric with its walls to avoid magnetic susceptibility effects that can influence signal integrals.[29,30] Susceptibility matching is also achieved by using the same solvent for the standard and sample solutions. Very high spinning speeds should be avoided as this causes vortexing of the sample solution resulting in degraded spectral resolution.[29]

1.4. STRATEGIES FOR QUANTITATIVE ANALYSIS OF SOLUTION-PHASE COMBINATORIAL LIBRARIES

External quantitation of reaction intermediates or products is probably impractical when parallel synthetic strategies are used to generate solution-phase libraries, especially if the products are contained in small volumes that would not be analyzed in a standard NMR tube. Internal standards can contaminate and dilute samples, which is undesirable where quantitation of limited sample quantities is necessary, as in the middle of a multiple-step synthesis. Thus, several quantitation alternatives have emerged to overcome these challenges. Chart 1.1 shows the structures of some unique internal standards that may be useful in quantitating solution-phase combinatorial products.

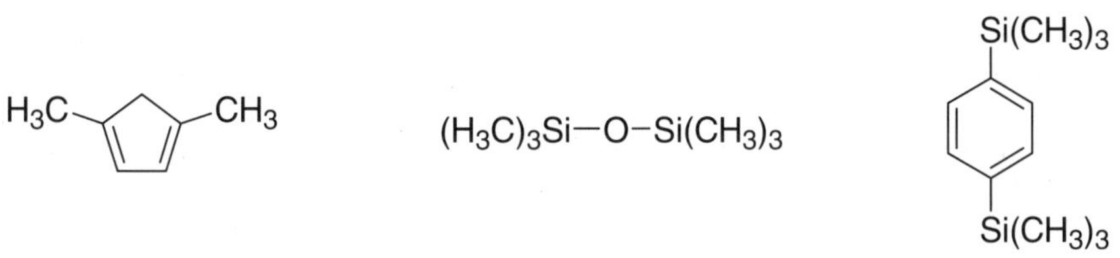

| 1 | 2 | 3 |

Chart 1.1. Useful internal standards for analysis of solution-phase combinatorial libraries. 2,5-dimethylfuran (DMFu, **1**) and hexamethyldisiloxane (HMDS, **2**) are volatile standards that can easily be evaporated from solution-phase samples after analysis. 1,4-bis (trimethylsilyl)benzene (BTMSB, **3**) is stable in DMSO for up to one month and is transparent in HPLC-UV and MS analyses.

1.4.1. Volatile Standards

Volatile internal standards can be evaporated from the sample after analysis, given that their boiling points are significantly lower than that of the analyte of interest. Such traceless NMR standards as 2,5-dimethylfuran (DMFu, **1**) and hexamethyldisiloxane (HMDS, **2**) are thus useful in quantitating reaction intermediates or biologically active samples that must undergo further testing.[31,32] After quantitation the standard can be removed from the sample by evaporation. This is especially convenient in a flow-probe format since nitrogen gas is used to flush the probe between samples. HMDS was used as an internal standard in an automated flow-probe analysis for determining product yields for a 96-member substituted methylene malonamic acid library.[32]

Quantitation by NMR requires time and skill that the typical organic chemist may lack.[31] Therefore it is desirable to have an internal standard indicator to reveal when proper experimental parameters for quantitation have been employed. The ^{1}H NMR spectrum of DMFu contains two singlets at 5.80 and 2.20 ppm. Gerritz et al. measured the ratio of these two signals as a function of the relaxation delay to provide an "internal relaxation standard" by which to estimate a reasonable experimental time for quantitation.[31] As shown in Table 1.4, quantitation accuracy was within 5% error for several analytes at varying concentrations.

The robustness of traceless NMR standards is questionable, considering their high volatility. Evaporative losses (which may vary according to the solvent used) over time can compromise results. Pinciroli et al. reported the use of 1,4-bis(trimethylsilyl) benzene (BTMSB, **3**) as a generic internal quantitation standard for library compounds soluble in dimethylsulfoxide (DMSO).[33] This standard does not have significant ultraviolet (UV) absorbance above 230 nm and is not ionized by electrospray. Therefore it

Table 1.4. Quantation Accuracy of "Traceless" NMR Using DMFu as an Internal Standard[a]

Analyte	Known Concentration	Calculated Concentration	% Error
1	0.86 mM	0.85 mM	0.7
1	13.09	12.38	5.4
1	70.26	72.63	3.4
2	4.14	4.23	2.3
2	28.47	28.42	0.2
3	22.98	22.06	4.0

[a] Data from reference 31.

provides no interference for samples that must also be analyzed by HPLC-UV-MS to confirm structural and purity data obtained from NMR. Once stability of BTMSB was demonstrated (about 1 month in solution) as well as the precision, accuracy, and linearity of the quantitation method validated, 314 combinatorial products were quantitated individually with this standard.[33]

1.4.2. Residual Protonated Solvent

Another convenient alternative is to simply quantitate using the residual protonated solvent signal as an internal standard. This can be challenging in aqueous (D_2O) or hygroscopic (DMSO) solvents since environmental instabilities (e.g., humidity) make it difficult to know the exact concentration of the solvent. Furthermore solvents like DMSO give rise to multiplets in a spectral region (~2.5 ppm) where sample signals may exist. Chloroform, which exhibits a singlet at about 7.2 ppm, downfield of some aromatic protons, is a better solvent standard and solubilizes many organic compounds. However, since expensive deuterated solvents are required, this method may be less useful for routine analyses.

1.4.3. ERETIC Method

For complex analytes or mixtures, it may be difficult to select a standard that contains resonances resolved from those of the sample. The ERETIC (Electronic Reference To access In vivo Concentrations) method overcomes this challenge by using an electronic signal as a reference.[34] Nothing is physically added to the sample or NMR tube, in contrast to traditional internal or external standards. The ^{13}C coil generates a pseudo-FID producing a signal that can be placed in a transparent region of the spectrum.[34] The ERETIC signal must be calibrated monthly against a sample of known concentration to provide the most accurate results. Table 1.5 shows the precision and accuracy of this method, as reflected by similar lactate concentrations determined using the ERETIC method and quantitation using trimethylamine hydrochloride (TMA) as an internal reference.

1.4.4. Special Issues Related to the Flow-Probe Format

The ERETIC method may be most useful for split-and-pool syntheses where it is undesirable to further complicate the NMR spectrum of complex mixtures by adding a standard. In both split-and-pool and parallel synthetic strategies, micro- to nanoscale reactions generate small amounts of product. For example, solution-phase libraries can be generated in 96-well microtitre

Table 1.5. Accuracy (Δ) and Precision (δ) of Lactate Concentrations Determined by the ERECTIC Method versus the Use of TMA as an Internal Standard[a]

Known Lactate Concentration	[Lactate] ± δ ERETIC	[Lactate] ± δ TMA	Δ (mM) ERETIC	Δ (mM) TMA
5.25 mM	5.19 ± 0.05 mM	5.46 ± 0.05	−0.06	0.21
10.92	11.01 ± 0.07	10.88 ± 0.05	0.04	−0.04
16.13	16.27 ± 0.07	16.56 ± 0.15	0.14	0.43
25.55	25.74 ± 0.14	25.51 ± 0.06	0.09	−0.14
54.11	54.55 ± 0.20	53.84 ± 0.22	0.23	−0.27

[a] Data from reference 34.

plates and the volumes contained in the individual wells are too small for analysis in standard NMR tubes. Hyphenation of HPLC with NMR (HPLC-NMR) has revealed the potential to analyze compounds such as peptides in a flowing system.[7] Commercially designed flow probes typically contain a total volume of about 120 μL while the active region is roughly half that value, making it possible to analyze library compounds dissolved in the volume of a single well. The development and commercialization of micro-coil NMR probes capable of measuring spectra for nL to μL volumes greatly reduces the sample volumes needed for NMR analysis and especially facilitates measurements for mass limited samples.[5,35,36] Using NMR flow probes, spectra can be acquired in on-flow (for concentrated samples) or stopped-flow (for minimal sample that requires signal averaging) formats. The low drift rate of most modern high-field magnets permits NMR flow analyses to be performed on-the-fly and unlocked, eliminating the need for expensive deuterated solvents to provide the deuterium lock. This is an advantage for quantitation of samples containing exchangeable protons (e.g., amide protons in peptides). In addition multiple components can be separated first by HPLC, stored individually in loops, and then flushed to the NMR probe for analysis.

Unique challenges exist for quantitation in flowing systems. Band broadening dilutes the sample to varying degrees depending on the flow rate and length of tubing connecting the HPLC to the flow probe. Multiple solvents create additional spectral interferences, and although solvent signals may be suppressed from the spectrum, signals underneath or nearby will likely be affected. Solvent gradients also change the composition of the solution (and in some cases the solubility of analytes), which will affect integration if the changes are significant on the time scale of the measurement.

Direct-injection NMR (DI-NMR) capitalizes on the small-volume capacity of the flow probe and averts the disadvantages encountered in the flowing system. This method is a modification of HPLC-NMR utilizing an automated liquids handler to directly inject samples into an NMR flow probe.[16] A syringe injects a fixed volume of a sample into the probe and withdraws it after analysis, depositing it back into its original well. The flow cell is flushed with rinse solvent between analyses and dried with nitrogen, thus preventing sample dilution by residual liquids.[16] Integrals of internal standards for each well can be used to quantitate individual reaction products while comparison of standard integrals for all wells provides quality control for monitoring injection and acquisition precision.[16] In a recent report DI-NMR characterized 400 potential pharmaceutical compounds overnight and provided information about reaction yields by quantitating against an internal standard.[37]

1.5. STRATEGIES FOR QUANTITATING SOLID-PHASE COMBINATORIAL LIBRARIES

Solution-phase synthesis is the traditional and perhaps favored approach for synthesizing parallel libraries. Since most reactions can be carried out in solution, solubility is not an issue and synthetic optimization is relatively simple. However, a bottleneck exists with product purification as unreacted starting materials and by-products remain in solution with the desired product. Solid-phase synthesis, utilizing resins such as the one introduced by Merrifield in the early 1960s,[38] is an alternative designed to facilitate easier purification of library compounds. Small, reactive molecules are attached to resin beads via tethers of variable length. Solutions of reagents added in excess drive reactions to completion. The product remains attached to the insoluble polymer while other materials remain in solution. The product can be purified by simple filtration. Such phase trafficking makes split-and-pool synthesis of large libraries much easier although transferring solution-phase reactions to the solid-phase often requires additional optimization.[39] After synthesis, product purity and structure can be assessed by NMR using two strategies: (1) cleave and analyze or (2) on-bead analysis.

1.5.1. Cleave and Analyze

Quantitation using internal or external standards as described above requires that products be cleaved from the bead and dissolved in an appropriate solvent. Cleavage occurs at the linker moiety, which connects the small organic molecule to the variable-length tentacle (the tentacle and linker together comprising the tether).[40] Quantitation is then relatively straightforward as long as an appropriate standard is selected. However, the product could be chemically altered during cleavage, making it difficult to determine product purity and yield. The cleavage reaction also requires extra time and adds another step in the overall synthesis. The product aliquot subject to cleavage is destroyed and cannot be used in the next step in the reaction scheme. Despite these limitations the cleave and analyze method has been widely utilized since, until recently, there have been few analytical techniques capable of analyzing compounds directly on the bead.

Wells et al. reported that the structure of a compound from a single bead could be confirmed by ^{1}H NMR with in situ cleavage in the NMR tube.[41] However, HPLC was used for quantitation due to the relatively poor sensitivity of NMR at nmol quantities of material.[41] Similarly Gotfredsen et al. reported that only 1.6nmol loading of an octapeptide onto a resin was required to obtain a high-resolution ^{1}H spectrum of the peptide (still bound to the resin) within 20 minutes.[42] Quantitation was achieved by fluorescence. Three times greater loading was required for complete structural elucidation by NMR.[42] After cleaving the peptide and analyzing by NMR in a small-volume probe, the ^{1}H solution-state spectrum revealed the presence of isomers whose resonances were unresolved in the spectra acquired on the bead.[42] The isomers may have been detected in the ^{13}C spectrum, where greater chemical shift dispersion improves resolution. Such an observation was made by Anderson et al., who used GC analysis of cleaved product to confirm the *exo/endo* isomeric ratio of norbornane-2-carboxylic acid inferred by relative peak intensities in the on-bead ^{13}C NMR spectrum.[43] The tether carbons of the polystyrene resin were also observed in this example, which provided an internal standard for determining the extent of reaction completion.[43] Riedl et al. were able to determine the enantiomeric excess of diasteroemeric Mosher esters by ^{13}C HR-MAS NMR.[44] The results obtained using seven pairs of ^{13}C resonances agreed within <1% with results obtained by HPLC on the cleaved products.[44] Lacey et al. used a solenoidal microcoil probe with an active volume of 800nL to measure ^{1}H NMR spectra of a compound cleaved from a single solid-phase resin bead in one hour.[45] The amount of product cleaved from the bead was determined using an internal standard to be around 500pmol, in good agreement with the loading capacities provided by the manufacturer of the Tentagel resin.

1.5.2. On-Bead Analysis

High-resolution magic-angle spinning (HR-MAS) NMR permits analysis of products still attached to solid-phase resin beads.[46] This makes quantitation possible at any step in the synthesis, without losing material to cleave-and-analyze methods that destroy the sample. Resin beads are swollen with solvent in a small rotor, and this rotor is entirely contained within the active region of the NMR coil. How much the resin swells, and hence the line widths in the NMR spectrum, depends on the resin structure and tether length.[47,48] The probe hardware is engineered with special materials to reduce magnetic susceptibility effects that contribute to broad lines. To further improve resolution, the sample is spun at the magic angle (54.7°) to minimize chemical shift anisotropy effects. Spinning at the magic angle helps average the magnetic susceptibility effects between the resin and solvent that also cause broad lines. These advances make it possible to obtain high-resolution liquid like ^1H and ^{13}C spectra for combinatorial products still attached to the bead. For example, Keifer et al. compared the line widths for small organic molecules attached to solid supports obtained in conventional liquids, solid-state, and HR-MAS NMR probes and showed that line widths can be reduced from several tens of Hz to a few Hz.[40]

1.6. ADVANTAGES AND DISADVANTAGES

HR-MAS NMR analysis of products attached to solid-phase resin beads is advantageous for several reasons. It is nondestructive, requires minimal sample preparation, and permits rapid analysis of small quantities (e.g., 1.6 nmol in 20 minutes).[42] Special NMR hardware is required for the analysis, but the experimental setup is equivalent to a traditional liquids analysis. The polymer portion of the resin contributes a broad spectral background that can be minimized by detecting ^{13}C rather than ^1H.[49] For maximal sensitivity this requires a probe optimized for ^{13}C detection and labeled compounds. A more practical approach is to suppress the polymer background with spectral editing techniques.

1.7. T_2 RELAXATION AND DIFFUSION EDITING

Spectral editing capitalizes on the differential behavior of specific nuclei in the magnetic field so that certain resonances can be suppressed in the recorded spectra. Such selectivity is important in mixture analysis where detection of one component over another may be preferred. The theory and practice of spectral editing for complex mixture analysis by NMR has recently been reviewed.[50] With respect to solid-phase synthesis, it is desirable to selectively detect the NMR resonances of the organic moiety attached to the resin without interference from the tentacle or polymer, which should be unaltered by the chemical reaction(s) performed.

The T_2 or transverse relaxation time roughly scales with the inverse of molecular mass. Polymers, which are typically bulky and have restricted tumbling in solution due to poor solubility, therefore have slower T_2 relaxation rates and shorter T_2 relaxation times. This leads to broad lines, because the T_2 time is also inversely proportional to the NMR line width at half height. It may be necessary to suppress the spectral background of the polymeric resin so that the resonances of the resin-bound combinatorial product are sufficiently resolved to permit quantitation. The product itself usually has a longer T_2 relaxation time and greater mobility in the solvent since the tentacles separate the product from the resin bead. This differential T_2 relaxation between the small organic moiety and the resin can be utilized to design NMR experiments permitting selective detection of the small molecule only.

Rather than acquiring data after the 90° pulse as in a typical one-dimensional NMR experiment [90° – Acquire], a 180° pulse train is inserted in the CPMG (Carr-Purcell Meiboom-Gill) method [90° – (τ – 180° – τ)$_n$-Acquire] to accomplish T_2 filtering.[51,52] After the initial 90° pulse, which tips the magnetization into the $x'y'$ plane, the magnetization will decay due to T_2 relaxation during the τ delay. The 180° pulse flips the magnetization, which is refocused after another τ delay. The τ delay and number of cycles through the CPMG train (n) are optimized to achieve the desired suppression of the resin signals. The improved resolution achievable with a CPMG-modified experiment is exemplified in Figure 1.9 for Fmoc–Lys–Boc bound to Wang resin. It is important to optimize τ and n to achieve a total pulse train length ($2\tau n$) that is both long enough to suppress the resin signals and short enough to preserve the remaining spectral information of the resin-bound small molecule. Some phase anomalies may appear near sharp resonances (e.g., near 7.5 ppm) due to rf pulse inhomogeities that are exacerbated by microcoil probes.

Diffusion-based methods also utilize T_2 filtering to discriminate against polymer resonances. A CPMG train at the end of the diffusion pulse sequence can selectively attenuate polymer resonances.[53] Alternatively, polymer signals are suppressed by T_2 relaxation during the relatively long gradient recovery delay times (tens of milliseconds) in the diffusion experiment used without CPMG filtering.[54] These methods have the added benefit that quickly diffusing molecules (like solvents) are suppressed with magnetic field gradient pulses.[53,55] Since any solvent or soluble component will be suppressed by the diffusion filter, these methods permit the use of protonated solvents (allowing beads to be analyzed directly from the reaction vessel) and discriminate against unreacted reagents.[56] Low-power presaturation of

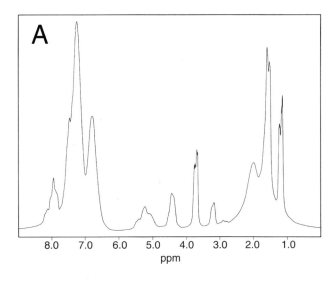

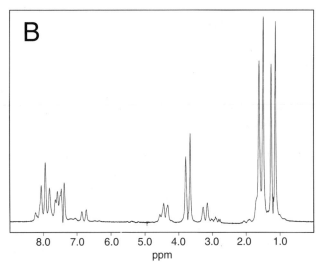

Figure 1.9. 500 MHz ^{1}H HR-MAS NMR spectra of 33.5 mg of Fmoc-Lys-Boc Wang resin swollen in d$_7$-DMF acquired without (A) and with (B) CPMG filtering. For both spectra, 16 scans were acquired and the sample was spun at 5 kHz. Sample heterogeneity and the restricted mobility of the polymer contribute to the broad lines observed in (A) for the one-dimensional spectrum. In (B) the t delay was 0.4 ms and the total length of the CPMG filter was 80 ms.

polymer resonances is another alternative to CPMG-filtering, although the success of this method may depend on the type of resin used.[40]

For library compounds with complex structures, quantitation on the bead may be impossible without modified NMR experiments to suppress the polymer spectral background. When analyzing data acquired with an edited NMR experiment, it must be assumed or known that there are not significant signal intensity losses for the resin-bound product, especially relative to the internal standard, which may or may not be resin bound. If T_2 losses for the resonances of interest are significant, the observed signal intensities must be corrected. The resonances of the resin-bound product used for quantitation must also be carefully selected. Slow resin dynamics can affect the observed NMR line widths for the product, and functional groups closer to the tentacle will exhibit greater line broadening (i.e., faster T_2 relaxation times) than those farther away from the tentacle.[57]

1.8.   QUANTITATION METHODS

NMR quantitation of materials synthesized on solid-phase beads is challenging, since resin dynamics and susceptibility differences within the sample contribute to broad lines in the spectrum. Spectral resolution of the desired analyte resonances from those of the resin as well as the reference material is important for both precise and accurate quantitation of the combinatorial product. Several quantitation methods, each of which will be addressed individually, have emerged to meet these challenges.

1.8.1.   Quantitation against a Soluble Internal Standard

Reaction kinetics can be monitored by comparing the HR-MAS NMR signal intensities of the resin-bound product and soluble reagent. If the amount of starting material is known, the amount of product can be calculated in an analogous manner to that outlined in Section 1.1. Warrass and Lippens applied this method to monitor Horner-Emmons reactions performed directly in the sample rotor.[58] The NMR signal integrals from a soluble dialdehyde reagent and resin-bound terminal aldehyde, which have ^1H chemical shifts well-resolved from the resin, were compared to determine reaction kinetics.[58] They also demonstrated the method on a larger reaction scale using protonated solvents followed by drying the resin and reswelling in deuterated solvent for NMR analysis. An inert soluble aldehyde was added as an internal standard. On either reaction scale, the number of product equivalents y was calculated from its signal integral I_2 in the following way:

$$y = x\left[\frac{I_2}{I_2 + 0.5I_1}\right], \tag{1.8}$$

where x is the number of equivalents of soluble reagent yielding an integral I_1. The integral I_1 is normalized for the two protons of the dialdehyde. The yield obtained by HR-MAS NMR was in good agreement with the value obtained independently by HPLC.[58] However, the absolute quantitation of product depends on knowing the density of reaction sites on the resin (i.e., the loading capacity, or how many equivalents of product can be attached per mass of resin). The accuracy and precision of the final results may then depend on the resin manufacturer's specifications for the lot of resin used and/or independent measurements to quantitate resin loading capacity.

Soluble inert materials have also been used as standards for quantitating solid-phase reactions. Examples include a traditional internal standard such as tetrakis(trimethylsilyl)silane (TKS) or the solvent used to swell the resin.[59] The mass of the resin (m_{resin}) and reference standard (m_{ref}) added to the rotor must be known to determine the loading capacity (l_i, in mmol/g) of the resin:

$$l_i = 1000\left[\frac{m_{ref} \times n_{C(ref)} \times I_i}{MW_{ref} \times I_{ref} \times n_{C(i)} \times m_{resin}}\right]. \tag{1.9}$$

In (1.9) the molecular weight of the reference compound (MW_{ref}, in g/mol) must also be known, and the integrals of the reference (I_{ref}) and compound of interest (I_i) normalized for the number of carbons giving rise to the signals if necessary. As demonstrated by Hany et al. and shown in Table 1.6, the ^{13}C HR-MAS NMR quantitation results obtained with either TKS or solvent (CDCl$_3$) agree with each other and those obtained by other chemical methods.[59] Accurate resin loading capacities were demonstrated for a variety of products and resins. However, the results depend on adequate swelling of the resin (3–4 mL/g).[59] This is illustrated by compound 1c in Table 1.6, which yielded a low loading capacity due to poor swelling. A relaxation reagent was also added to decrease the T_1 relaxation time of the TKS and hence shorten the analysis. Reagent addition is a potential disadvantage for highly reactive compounds (e.g., 2a, which could not be analyzed using a soluble standard, and 3a–c which had to be derivatized prior to quantitation to prevent reactivity). Thus this method is not a universal one for quantitating all solid-phase reactions. However, it was useful for confirming that the loading capacity of unfunctionalized polystyrene resin agreed with the calculated theoretical value, suggesting that the polystyrene resin itself could be used as an internal standard.[59]

Table 1.6. Quantitation of Resin Loading Using Soluble and Resin-Bound Standards [mmol/g]^a

Standard/ Compound^b	TKS	CDCl$_3$	Resin	Data from Chemical Methods
1a	2.43 ± 0.09	2.40 ± 0.08	2.51 ± 0.12	2.5
1b	1.05 ± 0.04	1.04 ± 0.05	1.03 ± 0.07	0.8
1c	1.90 ± 0.20	2.11 ± 0.24	3.21 ± 0.35	3.3
2a			1.09	1.0
3a			0.88	0.96
3b	1.97 ± 0.16	2.03 ± 0.11	2.00 ± 0.08	1.96
3c			2.07 ± 0.09	2.02
4	2.46	2.36	2.49	2.0
5	3.72 ± 0.25	3.70 ± 0.14	3.80 ± 0.28	3.0
6	0.60 ± 0.05	0.56 ± 0.03	0.58 ± 0.09	0.53

^{a} Resin Data from reference 59.
^{b} Each number represents a different resin; letters stand for individual products.

1.8.2.   Quantitation against Resin Components

Quantitation difficulties arising from differential molecular dynamics of soluble versus insoluble materials can be avoided by using an inert resin-bound moiety as an internal standard. The signal intensities of product to resin can be used to calculate the mole fraction of the product (x_i) and resin (x_j). By knowing the molecular mass of the resin (MW_j), the loading capacity of the resin can be determined according to Eq. (1.10):[59]

$$l_i = 1000 \left[\frac{x_i}{\sum_{j=1}^{n} x_j \times MW_j} \right].$$

(1.10)

The sample and solvent masses added to the rotor do not need to be measured. Hany et al. compared this method to quantitation against soluble standards.[59] Table 1.6 reveals that most resin loadings quantitated against the internal resin standard agreed with both the data obtained using soluble standards and independent quantitation methods. As shown in Eq. (1.10), this method can be used to simultaneously quantitate any number of organic molecules linked to the resin, as long as there are resolved NMR signals for each molecule. Resin-bound impurities (which are not always detectable by NMR) can also inflate the calculated loading capacities since they are not accounted for in the summation of Eq. (1.10).

Comparing resin-bound starting material and product signals in the HR-MAS NMR spectrum can also reveal reaction progress.[58] The signal selected from the starting material is usually one that is not changed by the reaction. This potentially removes another experimental variable, since its signal integral should remain constant during the reaction. Extensive sample preparation (e.g., drying, which could chemically alter the resin, and reswelling in deuterated solvent) is not required. Aliquots of the reaction mixture can be placed in the rotor and a small volume of deuterated solvent added for the NMR lock signal. Results do not depend on the theoretical loading capacity of the resin, which may change during the synthesis.[58] Warrass and Lippens utilized diffusion filtering to suppress the protonated solvent signal.[58] This prevented the use of a soluble standard, since its fast diffusion would also cause it to be attenuated in the final spectrum.

As indicated by the authors, diffusion filtering used in quantitation introduces potential error sources based on losses due to both T_1 and T_2 relaxation.[58] During the experimental diffusion time (ms scale), the magnetization is stored in the longitudinal (z') direction. If the nuclei whose signals are used for quantitation have significantly different T_1 relaxation times, the resulting signal intensities could be affected such that the quantitative result is skewed systematically high or low. As in a liquids NMR experiment, knowledge of the T_1 relaxation times for the nuclei of interest is important for properly selecting the parameters in the quantitative experiment. Differences in T_2 relaxation times can also affect quantitation, since the gradient recovery delay times are relatively long to achieve suppression of broad resin components. Nuclei in chemical exchange, such as amide protons in peptides, should not be used for quantitation, since exchange occurring on the time scales of the gradient recovery and/or diffusion times could affect the detected signals.[58,60]

1.8.3. Quantitation against an Independent Resin-Bound Internal Standard

A final alternative is to use an independent resin-bound standard that is added to the rotor with the resin containing the compound of interest. Once functionalized with the standard moiety, the resin standard undergoes no additional chemical reactions. After the loading capacity of the resin standard has been determined, this resin can be used as a universal reference for quantitating the loading capacity of other resins. Ideally the resin standard should be synthesized on the same type of matrix as the resin samples to be analyzed so that the swelling properties and resin dynamics, and hence NMR line widths, are similar for both the standard and the analyte resins.

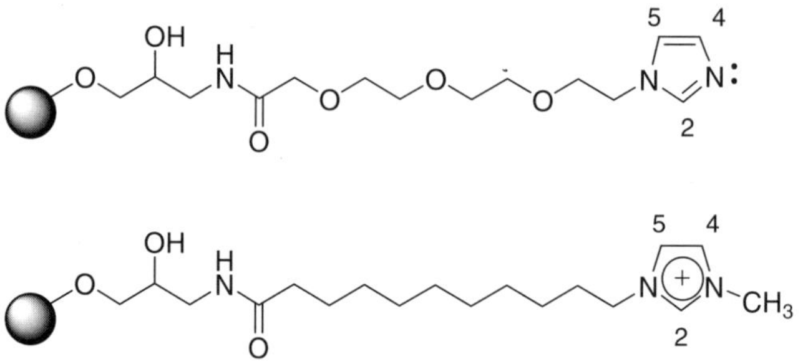

Chart 1.2. Structures of ImPEG resin (**1**) and ImQ resin standard (**2**). The shaded circle represents the resin backbone. The protons important for NMR quantitation are numbered on the structures.

Of course, the standard and the analyte resins must also yield unique NMR signals in spectral regions free from other sample or solvent signals.

This method has recently been explored in our lab for determining the loading capacities of various batches of resin prepared by carbodiimide coupling of 11-(1-imidazolyl)-3,6,9-trioxaundecanoic acid to Toyopearl AF–Amino 650M resin, a poly(methacrylate) having a hydrophilic aminoalkyl side chain (Chart 1.2, ImPEG resin). This resin was designed to serve as a stationary phase for affinity chromatography of cytochrome P450 4A1 (CYP4A1), a member of a family of cytochrome P450 enzymes that selectively hydroxylates the terminal methyl group of medium chain fatty acids. CYP4A1 is expressed along with other P450 isoforms in rat liver microsomes and is consequently difficult to purify by conventional protein chromatography. ω-Imidazolyl-fatty acids are known to be tight binding inhibitors of CYP4A1 due to a combination of strong hydrophobic interactions augmented by coordination of the imidazole nitrogen to the heme iron.[61] To quantitate ligand density for different preparations of resin, a resin standard containing a quaternized imidazolium moiety (ImQ resin in Chart 1.2) was designed and synthesized.

The density of the ligands on affinity resins can affect resin performance. In our case preliminary results suggested that if the ImPEG ligand density is too high, the ability of the column to adsorb enzyme was decreased significantly. Thus, to optimize the performance of the affinity support, several resin batches were made at various ligand : binding–site ratios. The number of mmol of reactive resin sites was calculated from the volume of settled resin used, the manufacturer's stated site density ($100 \pm 30\,\mu\text{mol/mL}$ settled wet resin), and our determination that 1 mL of wet settled resin yields 0.192 g dry resin (dried at high vacuum). Aliquots of each resin batch were mixed with aliquots of the resin standard (in approximately a 1:1 mass

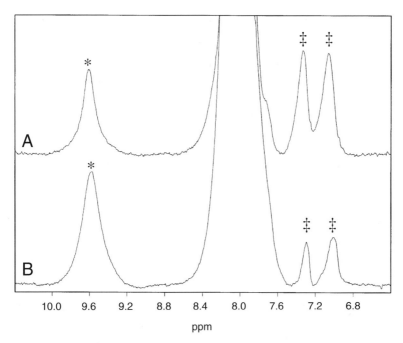

Figure 1.10. 500 MHz ^{1}H NMR spectra (in d_7-DMF) of imidazole-PEG (ImPEG) resins prepared with 1.26 (*A*) or 0.15 (*B*) equivalents of ligand. The C-2 proton from the ImQ resin is indicated with an asterisk (*), while the C-4/5 protons from the ImPEG resin (nonquaternized and therefore more upfield) are indicated with a double dagger (‡). The mass ratio of ImPEG resin to resin standard was approximately 1:1 in each mixture although different absolute amounts were used. The samples were spun at 5 kHz and 512 FIDs were coadded to yield each spectrum.

ratio, accurately weighed). The ligand density of the ImPEG analyte resin (LC_I) was then quantitated against the known ligand density of the internal resin standard (LC_{RS}) by HR-MAS NMR:

$$LC_i = \frac{A_I \times n_{H(RS)} \times LC_{RS} \times m_{RS}}{A_{RS} \times n_{H(I)} \times m_I}. \tag{1.11}$$

This equation is analogous to Eq.(1.9) presented by Hany et al., except the loading capacity, rather than the molecular weight, of the standard must be known. The ligand density of the resin standard in this example was independently determined using an inert soluble internal standard (TSP) and found to be 89.9 ± 4.1 nmol/mg.

Figure 1.10 shows the ^{1}H HR-MAS NMR spectra of two ImPEG/resin standard mixtures. The imidazole singlet at 9.6 ppm from the standard (*)

is well resolved from the ImPEG resonances (‡), and its area (A_{RS}, $n_{H(RS)}$ = 1) was compared to the combined integrals (A_I, $n_{H(I)}$ = 2) of the two unresolved ImPEG imidazole peaks at 7.0 and 7.4 ppm. (The other imidazole proton signals for the resin standard and ImPEG resins lie underneath the DMF signal at 8.0 ppm and thus contribute to both its intensity and line width.) The lines are broad for these resins, as stationary phase supports would be expected to have poor mobility even when solvated. Despite this fact the peaks are sufficiently resolved to permit quantitative integration.

The ImPEG resin prepared using 1.26 mol ImPEG ligand per mol resin sites (i.e., amino side chains) is shown in Figure 1.10A while the spectrum for the resin prepared with 0.15 equivalents of ligand is shown in Figure 1.10B. Even though there was more resin standard in the mixture shown in Figure 1.10B, the relative ImPEG integrals between spectra A and B reveal that the resin prepared with more equivalents of ImPEG ligand has a higher loading, as expected. Besides validating the use of the internal resin standard, this result illustrates that the loading capacities should be relatively predictable based on the reaction stoichiometry, especially for lower ligand densities which are more favorable for affinity column performance.

The ^{1}H HR-MAS NMR data and Eq. (1.11) were used to calculate the loading capacities of several resin batches prepared with various ligand:binding site ratios. The loading capacity of the resin prepared with 1.26 equivalents ImPEG ligand (81.0 ± 8.0 nmol/mg, Figure 1.10A) is approximately 5 times that of the resin prepared with 0.15 equivalents ligand (15.4 ± 1.3 nmol/mg, Figure 1.10B). This resin would be expected to give a loading capacity 8 to 9 times greater based on the amount of ligand used, suggesting that coupling is less efficient in the presence of excess ligand. Our collective results for several resin batches have confirmed this observation, revealing that higher coupling efficiencies are obtained when the number of reactive resin sites exceed the available amount of ligand and excess coupling reagent is used to drive the reactions to completion. The correlation of these results with the chromatographic performance of the ImPEG resins for purifying CYP4A1 and other P450 isoforms is currently being investigated further.

1.9. CONCLUSIONS

NMR spectroscopy provides a noninvasive way to quantitate solution- and solid-phase library compounds while also permitting total structural elucidation, which aids in determining structure-activity relationships for rational drug design. The mass sensitivity and resolution of NMR continue to improve with the development of small-volume flow probes for solution-phase libraries and HR-MAS techniques for analyzing solid-phase samples. In contrast to other analytical methods, no pure standards of the analytes are required for quantitation by NMR. This is a particular advantage in the combinatorial arena, since both parallel and split-and-pool synthetic strategies can lead to an enormous number of new compounds for which no standards exist. Automated devices for sample handling have greatly increased the throughput of NMR, helping to alleviate the analytical bottleneck created by combinatorial synthesis. The variety of standards and methods that have emerged illustrate the utility of NMR as a quantitative analytical tool. To appropriately quantify library compounds, many factors must be considered, including the instrumentation available, the chemical and physical properties of the analytes, and the time required for analysis. If the experimental parameters are carefully chosen based on the measured or known relaxation properties of the analytes and standards, reliable results can be obtained. These results are useful for assessing the utility of the synthetic strategy and play an important part of the detailed characterization of individual compounds for library archives.

ACKNOWLEDGMENTS

The authors thank Matthew A. Cerny, Yakov M. Koen, and Robert P. Hanzlik for the synthesis of the methacrylate resins used for HR-MAS NMR quantitation studies with an independent resin standard. NSF Grant CHE-9977422 provided partial funding for the 500 MHz spectrometer used in this research. L. H. Lucas acknowledges the support of NIH training grant 2 T32 GM08545 in the Dynamic Aspects of Chemical Biology.

REFERENCES

1. W. L. Fitch, *Mol. Diversity*, *4*, 39–45 (1999).

2. M. J. Shapiro and J. S. Gounarides, *Progr. Nucl. Magn. Reson. Spectrosc.* *35*, 153–200 (1999).

3. C. Marchioro, Analytical methods for solid and solution-phase combinatorial chemistry. Presented at Seminars in Organic Synthesis, Summer School "A. Corbella" 24th, Gargnano, Italy, June 14–18, pp. 149–170 (1999).

4. B. Yan, Y.-H. Chu, M. J. Shapiro, R. Richmond, J. Chin, L. Liu, A. Yu, *Analytical Methods in Combinatorial Chemistry*, Fenniri, H., ed., Combinatorial Chemistry, Oxford University Press, Oxford, pp. 263–286 (2000).

5. R. Subramanian, W. P. Kelley, P. D. Floyd, A. J. Tan, A. G. Webb, J. V. Sweedler, *Anal. Chem.* *71*, 5335–5339 (1999).

6. J. Chin, J. B. Fell, M. Jarosinski, M. J. Shapiro, J. R. Wareing, *J. Org. Chem.* **63**, 386–390 (1998).

7. J. C. Lindon, R. D. Farrant, P. N. Sanderson, P. M. Doyle, S. L. Gough, M. Spraul, M. Hoffman, J. K. Nicholson, *Magn. Reson. Chem.* **33**, 857–863 (1995).

8. G. F. Pauli, *Phytochem. Anal.* **12**, 28–42 (2001).

9. P. A. Keifer, *Drugs Fut.* **23**, 301–317 (1998).

10. D. Monleón, K. Colson, H. N. B. Moseley, C. Anklin, R. Oswald, T. Szyperski, G. T. Montelione, *J. Struct. Funct. Geonomics* **2**, 93–101 (2002).

11. D. J. Russell, C. E. Hadden, G. E. Martin, A. A. Gibson, A. P. Zens, J. L. Carolan, *J. Nat. Prod.* **63**, 1047–1049 (2000).

12. P. J. Hajduk, T. Gerfin, J.-M. Boehlen, M. Häberli, D. Marek, S. W. Fesik, *J. Med. Chem.* **42**, 2315–2317 (1999).

13. T. M. Logan, N. Murali, G. Wang, C. Jolivet, *Magn. Reson. Chem.* **37**, 512–515 (1999).

14. G. Lippens, R. Warrass, J.-M. Wieruszeski, P. Rousselot-Pailley, G. Chessari, *Comb. Chem. High Throughput Screening*, **4**, 333–351 (2001).

15. M. J. Shapiro, J. S. Gounarides, *Biotechnol. Bioeng.* **71**, 130–148 (2001).

16. P. A. Keifer, S. H. Smallcombe, E. H. Williams, K. E. Salomon, G. Mendez, J. L. Belletire, C. D. Moore, *J. Comb. Chem.* **2**, 151–171 (2000).

17. T. D. W. Claridge, *High-Resolution NMR Techniques in Organic Chemistry*, J. E. Baldwin, R. M. Williams, eds., Tetrahedron Organic Chemistry Series, Pergamon Press, Elmsford, NY, Vol. 19 (1999).

18. R. Freeman, *A Handbook of Nuclear Magnetic Resonance*, 2nd ed., Addison Wesley Longman, Singapore (1977).

19. J. K. M. Sanders, B. K. Hunter, *Modern NMR Spectroscopy A Guide for Chemists*, 2nd ed., Oxford University Press, New York (1993).

20. R. K. Harris, *Nuclear Magnetic Resonance Spectroscopy A Physicochemical View*, Pitman Publishing, Marshfield, MA (1983).

21. R. L. Vold, J. S. Waugh, M. P. Klein, D. E. Phelps, *J. Chem. Phys.* **48**, 3831–3832 (1968).

22. D. E. Jones, *J. Magn. Reson.* **6**, 191–196 (1972).

23. C. J. Galbán, G. S. Spencer, *J. Magn. Reson.* **156**, 161–170 (2002).

24. D. L. Rabenstein, *J. Chem. Ed.* **61**, 909–913 (1984).

25. E. D. Becker, J. A. Ferretti, P. N. Gambhir, *Anal. Chem.* **51**, 1413–1420 (1979).

26. D. L. Rabenstein, D. A. Keire, in *Modern NMR Techniques and Their Application in Chemistry*, A. I. Popov, K. Hallenga, eds., Marcel Dekker, New York, pp. 323–369 (1991).

27. F. Kasler, *Quantitative Analysis by NMR Spectroscopy*, Academic Press, New York (1973).

28. C. K. Larive, D. Jayawickrama, L. Orfi, *Appl. Spectrosc.* **51**, 1531–1536 (1997).

29. K. Hatada, Y. Terawaki, H. Okuda, *Org. Magn. Reson.* **9**, 518–522 (1977).

30. K. Hatada, Y. Terawaki, H. Okuda, K. Nagata, H. Yuki, *Anal. Chem.* **41**, 1518–1520 (1969).

31. S. W. Gerritz, A. M. Sefler, *J. Comb. Chem.* **2**, 39–41 (2000).

32. B. C. Hamper, D. M. Snyderman, T. J. Owen, A. M. Scates, D. C. Owsley, A. S. Kesselring, R. C. Chott, *J. Comb. Chem.* **1**, 140–150 (1999).

33. V. Pinciroli, R. Biancardi, N. Colombo, M. Colombo, V. Rizzo, *J. Comb. Chem.* **3**, 434–440 (2001).

34. S. Akoka, L. Barantin, M. Trierweiler, *Anal. Chem.* **71**, 2554–2557 (1999).

35. M. E. Lacey, R. Subramanian, D. L. Olson, A. G. Webb, J. V. Sweedler, *Chem. Rev.* **99**, 3133–3152 (1999).

36. D. L. Olson, M. E. Lacey, J. V. Sweedler, *Anal. Chem.* **70**, 645–650 (1998).

37. S. Borman, *Chem. Eng. News* **78** (May 15), 53–65 (2000).

38. R. B. Merrifield, *J. Am. Chem. Soc.* **85**, 2149–2154 (1963).

39. S. Borman, *Chem. Eng. News*, **80** (Nov, 11), 43–57 (2002).

40. P. A. Keifer, L. Baltusis, D. M. Rice, A. A. Tymiak, J. N. Shoolery, *J. Magn. Reson., Ser. A*, **119**, 65–75 (1996).

41. N. J. Wells, M. Davies, M. Bradley, *J. Org. Chem.* **63**, 6430–6431 (1998).

42. C. H. Gotfredsen, M. Grotli, M. Willert, M. Meldal, J. O. Duus, *J. Chem. Soc. Perkin Trans.* **17**, 1167–1171 (2000).

43. R. C. Anderson, M. A. Jarema, M. J. Shapiro, J. P. Stokes, M. Ziliox, *J. Org. Chem.* **60**, 2650–2651 (1995).

44. R. Riedl, R. Tappe, A. Berkessel, *J. Am. Chem. Soc.* **120**, 8994–9000 (1998).

45. M. E. Lacey, J. V. Sweedler, C. K. Larive, A. J. Pipe, R. D. Farrant, *J. Magn. Reson.* **153**, 215–222 (2001).

46. G. Lippens, M. Bourdonneau, C. Dhalluin, R. Warrass, T. Richert, C. Seetharaman, C. Boutillon, M. Piotto, *Curr. Org. Chem.* **3**, 147–169 (1999).

47. R. Santini, M. C. Griffith, M. Qi, *Tetrahedron Lett.* **39**, 8951–8954 (1998).

48. P. A. Keifer, *J. Org. Chem.* **61**, 1558–1559 (1996).

49. S. K. Sarkar, R. S. Garigipati, J. L. Adams, P. A. Keifer, *J. Am. Chem. Soc.* **118**, 2305–2306 (1996).

50. A. M. Dixon, C. K. Larive, *Appl. Spectrosc.* **53**, 426A–440A (1999).

51. H. Y. Carr, E. M. Purcell, *Phys. Rev.* **94**, 630–638 (1954).

52. S. Meiboom, D. Gill, *Rev. Sci. Inst.* **29**, 688–691 (1958).

53. J. A. Chin, A. Chen, M. J. Shapiro, *J. Comb. Chem.* **2**, 293–296 (2000).

54. J.-S. Fruchart, G. Lippens, C. Kuhn, H. Gras-Masse, O. Melnyk, *J. Org. Chem.* **67**, 526–532 (2002).

55. P. Rousselot-Pailley, N. J. Ede, G. Lippens, *J. Comb. Chem.* **3**, 559–563 (2001).

56. R. Warrass, J.-M. Wieruszeski, G. Lippens, *J. Am. Chem. Soc.* **121**, 3787–3788 (1999).

57. G. Lippens, G. Chessari, J.-M. Wieruszeski, *J. Magn. Reson.* **156**, 242–248 (2002).

58. R. Warrass, G. Lippens, *J. Org. Chem.* **65**, 2946–2950 (2000).

59. R. Hany, D. Rentsch, B. Dhanapal, D. Obrecht, *J. Comb. Chem.* **3**, 85–89 (2001).

60. M. Liu, H. C. Toms, G. E. Hawkes, J. K. Nicholson, J. C. Lindon, *J. Biomol. NMR.* **13**, 25–30 (1999).

61. P. Lu, M. A. Alterman, C. S. Chaurasia, R. B. Bambal, R. P. Hanzlik, *Arch. Biochem. Biophys.* **337**, 1–7 (1997).

CHAPTER

2

^{19}F GEL-PHASE NMR SPECTROSCOPY FOR REACTION MONITORING AND QUANTIFICATION OF RESIN LOADING

JOSEPH M. SALVINO

2.1. INTRODUCTION

Solid-phase organic synthesis has emerged as a powerful tool for the synthesis of chemical libraries.[1-4] A major drawback to solid-phase chemistry is that it is difficult to directly monitor the desired chemical reaction on resin. Standard analytical techniques for reaction optimization are available after the reaction product is cleaved from the solid support. However, the typically harsh conditions necessary to remove the reaction product from the solid support may introduce impurities and undesired side products, thus masking the true nature of the reaction being monitored. Both IR and NMR[5-15] spectroscopy have been used to monitor the progress of reactions on the solid phase. This chapter reviews the use of ^{19}F NMR spectroscopy as a tool to monitor solid-phase reactions directly, without having to cleave the product from the resin prior to analysis.

2.1.1. Fluorinated Linkers: Fluoro-Wang and Derivatives

A key aspect of a solid-phase synthesis is the choice of linker. The linker should be orthogonal to the required reaction conditions and allow quantitative cleavage of the product under mild reaction conditions. In addition a linker that incorporates a fluorine atom may serve as a powerful analytical handle as well as a synthetic handle. A fluorine atom in the linker allows the use of ^{19}F NMR spectroscopy to monitor solid-phase transformations. Advantages of using ^{19}F NMR to monitor solid-phase reactions include high sensitivity, due to the 100% natural abundance of the spin $^1/_2$ nucleus and the high gyromagnetic ratio of ^{19}F. A wide range of chemical shifts, approx-

Analysis and Purification Methods in Combinatorial Chemistry, Edited by Bing Yan.
ISBN 0-471-26929-8   Copyright © 2004 by John Wiley & Sons, Inc.

imately 200 ppm, is observed for closely related fluorine-containing compounds. Fluorine is sensitive to its chemical environment due to the large polarizability of the fluorine electron cloud. Remote changes in electron density thereby spread the ^{19}F resonances over a wide range of chemical shifts. Hence structural transformations even quite remote from the fluorine atom will affect the position of the ^{19}F signal. ^{19}F NMR is a nondestructive analysis. Nonfluorinated residual solvents or reagents are transparent to ^{19}F NMR. Therefore products can be analyzed immediately after washing to remove any excess fluorine-containing reagents, and the time-consuming exhaustive drying of the resin prior to analysis is unnecessary. ^{19}F NMR allows direct monitoring of a reaction without first having to cleave the product from the resin using harsh reaction conditions such as extended exposure to trifluoroacetic acid prior to analysis. Thus any impurities or decomposition can be avoided, and the true progress of a reaction observed.

Three fluorinated analogues of linkers are commonly used in solid-phase peptide synthesis, as has been recently reported.[12,14] These linkers are shown in Figure 2.1. 3-Fluoro-4-hydroxymethyl-benzoic acid linker is acid-stable, but cleavable under basic or nucleophilic conditions. 3-(3-Fluoro-4-hydroxymethyl-phenyl)-propionic acid linker requires strongly acidic conditions for cleavage (i.e., liquid hydrogen fluoride), while (3-fluoro-4-hydroxymethyl-phenoxy)-acetic acid linker requires cleavage under milder conditions such as trifluoroacetic acid. The latter fluorinated linker was utilized in the synthesis of Pilicide libraries, peptoids containing a coumarin moiety that bind to the active site of the periplasmic chaperones PapD and FimC.[12,14] Several of the steps leading to **8** (see Schemes 2.1 and 2.2) could be efficiently monitored and then optimized by use of ^{19}F NMR spectroscopy. High-quality spectra could be obtained within minutes for samples of resin in an ordinary NMR tube using a standard NMR spectrometer. Loading of the (3-fluoro-4-hydroxymethyl-phenoxy)-acetic acid linker to TentaGel S-NH2 resin was monitored particularly for optimization of the amide bond formation over ester formation, which resulted in a twofold coupling of the linker to the resin. Acylation of **2** to **3** was also monitored. The ^{19}F NMR signal for **2** appeared at −117.3 ppm. It was necessary to run the coupling

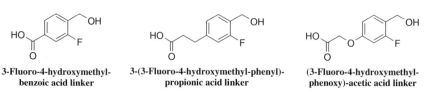

3-Fluoro-4-hydroxymethyl-benzoic acid linker   **3-(3-Fluoro-4-hydroxymethyl-phenyl)-propionic acid linker**   **(3-Fluoro-4-hydroxymethyl-phenoxy)-acetic acid linker**

Figure 2.1. Fluorinated analogues of linkers commonly used in solid-phase peptide synthesis.

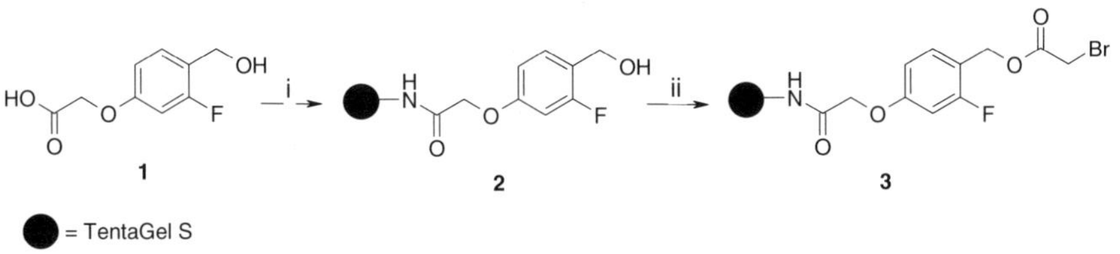

Scheme 2.1. Reagents and conditions: (i) Pentafluorophenol, DIC, TentaGel S HN_2, EtOAc; (ii) $BrCH_2CO_2H$, DIC, HOBt, DMAP, THF.

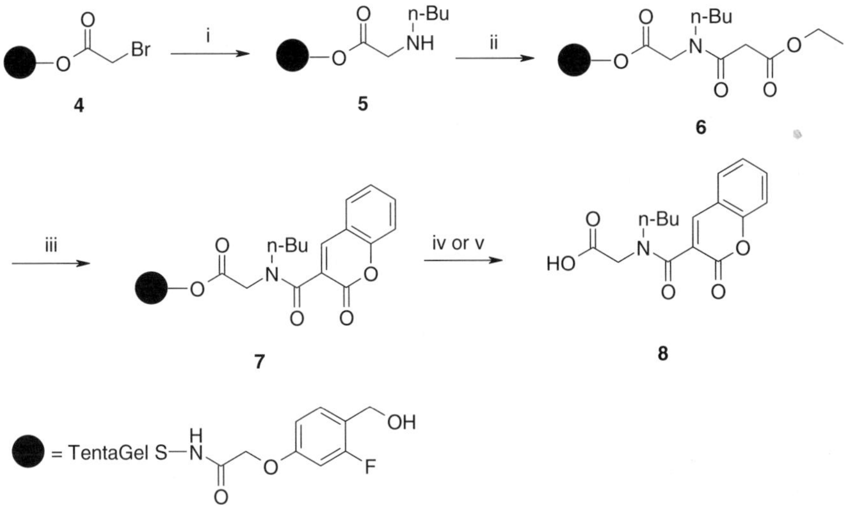

Scheme 2.2. Reagents and conditions: (i) n-Butylamine, CH_3CN, 0°C; (ii) $ClCOCH_2CO_2C_2H_5$, DIPEA, CH_2Cl_2, 0°C; (iii) salicyclaldehyde, piperidine, CH_3CN, reflux; (iv) TFA : H_2O (2 : 1); and (v) 1M LiOH, THF : MeOH : H_2O (3 : 1).

reaction twice in order to obtain complete conversion to **3** by observing the signal at −117.3 ppm disappear and a new signal at −115.1 appear that corresponds to **3**. (see Scheme 2.1 and Figure 2.2). Premature cleavage was observed during the transformation of **6** to **7** evident by the formation of a ^{19}F NMR signal at −117.3 for resin **2** (Scheme 2.2).

Another Fluoro-Wang type resin that was reported[15] is devoid of an amide bond in the attachment of the linker to the polymer support. This resin is readily available from 2-fluoro-4-hydroxy-acetophenone, and offers the advantage of being more chemically robust than the resins previously discussed. For example, the resin linkage is stable to lithium aluminum hydride (LAH) and inert to strongly basic conditions, such as treatment

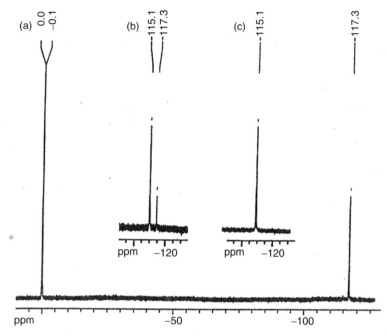

Figure 2.2. Monitoring the transformation of **2** into **3** by ^{19}F NMR spectroscopy; (*a*) **2**, (*b*) acylation of **2** with 3 equivalents of bromoacetic acid gave only partial conversion into **3**, and (*c*) full conversion was obtained after a second acylation.

with lithium bis (trimethylsilyl)amide (Scheme 2.3). ^{19}F NMR proved to be a powerful analytical technique to monitor the progress of these solid-phase reactions. Resin **11** (Scheme 2.3) displays a signal at −106 ppm corresponding to the fluorine on the aromatic ring of the linker. LAH reduction of the ester in resin **11** to give the alcohol resin **12** was evident by the shift of the fluorine signal from −106 to −118 ppm. The complete loss of the signal at −106 ppm confirms the completion of the reaction. ^{19}F NMR was used in this case to monitor the rate of loading diethylphosphonoacetic acid to resin **12**, to provide the phosphonate resin **13** (Scheme 2.4). The fluorine signal shifts from −118 ppm for resin **12** to −116 ppm for resin **13**. Optimization of the reaction, at least with respect to reaction time, is easily monitored. Small aliquots of resin are sampled, filtered, and washed at various time points during the reaction. This stops the progress of the reaction. ^{19}F NMR spectroscopic analysis of the samples then reveals the progress of the reaction for each sample. Integration of the fluorine signals determines the progress of the reaction (see Figure 2.3). The ability to observe the chemical shift differences in ^{19}F NMR spectra drops off the

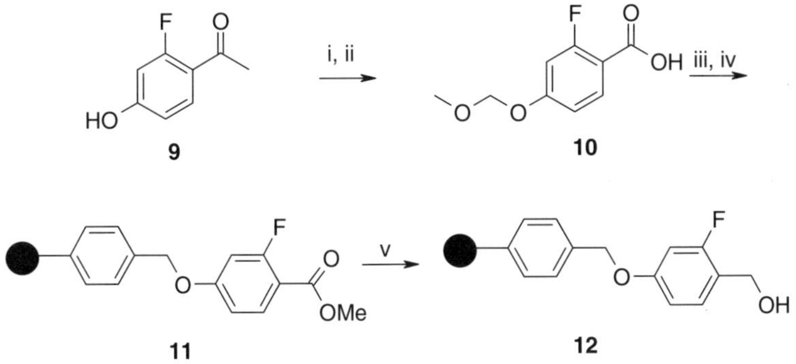

Scheme 2.3. Reagents and conditions: (i) MOM-Cl, NaH, THF, rt to 80°C, 1 h 93%; (ii) NaOCl, H$_2$O, dioxane, 70°C, 15 h, 99%; (iii) MeOH, HCl, reflux 12 h, 100%; (iv) Merrifield resin, Cs$_2$CO$_3$, DMF, 80°C, 12 h; and (v) LAH, THF, 25°C, 2 h.

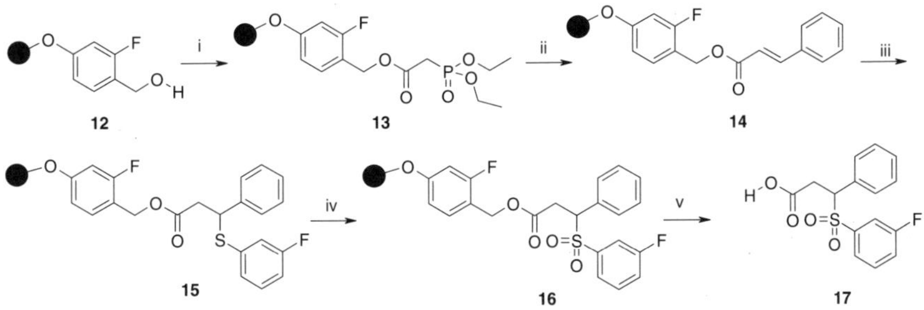

Scheme 2.4. Reagents and conditions: (i) Diethylphosphonoacetic acid, 2,6-dichlorobenzoyl chloride, anhydrous pyridine, DMF, 25°C, 8 h; (ii) lithium bis(trimethylsilyl)amide, THF, 0°C to 25°C, 60 min, then filter under argon, add benzaldehyde (4 equiv.), 60% cyclohexane in THF 25°C 24 h; (iii) 3-fluoro-phenylthiol, DMF, DBU (cat.), 65°C, 12 h; (iv) mCPBA, dioxane, 25°C, 12 h; and (v) 30% TFA in CH$_2$Cl$_2$, 25°C, 1 h.

farther away the chemical transformation is for the fluorine atom that is being monitored. Thus in the transformation of **13** to **14** (Scheme 2.4) the fluorine signal did not show any significant shift. In this case FT-IR was used as a complimentary technique, observing the shift in the carbonyl stretch for **13** and **14** to determine the progress of the reaction. The examples above demonstrate the power of ^{19}F NMR combined with these Fluoro-Wang type resins, **2** and **12**, for monitoring the loading directly on the solid phase. This methodology will facilitate the analysis of solid-phase reactions, particularly if the resin bound intermediate is unstable to the cleavage conditions.

^{19}F NMR spectroscopy has proved to be a powerful technique to quantify the formation of carboxylic acid esters and sulfonate esters to polymer-

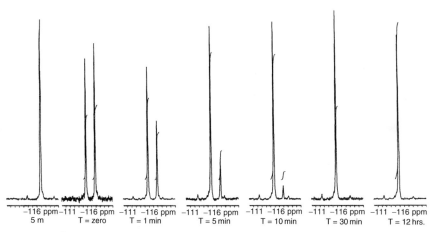

Figure 2.3. ^{19}F NMR of phosphonate loading to resin **12** (Scheme 2.4) at various time points.

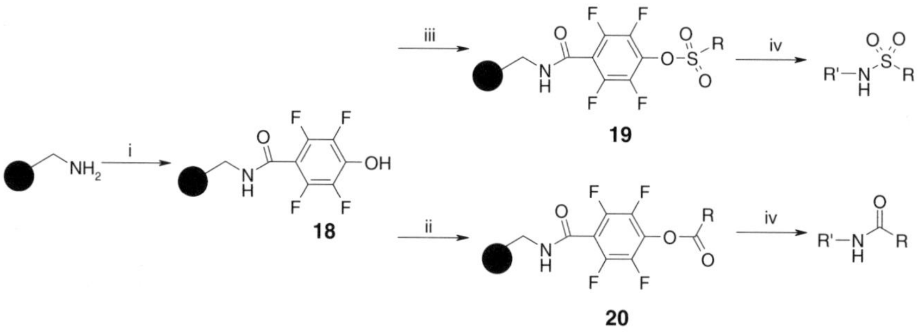

Scheme 2.5. Reagents and conditions: (i) 2,3,5,6,-tetrafluoro-4-hydroxybenzoic acid (1.7 equiv.), DIC (1.5 equiv.), DMF, HOBt (1.5 equiv.), 25°C, 16 h; (ii) sulfonyl chloride (2.0 equiv.), DIEA (3.0 equiv.), DMF, 25°C, 2 h; (iv) amine (0.8 equiv.), DMF, 25°C, 1–12 h; (iii) acid (2 equiv.), DIC (2 equiv.), DMAP (0.2 equiv.), DMF, 25°C, 3–22 h.

supported tetrafluorophenol (TFP)[16] resin **18** (Scheme 2.5). Only a limited amount of amine is needed to displace the carboxylate or sulfonate ester to form amides or sulfonamides, respectively, from this activated resin. Thus it is critical to be able to quantify the resin loading of the activated resin before library production to ensure synthesis of products devoid of amine starting material. The quality of the base resin **18** and the resulting polymeric activated resins, **19** and **20**, (Scheme 2.5) may be quantitatively determined in a nondestructive analysis using ^{19}F NMR spectroscopy. The ^{19}F NMR spectrum of the TFP resin, **18**, shows two resonances at −148 and

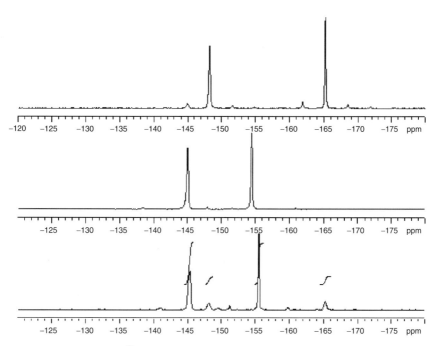

Figure 2.4. *Top spectrum*: ¹⁹F NMR spectra of tetrafluorophenol resin, **18**, Scheme 2.5. *Middle spectrum*: ¹⁹F NMR spectra of tetrafluorophenol sulfonate ester resin, **19**, Scheme 2.5. *Bottom spectrum*: ¹⁹F NMR spectra of tetrafluorophenol carboxylic acid ester, **20**, Scheme 2.5. Note the presence of resonances representing the residual unloaded resin in the bottom spectrum. All the spectra were acquired using the Varian's nanoprobe with magic angle spinning (1–2 kHz).

−165 ppm, each corresponding to two equivalent fluorine nuclei (top spectra of Figure 2.4). Analysis of the integration of the signals resulting from free and activated forms of the polymeric TFP resin determines the relative extent of resin loading (bottom spectra of Figure 2.4).

2.1.2. Quantitation Using ¹⁹F NMR Spectroscopy

There have been relatively few reports on techniques for quantifying chemical reactions on polymer-supported moieties. Thus any development of new methodologies to accelerate the analysis of solid-phase chemistry is important. One of the first uses of ¹⁹F NMR to quantitate loading of polymeric resin depended on internal concentration references derived by addition of a noncovalently bonded internal standard, such as the addition of C_6H_5F to the resin beads in an NMR tube, or addition of a solid-phase inter-

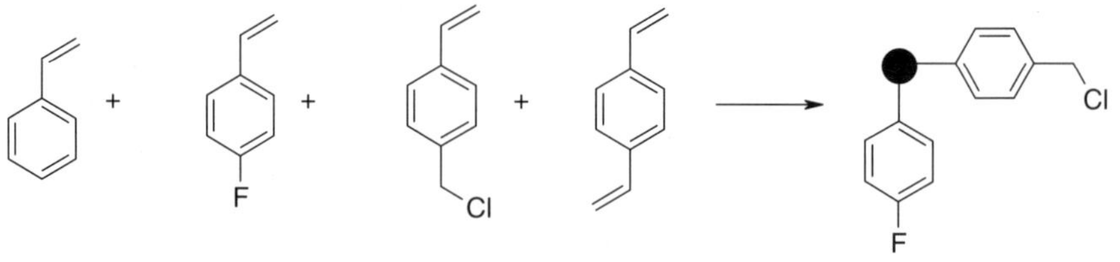

Figure 2.5. Suspension polymerization reaction incorporating an internal standard into the polymer support.

nal reference to the resin beads contained in an NMR tube. C_6H_5F was used as an internal standard and added as a solution, to determine the loading of amino acids with fluorine containing protecting groups into Merrifield resin.[17]

A recent paper describes a simple approach to quantify resin-bound chemistry using ^{19}F NMR spectroscopy in conjunction with a polymeric support bearing ^{19}F as an internal standard.[18] The internal standard renders the analysis independent of the resin sample mass. Solid-phase synthesis quantitation using this methodology requires a solid support having a well-characterized, chemically robust fluorine internal standard and an additional functional group or linker for covalent attachment of the desired reactants. An internal standard was incorporated into the solid support by co-polymerization with fluorinated monomer (Figure 2.5). The loading of the fluorine internal standard in this case was determined by combustion and subsequent fluoride ion chromatography. When a fluorine-containing reactant is bound to the solid-phase during a synthetic step, the reaction yield is determined from the ^{19}F NMR peak integration values of the product and the internal standard on the resin. This method permits rapid yield measurements at any step in a reaction sequence when a fluorine-labeled reactant is incorporated in the synthetic step. The utility of this methodology was demonstrated by optimization of a solid-phase Michael reaction (see Scheme 2.6). The reaction was optimized under various conditions, particularly where the base was varied. The method was contrasted to FT-IR spectroscopy, which could detect an incomplete reaction but afforded no quantitative information. This method presents a versatile and facile approach to determining reaction yields of solid-supported chemistries.

^{19}F NMR spectroscopy has also been used to monitor and optimize cleavage from the resin.[14] The m-fluorophenylalanine residue linked directly to the polymer support in **23a** and **23b** serves as an internal reference while the p-fluorophenylalanine was used to determine the amount of cleavage under various reaction conditions (Scheme 2.7). This resin was developed

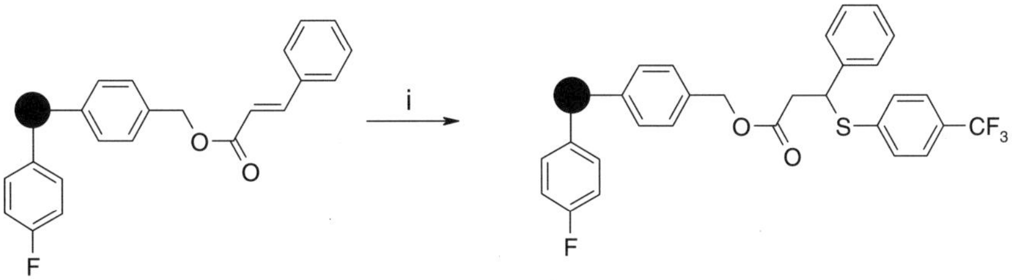

Scheme 2.6. Reagents and conditions: (i) Cs_2CO_3 or K_2CO_3 or $NaOCH_3$ or LiOH or $(C_2H_5)_3N$ or $LiN(Si(CH_3)_3)_2$ or DBU or n-Bu_4NOH, THF, 4-trifluoromethylthiophenol, 20°C, 72h.

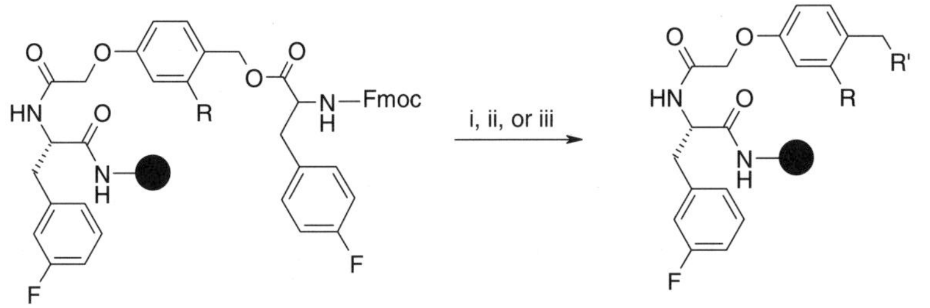

23a R=H

23b R=F

24a R=H R'=scavenger and OH

24b R=F R'=scavenger and OH

Scheme 2.7. Reagents and conditions: (i) TFA:water:thioanisole:ethanedithiol (87.5:5:5:2.5), 25°C, 2h; (ii) TFA:water:thioanisole:ethanedithiol (87.5:5:5:2.5), 60°C, 2h; (iii) 1M aq. LiOH in THF:MeOH:H_2O (3:1:1), 25°C, 2h.

to explore the conditions for cleavage of the fluorinated linker versus the corresponding nonfluorinated linker. The resin bound Fmoc-F-phenylalanines derivatives were prepared and subjected to cleavage under different conditions. The m-fluorophenyl alanine residue linked directly to the resin served as an internal reference while the p-fluorophenylalanine residue was used to determine the amount of cleavage.

2.1.3. ^{19}F NMR Reaction Monitoring

Several methods have been reported to monitor solid-phase reactions using ^{19}F NMR without a Fluoro-Wang type linker. For example, the monitoring of solid-phase nucleophilic aromatic substitution reactions via ^{19}F NMR through the loss of the aromatic fluorine atom has been reported.[13] ^{19}F NMR monitors the progress of the reaction, hence monitoring the reaction kinetics of product formation (Scheme 2.8).

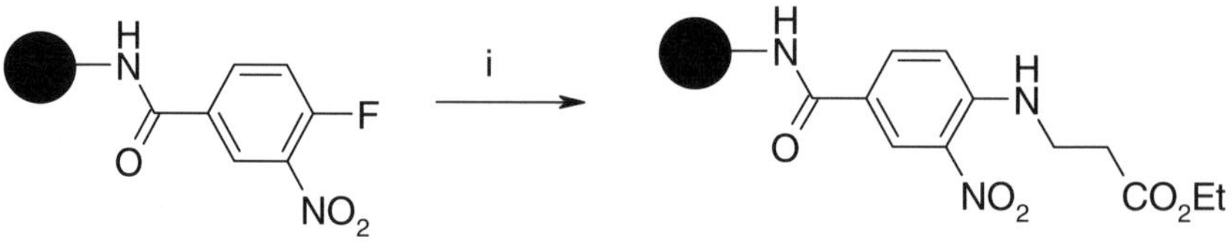

Scheme 2.8. Reagents and conditions: (i) DIEA, 3-amino-propionic acid ethyl ester, DMF, 25°C.

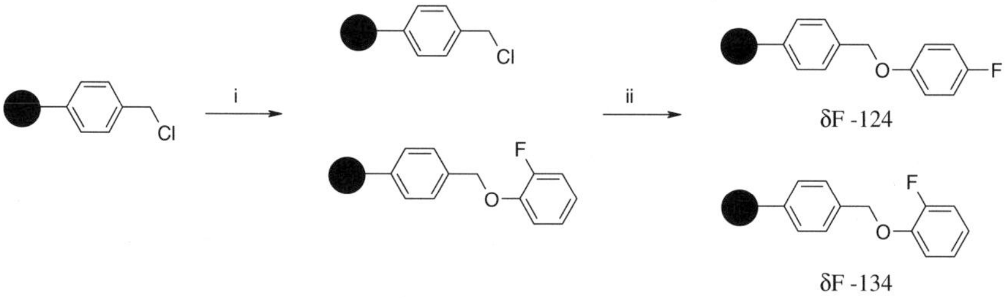

Scheme 2.9. Reagents and conditions: (i) NaH, DMF, 2-fluoro-phenol; (ii) NaH, DMF, 4-fluorophenol.

Another report[19] demonstrates the capability of providing quantitative information on the extent of conversion of a solid-phase material. This method may be applicable to a wide range of reaction chemistries and resin-supported materials. These authors use ^{19}F NMR to quantify the extent of the derivatization of functionalized polymeric material with a fluorinated substrate. Thus the rate of the reaction of Merrifield resin with 2-fluo-rophenol was examined. At various time intervals portions of the reacting resins were removed and were exhaustively reacted with excess 4-fluo-rophenol. Analysis of the resin samples shows two signals (Scheme 2.9), one for the 2-fluoro derivative at −134 ppm and the other for the 4-fluoro deriv-ative at −124 ppm. Quantitative information on the extent of the reaction was obtained by measuring the relative intensities of the resin bound fluo-rine and a known concentration of an internal standard in solution, such as fluorobenzene. It was noted that there needed to be a sufficient time delay between the RF pulses so that the solid-phase and solution-phase nuclei would be similarly relaxed in the ^{19}F NMR spectra. This method provided excellent correlation for resin-bound F atom content in samples containing 30 mg resin samples blended to various extents with 100% 2-fluorophenyl Merrifield ether and unfunctionalized Merrifield resin. This fluorophenol based assay was then used to optimize the attachment of an epoxide to the Merrifield resin (Scheme 2.10). The alkylation reaction was monitored at

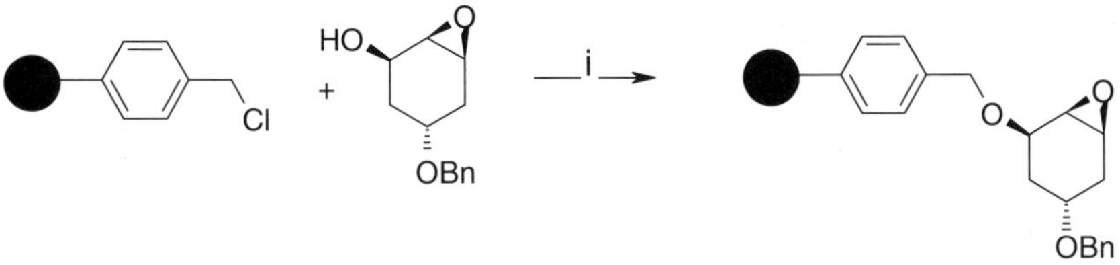

Scheme 2.10. Reagents and conditions: (i) NaH, DMF, 60°C, 3 h, then rt for 16 h.

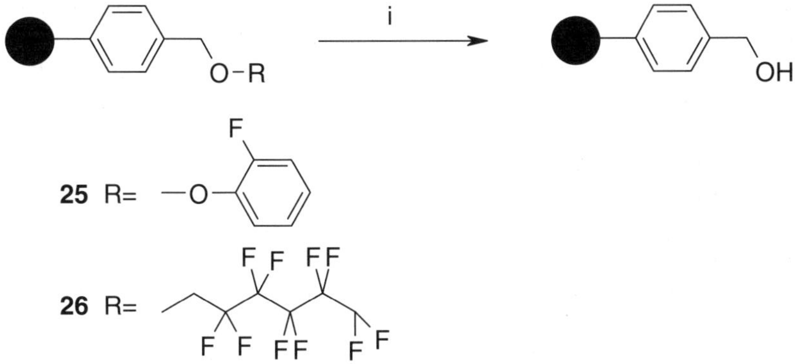

Scheme 2.11. Reagents and conditions: (i) SnCl$_4$ (10 equiv.), chloroform, H$_2$0, 30°C, 1.5–4 h.

various time points and then treated with excess 4-fluorophenolate. The ^{19}F NMR analysis determined the rate of formation of the Merrifield ether. The rate of the cleavage reaction was studied for the two Merrifield ethers **25** and **26** using ^{19}F NMR spectroscopy (Scheme 2.11). A 1.5-fold excess of stannic chloride was added to each resin sample, **25** and **26**, suspended in d-chloroform, and the increase in the intensities of the ^{19}F signals at −142 and −124 ppm, due to the cleaved products, respectively, were monitored over some time.

2.2. ^{19}F NMR AS A TOOL FOR ENCODING LIBRARIES

^{19}F NMR is an excellent tool for generating the encoding method for single beads in combinatorial libraries produced by the mix and split synthetic method. The same advantages of using ^{19}F NMR spectroscopy to monitor reactions are true for using ^{19}F NMR as a tool for encoding libraries. Numerous fluorine tags including many from commercially available sources are available for use. The relative stability of the fluorine tag is also an important positive characteristic. Due to the wide range of chemical shifts observed for fluorinated molecules, even library compounds that contain a fluorine atom usually display a chemical shift different from any encoding tags used. This characteristic usually distinguishes the library compound from the tag. Also in most cases the library compound is cleaved from the resin before encoding so that overlap of the ^{19}F signal does not become an issue.

A recent report[20] uses a fluorine encoding method for the first position monomers in a mix and split library to assist in the mass spectral deconvolution of library hits. Mass spectroscopy is used as the primary structural deconvolution tool. However, due to mass redundancies in the screened pools, mass spectral analysis alone cannot unequivocally determine the structure of the hits. Thus, to avoid having to re-synthesize numerous compounds, ^{19}F encoding is used to provide additional structural information. A hybrid structural determination strategy is employed that relies on mass spectral analysis of the active bead coupled with on-bead analysis of the first position nonreleasable code. This strategy reduces the number of structural possibilities to a single structure in every case. In practice, a series of commercially available fluorine carboxylic acids are used to generate an encoding system for mix and split libraries. The tags are attached to the resin via an amide bond, and each tag displays a unique distinguishable fluorine chemical shift. A large number of codes may be generated from a small set of differentiable fluorine carboxylic acid tags. Combinations of two fluorine tags are used at a time, along with 50 to 100 first-position encoded resin pools that require 11 to 15 individual tags. The large batches of encoded resin are synthesized as outlined in Scheme 2.12.

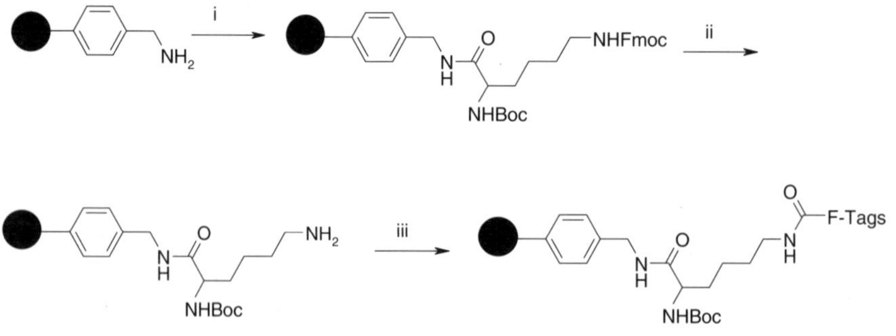

Scheme 2.12. Synthesis of bulk ^{19}F recoded resin. **Reagents and conditions:** (i) N-Boc-ε-N-Fmoc-Lys-OH, DIC, HOBt, DCM; (ii) piperidine, DCM; (iii) F-tags-COOH, DIC, HOBt, DCM.

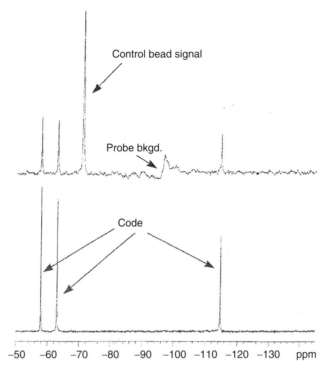

Figure 2.6. Decoding of a combinatorial chemistry library bead by ^{19}F NMR.

A typical synthesis of a combinatorial mix and split library proceeds by first defining the required resin pools, followed by deprotection of the α-amine position, attachment of the appropriate linker, and then attachment of the library core and subsequent derivatization following a typical mix and split protocol. Deconvolution of the active bead proceeds by acquiring the ^{19}F NMR spectra of the bead. These investigations used an NMR equipped with a magic angle spinning probe. It took one to four hours to "read" the ^{19}F code from a single bead for a typical sample using 130 μm aminomethyl polystyrene resin with a library and tag loading capacity of 1.2 mmol/g each. Single-bead spectra of library hits are compared to known spectra of coded resin batches shown in Figure 2.6.

Another recent report[21,22] uses a set of chemically robust aryl fluorides as molecular tags. These molecules exhibit chemical shifts that range from −55 to −25 ppm relative to fluorobenzene, and are tolerant to a wide variety of chemical transformations in solid-phase chemistry. Poly-*N*-alkylglycines were chosen as a model system for the preparation of combinatorial

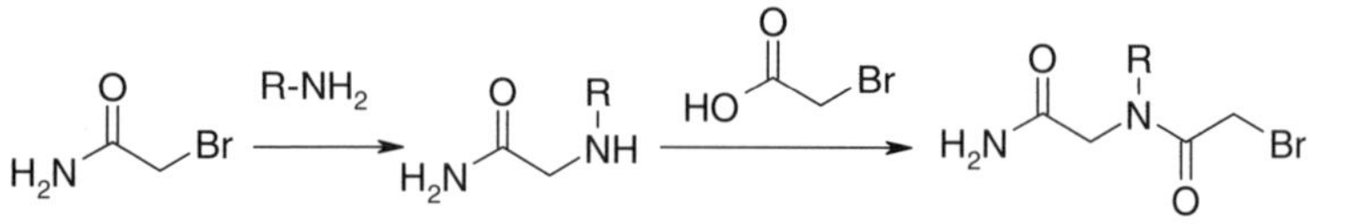

$$(2.1)$$

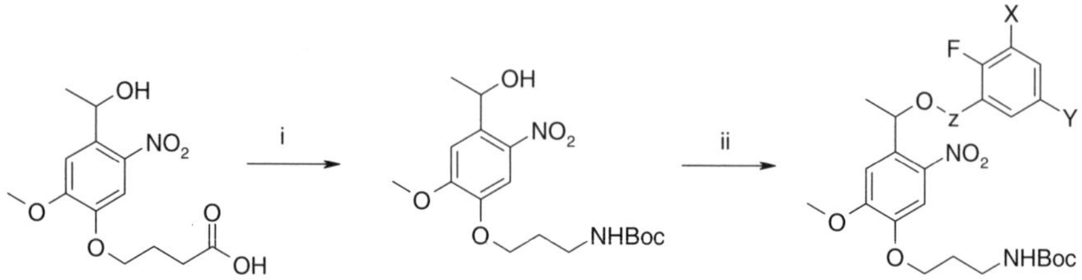

Scheme 2.13. Reagents and conditions: (i) $N_3PO(OPh)_2$, triethylamine, tert-butanol, reflux, 61%; (ii) alkyl bromide F-tags, NaH, DMF, 25°C, 2 h, 71–83%.

libraries using these fluorinated tags (Eq. 2.1). The detection sensitivity in ^{19}F NMR for fluorinated tag molecules was established to define the scope of the method. A series of fluorinated tags have been prepared that fall into two molecular classes, hydroxybutyl ethers of phenols and benzyl alcohols. The fluoroarene tags were attached to a linker that permits them to be released photochemically prior to reading, and bears a primary amine for chemical synthesis (Scheme 2.13). This method of decoding combinatorial libraries uses solution-phase methodology. The photo-linker to the tags enables them to be readily released for NMR analysis in solution, rather than requiring a solid-phase NMR method. In this application the fluoroaromatics perform as analogue codes in that they directly represent building blocks in the library. The frequency of the appearance of a tag is therefore a direct measure of the performance of the building block in the assay.

REFERENCES

1. L. A. Thompson, J. A. Ellman, *Chem. Rev.* **96**, 555 (1996).
2. D. Obrecht, J. M. Villalgordo, *Solid-Supported Combinatorial and Parallel Synthesis of Small-Molecular Weight Compound Libraries*, Elsevier Science, Amsterdam (1998).
3. B. A. Bunin, *The Combinatorial Index*, Academic Press, San Diego (1998).
4. F. Balkenhohl, C. von dem Bussche-Hunnefeld, A. Lansky, C. Zechel, *Angew. Chem. Int. Ed. Engl.* **35**, 2288 (1996).

5. B. Yan, G. Kumaravel, H. Anjara, A. Wu, R. C. Petter, C. F. Jewell, J. R. Wareing, *J. Org. Chem.* **60**, 5736 (1995).

6. B. Yan, Q. Sun, J. R. Wareing, C. F. Jewell, *J. Org. Chem.* **61**, 8765 (1996).

7. W. Li, B. J. Yan, *J. Org. Chem.* **63**, 4092 (1998).

8. J. E. Egner, M. Bradley, *Drug Discovery Today* **2** (3), 102 (1997).

9. P. A. Keifer, *Drug Discover Today* **2** (11), 468 (1997).

10. Y. Luo, X. Ouyang, R. W. Armstrong, M. M. Murphy, *J. Org. Chem.* **63**, 8719 (1998).

11. A. Svensson, T. Fex, J. Kihlberg, *Tetrahedron Lett.* **37**, 7649 (1996).

12. A. Svensson, K.-E. Berquist, T. Fex, J. Kihlberg, *Tetrahedron Lett.* **39**, 7193 (1998).

13. M. J. Shapiro, G. Kumaravel, R. C. Petter, R. Beveridge, *Tetrahedron Lett.* **37**, 4671 (1996).

14. A. Svensson, T. Fex, J. Kihlberg, *J. Comb. Chem.* **2**, 736 (2000).

15. J. M. Salvino, S. Patel, M. Drew, P. Krowlikowski, E. Orton, N. V. Kumar, T. Caulfield, R. Labaudiniere, *J. Comb. Chem.* **3**, 177 (2001).

16. J. M. Salvino, N. V. Kumar, E. Orton, J. Airey, T. Kiesow, K. Crawford, R. Mathew, P. Krolikowski, M. Drew, D. Engers, D. Krolikowski, T. Herpin, M. Gardyan, G. McGeehan, R. Labaudiniere, *J. Comb. Chem.* **2**, 691 (2000).

17. S. L. Nanatt, W. T. Frazer, J. T. Cudman, B. E. Lenk, J. F. Lubetich, E. A. McNelly, S. C. Smith, D. J. Templeton, R. P. Pinnell, *Tetrahedron Lett.* **21**, 1397 (1980).

18. M. Drew, E. Orton, P. Krolikowski, J. M. Salvino, N. V. Kumar, *J. Comb. Chem.* **2**, 8 (2000).

19. D. Stones, D. J. Miller, M. W. Beaton, T. J. Rutherford, D. Gani, *Tetrahedron Lett.* **39**, 4875 (1998).

20. J. E. Hochlowski, D. N. Whitten, T. J. Sowin, *J. Comb. Chem.* **1**, 291 (1999).

21. M. C. Pirrung, K. Park, L. N. Tumey, *J. Comb. Chem.* **4**, 329 (2002).

22. M. C. Pirrung, K. Park, *Bioorg Med. Chem. Lett.* **10**, 2115 (2000).

CHAPTER

3

THE APPLICATION OF SINGLE-BEAD FTIR AND COLOR TEST FOR REACTION MONITORING AND BUILDING BLOCK VALIDATION IN COMBINATORIAL LIBRARY SYNTHESIS

JASON J. COURNOYER, CLINTON A. KRUEGER,
JANICE V. WADE, and BING YAN

3.1. INTRODUCTION

Combinatorial library development in solution[1] or on solid phase[2,3] requires optimizing reaction conditions and evaluating the compatibility of building blocks under such synthesis conditions. A synthesis scheme must first be proved to yield the desired product with high yield and purity. Then all building blocks are evaluated in one of two ways: in parallel validation as one site of diversity is varied, and the remaining sites of diversity are held constant, or in matrix validation as all sites of diversity are varied. Whereas a parallel validation helps determine how one set of precursors performs with a specific substrate, a matrix library provides information of cross-compatibility of all sites of diversity. The analysis of these products remains a major challenge.[4] This chapter will discuss combinatorial synthesis using solid-phase organic synthesis (SPOS) methods.

A possible way to monitor solid-phase synthesis is to cleave compounds from the solid support and analyze them by liquid chromatography mass spectrometry (LC/MS). However, compounds are subject to chemical changes under cleavage conditions. In addition this process is time-consuming and labor intensive. Alternatively, more appropriate methods were developed for feasibility studies and building block validation. The organic reactions used in combinatorial library synthesis are usually well optimized in solution so that the corresponding SPOS reactions will lead to the expected resin-bound products. Consequently the confirmation of the presence or absence of the desired product or starting material becomes

Analysis and Purification Methods in Combinatorial Chemistry, Edited by Bing Yan.
ISBN 0-471-26929-8   Copyright © 2004 by John Wiley & Sons, Inc.

the major goal of analysis. Single-bead FTIR[4] is an effective tool for this type of analysis because most organic reactions involve functional group transformations. This analysis has the advantage of monitoring organic reactions from only one bead without sample preparation, and it can be rapidly performed at any time during synthesis. Additionally color tests for the solution-phase reaction monitoring are transferred to solid-phase chemistry whenever possible in our laboratories.[5,6] Some color tests have been developed especially for solid-phase reaction monitoring such as fluorescent tagging for various resin-bound functional groups.[7,8]

Color tests can be conducted in the synthesis laboratories as a quick check for the presence or absence of certain functional groups. They also indicate the extent of reaction completion based on the disappearance of the functional group in the starting material. Therefore a combination of color tests and FTIR analysis can provide a clear picture of progress in solid-phase reactions. Ensuring complete reaction for every step is a more logical way to increase reaction yield than adding excess materials to an inefficient synthesis that still must confront an unavoidable loss in product purification. In the following, we show that the monitoring of each reaction during development and subsequent production by FTIR and color test is useful and indispensable. Scaffold loading, the addition of diversity side chains onto the scaffold, and cleavage of the product from resin can all be monitored based on the presence or lack of organic functional groups.

3.2. REACTION MONITORING FOR FEASIBILITY AND SYNTHESIS OPTIMIZATION

A 90% yield for each reaction step in a four-step synthesis sequence would only give a product with 65% final yield and a low purity. Therefore, during feasibility and optimization studies, it is important to know that each reaction step is complete. An individual test, FTIR or color test, can reveal whether a reaction is occurring and, in some cases, whether it is complete. The combination of different tests can provide more conclusive information on reaction completion.

The coupling of carboxylic acid **2** to Marshall resin **1** to form **3** occurred in the presence of coupling catalysts and base in dichloromethane (Scheme 3.1). To ensure reaction completion, both the phenol test[9] and single-bead FTIR are used. Normally, a positive phenol test resulting in dark-purple resin beads indicates the presence of the phenol group. The color intensity is proportional to the phenol concentration (Figure 3.1). FTIR showed the disappearance of the IR band for phenol group at $3278\,\mathrm{cm}^{-1}$ and the appearance of the IR band for ester group at $1710\,\mathrm{cm}^{-1}$ (Figure 3.2). In combining

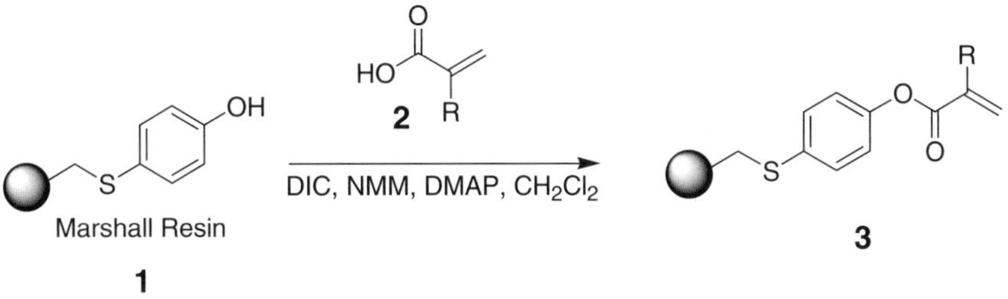

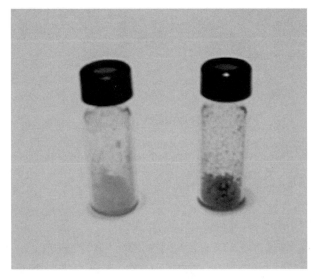

Scheme 3.1. Carbxoylic acid coupling to Marshall resin.

Figure 3.1. Phenol color test. Colorless beads on left indicate the absence of phenol, dark blue color beads on right indicate presence of phenol.

the FTIR results and the negative color test for phenol groups, we could prove that a complete carboxylic acid coupling on Marshall resin had occurred.

The coupling of Fmoc-protected amino acid **5** to Wang resin **4** to form **6** in the presence of coupling catalysts in DMF (Scheme 3.2) was also monitored. A positive test for hydroxy groups would result in the fluorescent resin beads,[8] indicating the presence of hydroxyl groups and an incomplete coupling to the Wang resin. Single-bead FTIR showed the disappearance of IR band for hydroxyl stretch at 3437 cm^{-1} and the appearance of the IR band for the ester carbonyl group at 1710 cm^{-1} (Figure 3.3). A negative test

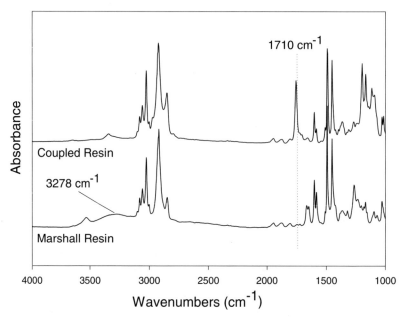

Figure 3.2. Single-bead FTIR spectra of Marshall resin **1** (*bottom*) and resin **3** (*top*). Disappearance of the phenol—OH group (broad stretch at $3278\,\text{cm}^{-1}$) and the appearance of the carbonyl group (strong stretch at $1710\,\text{cm}^{-1}$) indicate the coupling of the carboxylic acid **2** to the Marshall resin.

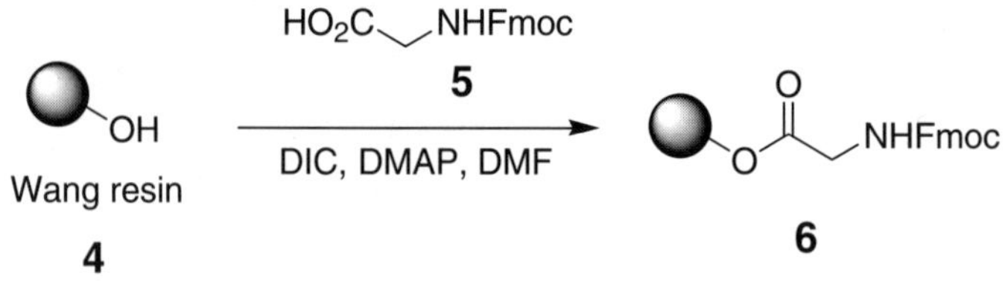

Scheme 3.2. FMOC coupling to Wang resin.

for the presence of hydroxy groups (Figure 3.4, nonfluorescent beads) and the single-bead FTIR results (Figure 3.3) indicated a complete coupling of the protected amino acids to the Wang resin.

3.3. PRECURSOR VALIDATION USING FTIR

Rigorous validation of all building blocks is required for a highly diverse combinatorial library to have good yield and purity. In the course of vali-

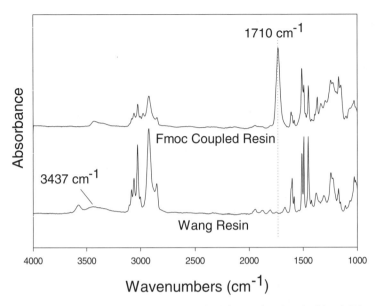

Figure 3.3 Single-bead FTIR spectra of Wang resin **4** (*bottom*) and resin **6** (*top*). Disappearance of the hydroxy group (broad stretch at $3437\,cm^{-1}$) and the appearance of the carbonyl group (strong stretch at $1710\,cm^{-1}$) indicate the coupling of the FMOC carboxylic acid **5** to the Wang resin.

Figure 3.4 Resin-bound alcohol fluorescent test. Nonfluorescent beads on left indicate the absence of hydroxyl group, and fluorescent beads on right indicate presence of hydroxyl group.

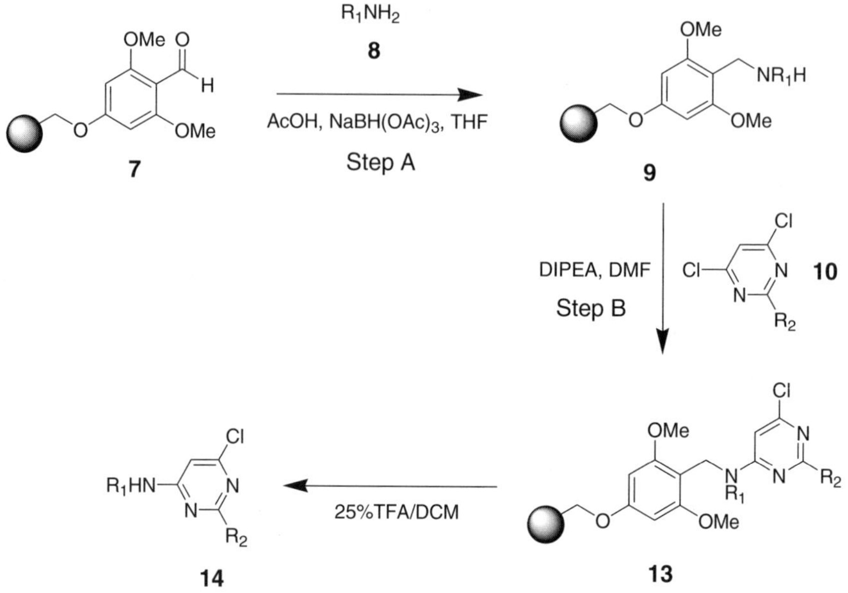

Scheme 3.3. Library 1 scheme with cleavage step.

dation studies this can be accomplished by the combination of color test and FTIR analysis without an extra cleavage step needed for LC/MS analysis. The first test case was to validate the amine precursors for an aryl pyrimidine library (Lib 1) on solid phase (Scheme 3.3).[10] Reductive amination was carried out on 4-formyl-3,5-dimethoxyphenoxymethyl resin **7** (step A), followed by the addition of an amines **8** to the resin (step B) to form **9**. An aldehyde functional group on the bead was transformed to a secondary amine in step A, which would result in the disappearance of IR bands for C=O stretch from aldehyde at $1683\,cm^{-1}$ and C—H stretch at $2775\,cm^{-1}$ (Figure 3.5). Step A progressed in two stages. The intermediate is an imine **11** that is further reduced to the secondary amine **9** (route 1 of Scheme 3.4). Route 2 of Scheme 3.4 illustrates that an alcohol **12** would be formed after reduction treatment if the conversion of **7** to **11** were not complete. To examine if route 2 occurred, a fluorescent test for resin bound hydroxyl[8] was used in addition to single-bead FTIR analysis. The fluorescence dye 9-anthroylnitrile specifically reacts with the hydroxyl group, and not the secondary amines.[8] After the reductive amination with an amine, the disappearance of the IR stretches associated with the aldehyde group and the negative result for the presence of hydroxyl group indicated that the reaction went to completion.

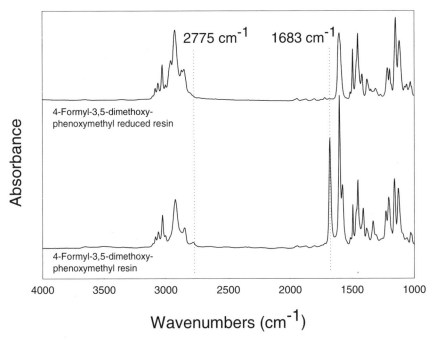

Figure 3.5. Single-bead FTIR spectra of 4-formyl-3,5-dimethoxyphenoxymethyl resin **7** (*bottom*) and reduced resin **9** (*top*). Disappearance of the aldehyde groups (strong stretch at 1679 cm⁻¹ and weak stretch at 2775 cm⁻¹) and the lack of formation of hydroxyl signal indicate the successful reductive amination of resin **7**.

A total of 59 amines (Chart 3.1) were examined by both fluorescence test and single-bead FTIR. A selection of FTIR spectra of resin **9** was shown in Figure 3.6. The IR spectrum of 4-formyl-3,5-dimethoxyphenoxymethyl resin showed an IR band for aldehyde $C=O$ stretch at 1683 cm⁻¹ (Figure 3.6A). The rest of the spectra in Figure 3.6 (from *B* to *H*) showed decreased intensity in this band. For quantitative analysis, an integration analysis of this band was carried out. All integration values were normalized to a polystyrene band at 1947 cm⁻¹. Assuming the integrated IR band for $C=O$ stretch from 4-formyl-3,5-dimethoxyphenoxymethyl resin was 100%, the relative percentage completion of each amine was calculated from their integrated bands at 1683 cm⁻¹. A summary of FTIR results for all 59 amines was listed in Table 3.1. Among the 59 amines evaluated, only amine **33** did not give desired product at all, while amines **34** and **44** showed less than 50% completion, the rest of them showed greater than 95% completion. The broad feature from O—H stretch of alcohol in the range of 3200 to

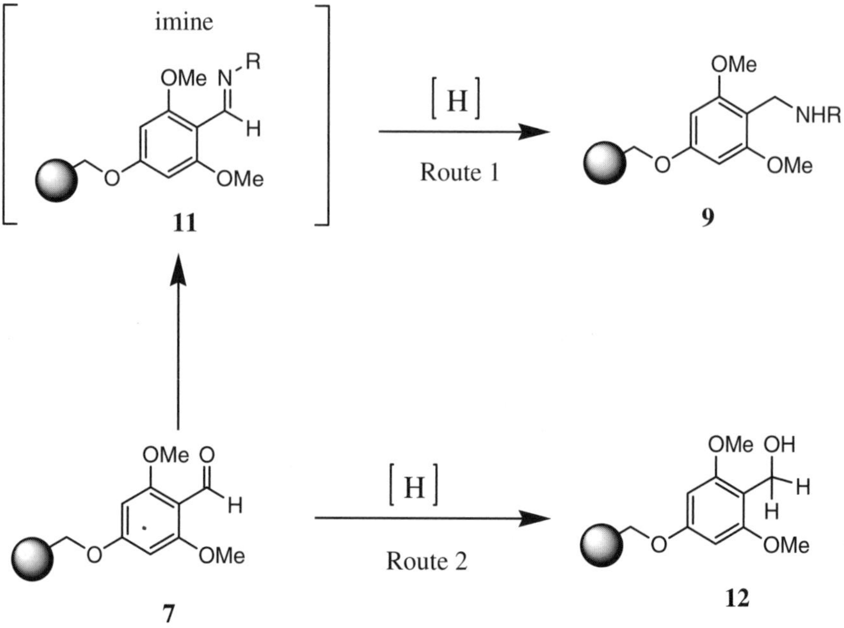

Scheme 3.4. Reductive amination of 4-formyl-3,5-dimethoxyphenoxymethyl resin.

$3600\,\mathrm{cm}^{-1}$ was observed in some of the amines such as H in Figure 3.6 indicating incomplete reaction.

In step B (Scheme 3.3) scaffold **10** was added to the resins after the reductive amination. The products on resin **13** were cleaved from the bead and analyzed by LC/MS (**14**). Experimental results were also summarized in Table 3.1. Twenty-six amines gave products with purity greater than 90% by UV214 detection. Six amines gave products with purity between 50% and 90%. Twenty-seven amines gave products with purity less than 40%. Only 22 out of 26 amines (>90%) were used in the final library production because four amines would produce compounds with a calculated C log P value larger than five (not druglike)[11] and were excluded from synthesis.

3.4. REACTIVITY COMPARISON

To capture chemical reactivity information as a guide for future library development, we compared the building block validations of various libraries. Two approaches were adopted: (1) the same set of building blocks

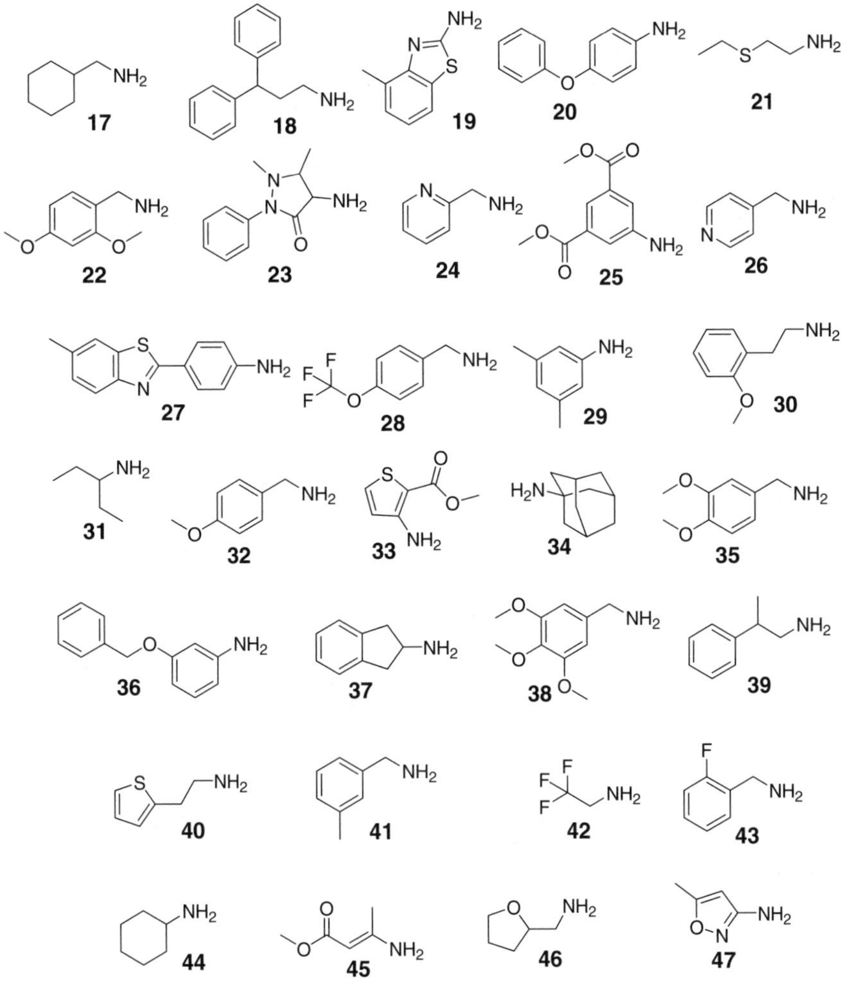

Chart 3.1. Amines used in validation of Library 1 (continued to next page).

for the same reaction on the same starting resin and (2) the same set of building blocks for the same reaction on different starting resins. In subsequent studies a set of amines was also used for the reductive amination reaction in two other library syntheses (Lib 2 is the same as step A in Scheme 3.3 and the reaction for Lib 3 is shown in Scheme 3.5). Results of comparisons were summarized in Table 3.2. Amines proved to work well for reductive amination in Lib 1 also worked well for Lib 2 and Lib 3. This

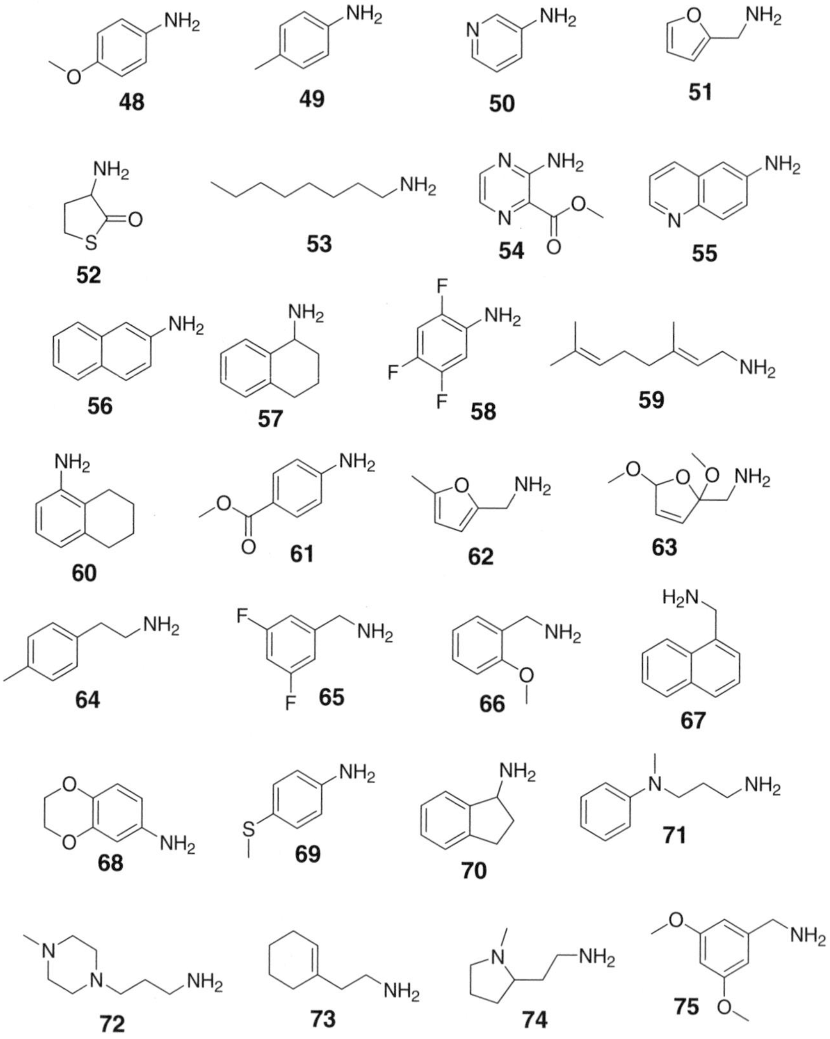

Chart 3.1. *Continued*

demonstrates the value of previous chemistry information in guiding future development of libraries when the same reaction or comparable reaction is performed. With accumulation of more data, such lists could decrease the development time of libraries by qualifying diversity building blocks shown to work previously and mitigating the search for diversity elements that may have already proved successful for other similar reactions.

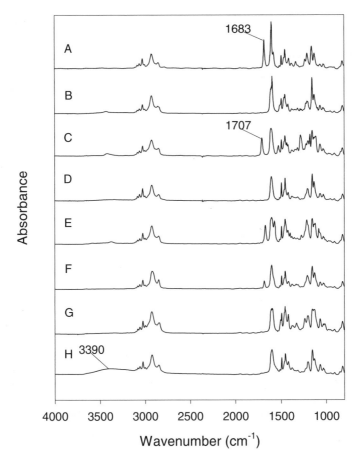

Figure 3.6 Single-bead FTIR spectra of resin **7** (*A*) and reduced resins **9** (*B–H*). The normalized area of the band at 1683 cm^{-1} is used to measure reaction completion. The FTIR stretch at 1707 cm^{-1} for resin *C* (used amine **25**) shows that the resultant resin contains an additional carbonyl group and resin *H* shows an incomplete imine formation from the aldehyde resulting in a hydroxyl formation after reduction (**12** of Scheme 3.4).

3.5. PRODUCT CLEAVAGE REACTIONS

Monitoring complete cleavage of product by single-bead FTIR is a simple yet significant test to enhance reaction yield. The optimization of this step proved to be very important because using too mild a cleavage method may result in only partial release of resin-bound compounds but too severe a condition may result in the breakdown of labile products. Fortunately,

Table 3.1. FTIR and LC/UV$_{214}$/MS Results from the Amine Validation

Diversity Element	-CHO Disappearance 1683 cm^{-1} (area%)	O–H Formation	LC/MS Purity Step B (%UV$_{214}$)	Library 1
17	>95		100	Used
18	>95		93	Not used
19	>95		0	Failed
20	>95		5	Failed
21	>95		78	Used
22	>95		0	Failed
23	>95		2	Failed
24	>95		89	Used
25	>95		0	Failed
26	>95		0	Failed
27	>95		0	Failed
28	>95		100	Used
29	>95	Yes	7	Failed
30	>95		79	Used
31	>95		100	Used
32	>95		100	Used
33	<20	Yes	0	Failed
34	<70		0	Failed
35	>95		100	Used
36	>95		0	Failed
37	>95		91	Used
38	>95		100	Used
39	>95		100	Used
40	>95	Yes	50	Not used
41	>95		97	Used
42	>95	Yes	0	Failed
43	>95		98	Used
44	<70		100	Used
45	>95		40	Not used
46	>95		100	Used
47	>95		0	Failed
48	>95	Yes	46	Not used
49	>95	Yes	76	Not used
50	>95	Yes	0	Failed
51	>95		0	Failed
52	>95		76	Used
53	>95		100	Not used
54	>95		0	Failed
55	>95		0.7	Failed
56	>95		0	Failed
57	>95		76	Not used
58	>95		0	Failed

Table 3.1. *Continued*

Diversity Element	-CHO Disappearance 1683 cm^{-1} (area%)	O–H Formation	LC/MS Purity Step B (%UV$_{214}$)	Library 1
59	>95		68	Not used
60	>95		0	Failed
61	>95		0	Failed
62	>95		63	Not used
63	>95	Yes	0	Failed
64	>95		100	Used
65	>95		100	Used
66	>95		96	Used
67	>95		95	Used
68	>95		23	Not used
69	>95		0	Failed
70	>95		94	Used
71	>95		100	Used
72	>95		0	Failed
73	>95		53	Not used
74	>95	Yes	53	Not used
75	>95	Yes	43	Not used

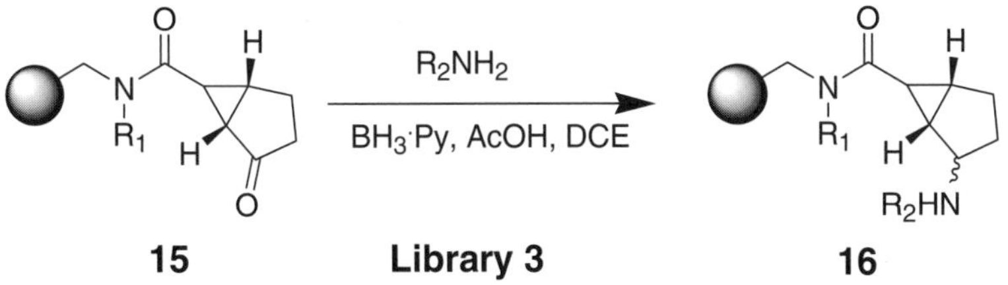

Scheme 3.5. Reductive amination step for Library 3.

different reaction conditions for product cleavage can be monitored by FTIR. An example of this type of study is illustrated by monitoring amine cleavage of product from Marshall linkers. Seven different resin-bound thiophenol esters were cleaved with three structurally diverse amines (Figure 3.7), and their cleavage efficiencies were measured by single-bead FTIR.[12] The resin beads (10.0 mg) were added to a reaction tube in anhydrous pyridine and allowed to swell for 30 minutes. The swelling solvent was filtered off. A solution of 5 equivalents of amine in 1.5 mL of anhydrous pyridine was added to the reaction tube and subsequently placed on a

Table 3.2. List of Amines Used for Reductive Amination in Libraries 1, 2, and 3

Amine	Library-1	Library-2	Library-3
Butylamine	y	y	y
2-Methoxyethylamine	y	y	y
Benzylamine	y	y	y
3-Ethoxypropylamine	y	y	y
Propylamine	y	y	y
Isopropylamine	y	y	y
4-Methoxybenzylamine	y	y	y
3-Methoxypropylamine	y	y	y
3-Isopropoxypropylamine	y	y	y
Cyclopropanemethylamine	y	nt	y
3-butoxypropylamine	y	y	y
Isobutylamine	y	y	y
4-Fluorobenzylamine	y	y	y
2-Fluorobenzylamine	y	y	y
2-Methoxybenzylamine	y	y	y
Cyclopropylamine	y	y	y
Isoamylamine	y	y	y
3-Methylbenzylamine	y	y	y
Cyclobutylamine	nt	y	y
Cyclopentylamine	y	y	nt
2-Aminopentane	y	y	y
3-Fluorobenzylamine	y	y	y
3-Aminopentane	y	y	nt
2-Ethoxyethylamine	nt	y	y
2-Fluorophenethylamine	y	nt	y
3-Fluorophenethylamine	y	nt	y
2,4-Difluorobenzylamine	y	nt	y
Phenethylamine	y	y	nt
2-(2-Aminoethyl)pyridine	y	y	nt

Note: y—validated and used for library synthesis; nt—not tested.

rotator. FTIR spectra were acquired on single beads at various times during the cleavage reaction. Normalized peak areas were measured for the diminishing peak at $1750 \, cm^{-1}$ corresponding to the ester group that is being cleaved off of the resin and the progression of the reaction was monitored at specific time intervals to confirm complete cleavage (Figure 3.8). The study showed that a less sterically hindered amine facilitates a faster cleavage reaction. The effect of temperature was also studied by monitoring the reaction at 20°C, 40°C, and 60°C for one of the resins. The FTIR showed

Resin-bound product

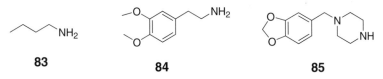

Figure 3.7. Structures of the seven resin-bound thiophenol esters and three cleavage amines.

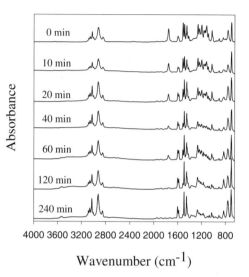

Figure 3.8. Single-bead FTIR spectra of resin **82** at various times during the cleavage reaction with amine **83**.

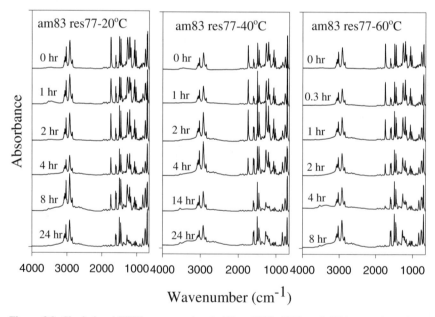

Figure 3.9. Single-bead FTIR spectra of resin **77** at 20°C, 40°C, and 60°C at various times during cleavage with amine **83**.

that the rate of cleavage reaction to increase twofold for every 10°C rise in temperature (Figure 3.9). Similar to this work, the cleavage of acid-labile products with TFA was also studied.[13]

3.6. CONCLUSION

In this chapter several examples were used in the validation of chemistry and precursors. They illustrate the importance of FTIR and color tests in aiding the development of solid-phase chemistry in the loading, diversification, and cleaving steps in combinatorial synthesis. The combined method provides crucial information on reactions carried out on solid supports and offers a fast way to monitor a range of reactions without the need to cleave the compound from resin at intermediate steps. Building block reactivity for similar reactions can also be achieved to guide future synthesis development, as these data may prove powerful for efficient and cost-effective combinatorial library development.

REFERENCES

1. C. M. Baldino, *J. Comb. Chem.* **2**, 89–103 (2000).
2. H. Fenniri, *Combinatorial Chemistry: A Practical Approach*, Oxford University Press, Oxford (2000).
3. P. Seneci, *Solid-Phase Synthesis and Combinatorial Technologies*, Wiley, New York, 2000.
4. B. Yan, *Analytical Methods in Combinatorial Chemistry*, CRC Press, Roca Baton, CL (1999)
5. B. Yan, G. Kumaravel, H. Anjaria, A. Wu, R. Petter, C. F. Jewell Jr., J. R. Wareing, *J. Org. Chem.* **60**, 5736–5738 (1995).
6. B. Yan, *Acct. Chem. Res.* **31**, 621–630 (1998).
7. J. J. Cournoyer, T. Kshirsagar, P. P. Fantauzzi, G. M. Figliozzi, T. Makdessian, B. Yan, *J. Comb. Chem.* **4**, 120–124 (2002).
8. B. Yan, L. Liu, C. A. Astor, Q. Tang, *An. Chem.* **71**, 4564–4571 (1999).
9. C. R. Johnson, B. Zhang, P. Fantauzzi, M. Hocker, K. M. Yager, *Tetrahedron* **54**, 4097 (1998).
10. J. V. Wade, C. A. Krueger, *J. Comb. Chem.* **5**, 267–272 (2003).
11. C. A. Lipinski, F. Lombardo, B. W. Dominy, P. J. Feeney, *Adv. Drug Delivery Rev.*, **23**, 3–25 (1997).
12. L. Fang, M. Demee, T. Sierra, T. Kshirsagar, A. Celebi, B. Yan, *J. Comb. Chem.* **4**, 362–368 (2002).
13. B. Yan, N. Nguyen, L. Liu, G. Holland, B. Raju, *J. Comb. Chem.* **2**, 66–74 (2000).

CHAPTER

4

HR-MAS NMR ANALYSIS OF COMPOUNDS ATTACHED TO POLYMER SUPPORTS

MERITXELL GUINÓ and YOLANDA R. DE MIGUEL

4.1. INTRODUCTION

In the last decade the great interest in the area of solid-phase synthesis sparked the need for development of analytical techniques for on-bead reaction monitoring and characterization of resin-bound molecules. For many years the traditional method of characterization for on-bead molecules consisted in the cleavage of the compound from the resin and its analysis by conventional solution-phase techniques such as NMR, IR, and MS. This cleave-and-characterize technique has the limitations of being destructive, time-consuming and not always applicable. Therefore other methods that do not require cleavage from the resin were used, such as color tests,[1] which are very simple and rapid despite their destructive nature, and microanalysis.[2] Combustion elemental analysis is widely used in the determination of solid-phase organic reaction yields because of its accuracy and reproducibility. Although this method is destructive, it requires only a small amount of resin (about 2 mg) for CHN analysis. It is also possible to analyze Cl, S, P, Br, I, and metals. The analysis can be automated to speed up the process; however, not all laboratories have access to this technique. The procedure actually turns out to be expensive and time-consuming because the samples must be sent to an external service. Another drawback of this method is that to quantify solid-supported reactions, the reaction has to undergo an increase or decrease in the amount of a particular heteroatom during the course of the reaction.

The development of single-bead FT-IR spectroscopy (for details, see Section 4.3) represented a great advancement for the monitoring of on-bead reactions. It has brought sensivity, speed, convenience, and effectiveness to functional group transformations.[3] Just one bead is required to carry out the analysis, without the need for sample preparation.

Analysis and Purification Methods in Combinatorial Chemistry, Edited by Bing Yan.
ISBN 0-471-26929-8   Copyright © 2004 by John Wiley & Sons, Inc.

Nevertheless, the primary tool used in solution-phase synthesis for structural elucidation is NMR spectroscopy. There has been some interest in developing NMR methods for analyzing resin-bound compounds and monitoring the progress of chemical reactions on solid-phase by on-bead analysis (without cleavage of the product from the resin).[4] Many improvements have been achieved in NMR for SPS, but limitations remain. On the one hand, the limited mobility of polymers, as well as the poor mobility of attached compounds, leads to broad bands with low spectral resolution. On the other hand, the broad signals due to the polymer matrix can mask or overlay with bands from the desired product. The main advantage of this technique is its nondestructive nature, as the sample can be easily recovered. The next section will focus on the most powerful technique developed to monitor and characterize polymer-supported compounds: gel-phase HR-MAS NMR spectroscopy.

4.2. BACKGROUND TO GEL-PHASE HR-MAS NMR SPECTROSCOPY

4.2.1. Gel-Phase NMR Spectroscopy

The first approach to monitor SPOS by gel-phase NMR was based on the swelling of polystyrene (PS) resins (cross-linked with divinylbenzene). This method uses an appropriate deuterated solvent to increase the mobility of the sample and obtain narrower peaks. The resulting swollen resin is neither fully liquid nor fully solid but gel-like. The linewidth is a function of the amount of divinylbenzene (DVB), which defines the level of polystyrene cross-linking. As a result resins with low DVB content (1–2%), which means low cross-linking, and resins with long mobile chains, such as polyethyleneglycol (PEG), give sharper resonances as the high mobility of the product simulates better the solutionlike state. For this reason PEG-containing resins like Tentagel or Argogel are very good for gel-phase NMR. An advantage of this technique is that the NMR analysis of samples is carried out with a conventional NMR spectrometer. Thus the resin is swollen as a heterogeneous slurry in a standard NMR tube. A large amount of beads is required (ca. 100 mg) because of the relatively low concentration of the analyte molecule, which leads to poor sensivity. However, the technique is not destructive, as the sample can be easily recovered.

The heterogeneity of the gel sample leads to differences in magnetic susceptibility throughout, resulting in signal line broadening. This generally limits gel-phase NMR to certain heteronuclei, which have larger chemical shift dispersions, such as ^{13}C NMR[5], ^{19}F NMR[6], and ^{31}P NMR[7]. ^{1}H gel-phase

NMR spectra are usually too broad to be used for purposes of structure elucidation.

4.2.2.   Gel-Phase Magic Angle Spinning NMR (MAS-NMR) Spectroscopy

NMR spectra obtained of resin-bound compounds are comparable to those obtained for soluble compounds. Fréchet et al.[8] first reported a method that reduces spectral peak width in the characterization of solids. This technique consists in spinning the sample at the magic angle of 54° 44′ relative to the static magnetic field, which removes substantially the line broadening (Figure 4.1). Line broadening is due to interactions between the nuclear dipoles of the molecules and the magnetic field direction, as well as variations in magnetic field susceptibility throughout the sample. The dipole-dipole coupling, which depends on the angle between the dipolar pair and the static field direction in the form of $(3 \cos^2 \theta - 1)$, and the broadening due to magnetic susceptibility, can be removed by MAS. Spinning the sample at the magic angle averages the dipolar couplings to zero (i.e., $3 \cos^2 \theta - 1 = 0$).

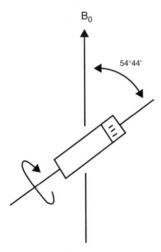

Figure 4.1

4.3. HR-MAS NMR EXPERIMENTS

The most powerful development in this area was the introduction of a high-resolution MAS (HR-MAS) NMR probe or Varian nano-probe.[9] This technique became a great tool for the characterization of resin-bound products, as their ^{1}H NMR spectra were comparable with standard solution spectra. However, the equipment required to carry out these HR-MAS ^{1}H NMR spectra, which is the special HR-MAS NMR probe, is expensive. Thus it is not available in all laboratories.

Keifer et al.[10] reported a comparison between the NMR spectra of N-Fmoc-L-aspartic-β-(t-butyl)ester Wang resin using conventional high-resolution MAS and HR-MAS probes. The conventional 5-mm liquid probe produces the poorest resolution spectrum with really broad signals. An improvement already is shown when using the 5-mm CP/MAS probe. In this case the MAS removes the susceptibility broadening, providing some line narrowing. However, the highest resolution spectrum is obtained using the ^{1}H HR-MAS probe where MAS and susceptibility matching are combined.

4.3.1. Sample Preparation

The experiments performed in the HR-MAS probe require special sample preparation as the resin has to be placed in 4-mm rotors instead of the normal NMR tube used for standard solution compounds. A few milligrams of resin are transferred into the 4-mm HR-MAS NMR rotor with a spatula. Some µL (≤40µL) of deuterated solvent is added to the rotor until the beads are completely swollen in the solvent, making a gel-like solution. (Some resin or some more solvent can be added to the rotor until the gel stage is acquired.) Finally the rotors are closed manually by using standard caps.

4.3.2. Types of Resin

To obtain high-quality HR-MAS ^{1}H NMR spectra, polymers are required to have low levels of cross-linking (1–2% polystyrene-divinylbenzene, PS-DVB) and good swelling properties. Thus traditional polystyrene resins with low cross-linking are suitable for HR-MAS ^{1}H NMR, although their swelling properties are not as good as PEG-polystyrene resins such as ArgoGel or TentaGel. The presence of a long polyethylene glycol chain between the rigid polymer backbone and the attached molecule increases the mobility of the mentioned molecule, thus acquiring a higher liquid-like mobility that is reflected in the resolution of the spectrum. The high mobility of the long and flexible tethered PEG chains results in high-quality and narrow ^{1}H NMR linewidths. The use of solvents where both the matrix and the resin-bound compound have a good swelling and solubility further improves the quality of the spectra. However, only the deuterated solvents that swell the beads, such as $CDCl_3$, can be used, for solvents like D_2O or MeOD are not suitable.

4.3.3. Types of Experiments

A problem occurs when the spectrum of a polymer-supported molecule as compared with the analogue compound in solution, shows broad signals. The broadness is due to the polystyrene backbone, and even the PEG signals (for resins that contain this chain), masking or complicating the spectral assignments. Techniques have been developed to decrease these peaks and thus improve the quality of the spectra. One is to pre-saturate at the frequencies of the polyethylene glycol (PEG) group and suppress the PEG signal with minimum distortion of the spectrum. Another is to apply the CPMG (Carr-Purcell-Meiboom-Gill) pulse sequence to the sample to remove the broad bands arising from the polystyrene backbone and reveal the hidden peaks in the sample. Also a combination of CPMG and pre-saturation can decrease the intensity of the PEG signal and the polystyrene backbone at the same time, thus obtaining a high-quality spectrum comparable to a solution-like spectrum.

Several NMR experiments have been carried out by our group on one sample supported on Argogel resin as shown in Figure 4.2. First, the gel-phase NMR spectrum was obtained on a 5-mm solution QNP probe (spectrum A). The spectrum provide essential structural information, although there was an extremely broad signal centered at 3.7 ppm arising from PEG. A remarkable improvement occurs with HR-MAS NMR (spectrum B), as sharp resonances appear for the bound-molecules. The CPMG pulse sequence was applied to the sample to eliminate the broad signals due to the polystyrene backbone (spectrum C). Pre-saturation to eliminate the PEG signal combined with the CPMG further improved the spectrum and gave a very high-quality spectrum (spectrum D).

In order to interpret the spectra and make all the assignments and characterizations of complex structures, 2D correlation spectra have been developed. The ^{1}H NMR spectra of the swollen beads have a linewidth of the order of 6 to 10 Hz and the direct measurement of the coupling constants is not possible.

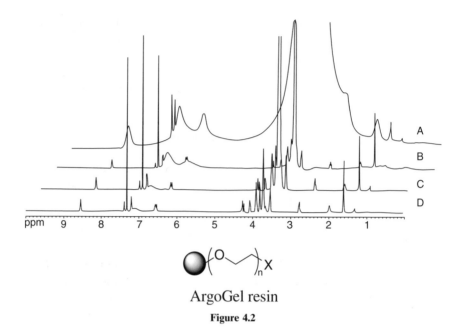

ArgoGel resin

Figure 4.2

4.4. REACTION MONITORING AND CHARACTERIZATION OF POLYMER-SUPPORTED COMPOUNDS

Advancements in solid-phase reactions have been facilitated by the introduction of HR-MAS ^{1}H NMR. The technique provides a powerful way to monitor and characterize compounds attached to resins, with a resolution comparable to solution-like spectra. An example of similar quality of ^{1}H NMR spectra obtained for a supported molecule by gel-phase HR-MAS NMR and for the same unsupported molecule in solution has been carried out in our group (Figure 4.3). This molecule is the methyl ester in solution phase, while it is the Argogel ester on solid phase.

Early use of HR-MAS NMR as a tool for monitoring the progress of a reaction on solid phase was reported by Wehler et al.[11] Spectra taken after 0.5, 1, and 1.5 hour showed peaks from the aromatic region of the Fmoc-protecting group during the intramolecular transformation from **1** to **2** in the NMR rotor (Scheme 4.1). The broad peaks due to compound **1** being substituted in the resin are less intense relative to the narrow peaks from compound **2**, which is free in solution.

Sarkar et al.[12] reported the combination of ^{1}H HR-MAS NMR technique with a spin echo sequence to follow the completion of the reduction of a

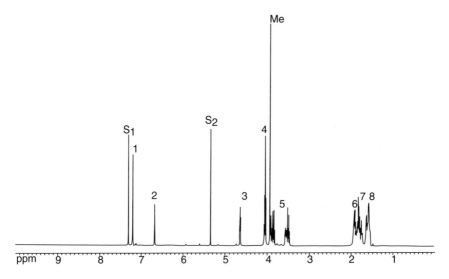

5 mm solution state spectrum
R = Me

Figure 4.3

resin-bound methyl benzoate **3** to the corresponding benzyl alcohol **4** (Scheme 4.2). The reaction made it easy to follow the disappearance of a peak at the chemical shift of 3.9 ppm (for the methyl ester) and the appearance of a peak at 4.7 ppm for the CH_2OH.

The HR-MAS ^1H NMR spectra of the reaction sequence shown in Scheme 4.3 can be used to monitor each reaction step (using a single macro bead) because of the high mobility of the long PEG spacer of the resin.[13]

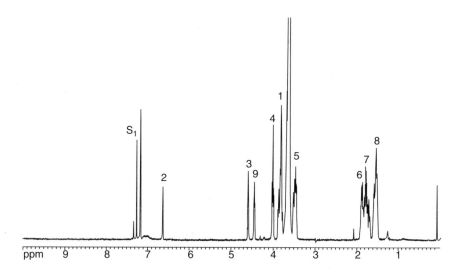

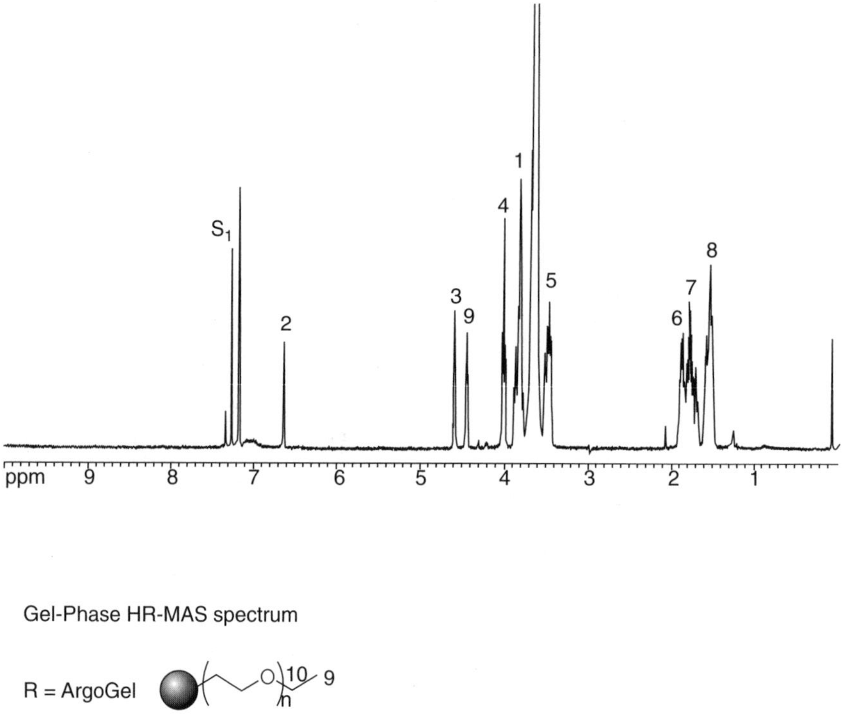

Gel-Phase HR-MAS spectrum

R = ArgoGel

Figure 4.3a

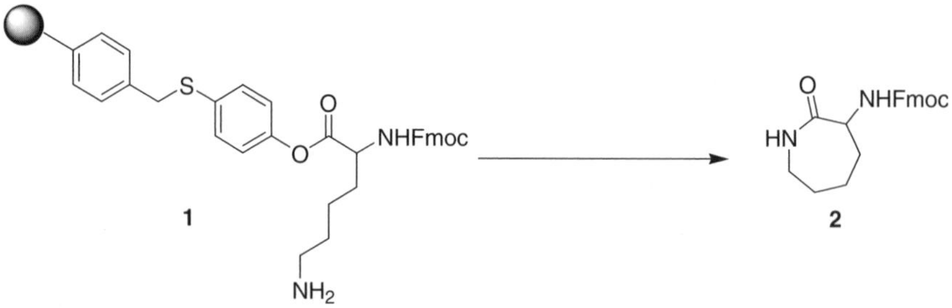

Scheme 4.1

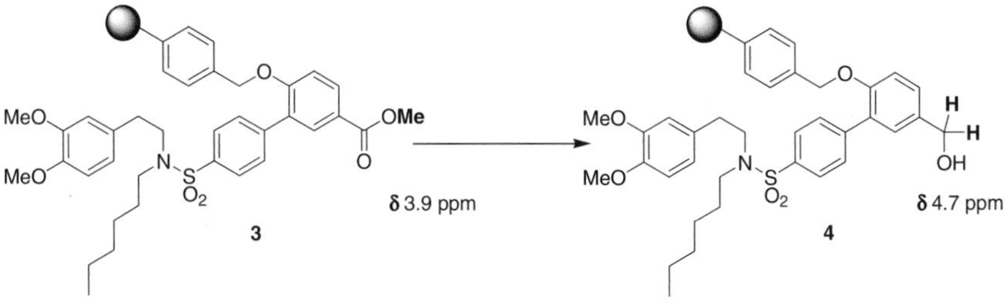

Scheme 4.2

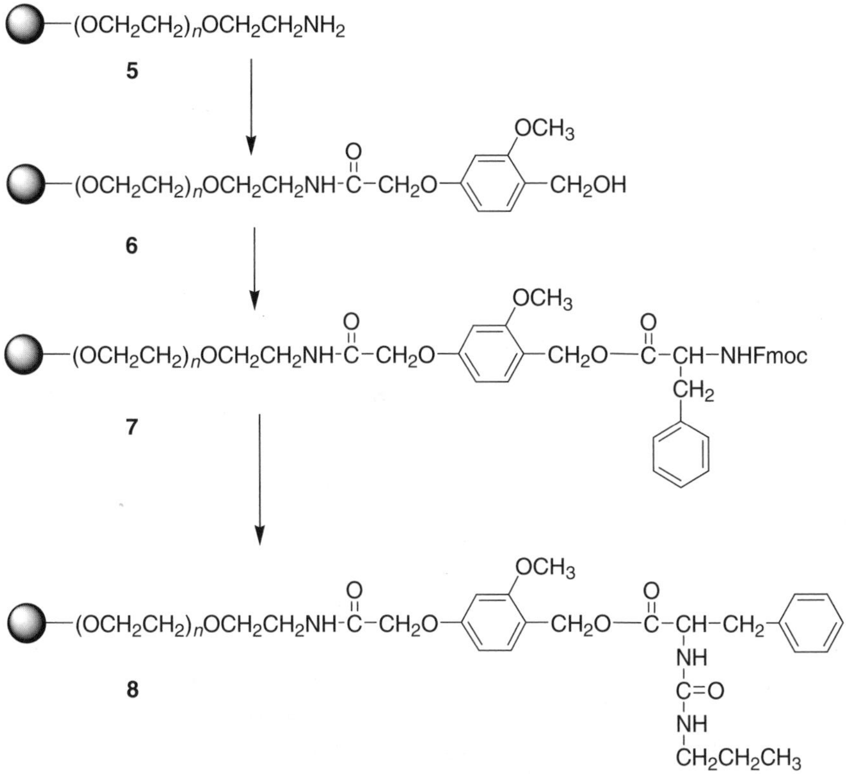

Scheme 4.3

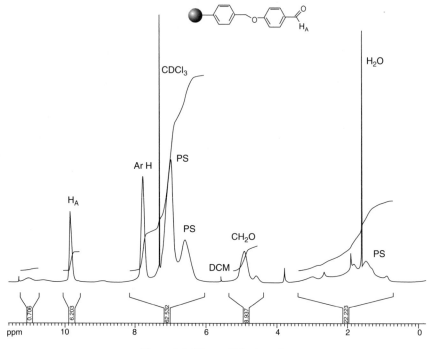

Figure 4.4. Wang aldehyde resin.

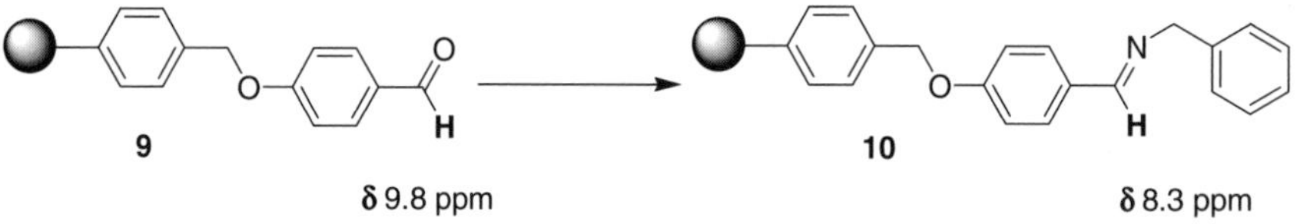

Scheme 4.4

The main signal of the spectrum of the OCH$_2$CH$_2$ units appears at 3.5 ppm. With every subsequent synthetic step, new characteristic peaks are observed. The binding of a linker to resin **5** to give the polymer-supported compound **6** is indicated by additional peaks in the aromatic region (δ = 6–7.5 ppm) and signals from the CH$_2$O (δ = 4.8 ppm) and the methoxy group (δ = 3.75 ppm). The next step involves the binding of Fmoc-phenylalanine to give **7** as is shown by proton signals in the aromatic region; this is due to

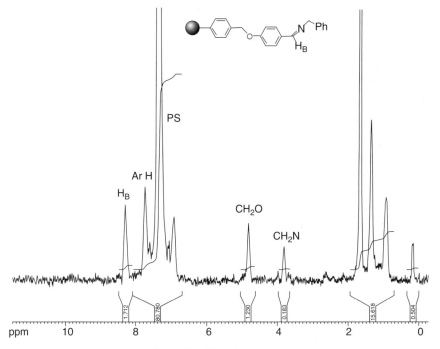

Figure 4.4a. Wang imine resin.

the Fmoc unit. The last step involves cleavage of the Fmoc protecting group and leads to a decrease of the intensity of the peaks in the aromatic region; new signals appear in the aliphatic region. Although the 1D experiment for the direct monitoring of the syntheses on the resin is successful, the assignment of spectra signals of complex resin-bound compounds such as **8**, two-dimensional NMR techniques are crucial.

Research was carried out by our group[14] on the use and regeneration of the Wang aldehyde scavenger **9** to eliminate excess primary amines in libraries of secondary amines prepared by reductive amination. The reaction in this process can be easily followed by gel-phase HR-MAS ^{1}H NMR. The signal of the aldehyde proton disappears at 9.8 ppm; a new peak for the imine proton appears at 8.3 ppm in the spectrum as shown in Figure 4.4 and Scheme 4.4.

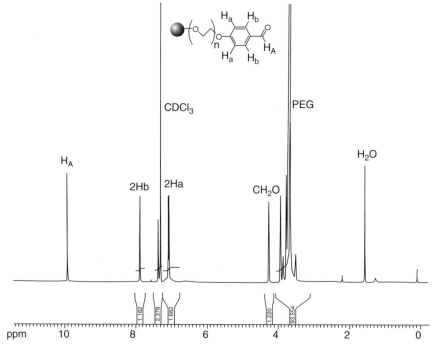

Figure 4.5. AG-CHO.

The spectra shown in Figure 4.4 for the Wang aldehyde resin **9** and the Wang imine resin **10** can be compared to the spectra of the analogous aldehyde and imine attached to the ArgoGel resin instead of the Wang resin. The reaction between ArgoGel aldehyde **11** (with a peak at 9.8 ppm) and benzylamine gives an ArgoGel imine **12** (with a peak at 8.3 ppm) as shown in Scheme 4.5 and Figure 4.5. In this case the peaks are sharper than before. This is because the long PEG chain of the ArgoGel resin gives better mobility to the attached molecule and results in a higher quality spectrum. As shown in the spectra, the reaction is not quantitative because the signal at 9.8 ppm has not completely disappeared.

As with resin beads, gel-phase HR-MAS NMR spectroscopy can be used to analyze systems where the resin is grafted onto the base polymer units, known as "crowns." The first example of HR-MAS NMR on crowns was demonstrated by Shapiro et al.[15] The spectrum of a crown with the aldehyde **13** attached to the graft polymer has the characteristic peak at 10.2 ppm (Scheme 4.6). A Wittig reaction was performed on this substrate, and its

11 δ 9.8 ppm   **12**   δ 8.3 ppm

Scheme 4.5

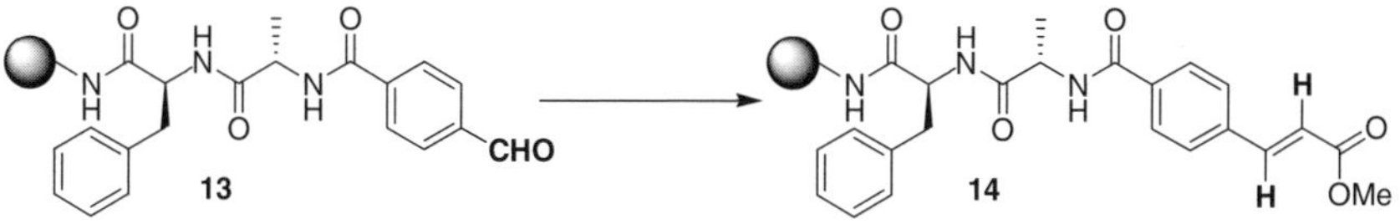

13 δ 10.2 ppm   **14**   δ 6.8, 7.7 ppm

Scheme 4.6

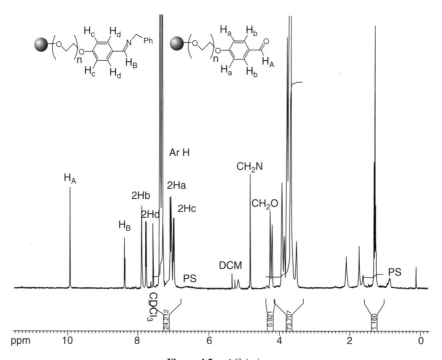

Figure 4.5a. AG-imine.

success was confirmed by the disappearance of the aldehyde peak and the appearance of two new peaks due to the olefin **14** at 6.8 and 7.7 ppm.

Other examples concerned the use of macroscopic systems such as Synphase lanterns. Lippens et al.[16] compared the results they obtained using Wittig-Horner condensation on a macroscopic support with earlier results[17] of the same reaction on a standard resin. As in all of the examples shown above, gel-phase HR-MAS NMR spectroscopy proved to characterize better the supported intermediates[18] and final products and provide better understanding of the interactions between the polymer and the reagents. This is an important development in supported organic chemistry.

4.5. CONCLUSION

In complex reactions on resins, one-dimensional gel-phase HR-MAS NMR spectroscopy does not always afford conclusive information to be obtained on the characterization of all the intermediates and final products. Two-dimensional HR-MAS NMR spectroscopy (COSY, TOCSY, HETCOR, and SECSY) provides crucial information on proton connectivity through J coupling information.[19,20] With 2D HR-MAS NMR complete assignments can be made of spectra of complex bound-molecules, such as oligosaccharides[21] and peptides,[22] and even of noncovalent interactions between compounds connected to resins and receptor molecules.[23] In sum, HR-MAS NMR spectroscopy has been successfully used for characterizing complex resin-bound products and intermediates. It has been used for analyzing biochemical compounds as well.[24] Future work will adapt this technique to reaction monitoring and to routine quantitation besides its use as a characterization tool.

REFERENCES

1. C. Kay, O. E. Lorthioir, N. J. Parr, M. Congreve, S. C. McKeown, J. J. Scicinski, S. V. Ley, *Biotechnol. Bioeng.* **71** (2) 110–118 (2000).
2. B. Yan, C. F. Jewell, S. W. Myers, *Tetrahedron* **54**, 11755–11766 (1998).
3. For a review of this technique: Y. R. de Miguel, A. S. Shearer, *Biotechnol. Bioeng.* **71**, 119–129 (2001).
4. (a) M. J. Shapiro, J. S. Gounarides, *Biotechnol. Bioeng.* **71**, 130–148 (2001); (b) P. A. Keifer, *Drug Discov. Today* **2** (11), 468–478 (1997); (c) M. J. Shapiro, J. R. Wareing, *Curr. Opin. Chem. Biol.* **2** (3), 372–375 (1998); (d) M. J. Shapiro, J. S. Gounarides, *Prog. Nucl. Magn. Reson. Spectrosc.* **35** (2), 153–200 (1999).

5. (a) H. Sternlicht, G. L. Kenyon, E. L. Packer, J. Sinclair, *J. Am. Chem. Soc.* **93** 199–208 (1971); (b) G. C. Look, C. P. Colmes, J. P. Chinn, M. A. Gallop, *J. Org. Chem.* **59** (25), 7588–7590 (1994).

6. (a) A. Svensson, T. Fex, J. Kilhberg, *Tetrahedron Lett.* **37**, 7649–7652 (1996); (b) M. J. Shapiro, G. Kumaravel, R. C. Setter, R. Beveridge, *Tetrahedron Lett.* **37**, 4671–4674 (1996).

7. (a) R. Quarrel, T. D. W. Claridge, G. W. Weaver, G. Lowe, *Mol. Divers.* **1**, 223–232 (1996); (b) C. M. G. Judkins, K. A. Knights, B. F. G. Johnson, Y. R. de Miguel, R. Raja, J. M. Thomas, *Chem. Commun.* 2624–2625 (2001).

8. H. D. Stöver, J. M. Fréchet, *Macromolecules* **22**, 1574–1576 (1989).

9. W. L. Fitch, G. Detre, C. P. Holmes, J. N. Shoolery, P. A. Keifer, *J. Org. Chem.* **59**, 7955–7956 (1994).

10. P. A. Keifer, L. Baltusis, D. M. Rice, A. A. Tymiak, J. N. Shoolery, *J. Magn. Reson.*, Ser. A, **119**, 65–75 (1996).

11. T. Wehler, J. Westman, *Tetrahedron Lett.* **37** (27), 4771–4774 (1996).

12. R. S. Garigipati, B. Adams, J. L. Adams, S. K. Sarkar, *J. Org. Chem.* **61**, 2911–2914 (1996).

13. M. Pursch, G. Schlotterbeck, L. Tseng, K. Albert, W. Rapp, *Angew. Chem. Int. Ed. Engl.* **35**, 2867–2869 (1996).

14. M. Guinó, E. Brulé, Y. R. de Miguel, *J. Comb. Chem.*, **5**, 161–165 (2003).

15. J. Chin, B. Fell, M. J. Shapiro, J. Tomesch, J. R. Wareing, A. M. Bray, *J. Org. Chem.* **62**, 538–539 (1997).

16. P. Rousselot-Pailley, N. J. Ede, G. Lippens, *J. Comb. Chem.* **3**, 559–563 (2001).

17. R. Warrass, G. Lippens, *J. Org. Chem.* **65**, 2946–2950 (2000).

18. J. Fruchart, G. Lippens, C. Kuhn, H. Gras-Masse, O. Melnyk, *J. Org. Chem.* **67**, 526–532 (2002).

19. (a) G. Lippens, M. Bourdonneau, C. Dhalluin, R. Warrass, T. Richert, C. Seetharaman, C. Boutillon, M. Piotto, *Curr. Org. Chem.* **3**, 147–169 (1999); (b) M. J. Shapiro, J. Chin, R. E. Marti, M. A. Jarosinski, *Tetrahedron Lett.* **38** (8) 1333–1336 (1997); (c) J. Chin, B. Fell, S. Pochapsky, M. J. Shapiro, J. R. Wareing, *J. Org. Chem.* **63**, 1309–1311 (1998).

20. A. Bianco, J. Ferrer, D. Limal, G. Guichard, K. Elbayed, J. Raya, M. Piotto, J. Briand, *J. Comb. Chem.* **2**, 681–690 (2000).

21. P. H. Seeberger, X. Beebe, G. D. Sukenick, S. Pochapsky, S. J. Danishefsky, *Angew. Chem. Int. Ed. Engl.* **36** (5), 491–493 (1997).

22. (a) C. Dhalluin, C. Boutillon, A. Tartar, G. Lippens, *J. Am. Chem. Soc.* **119**, 19494–10500 (1997); (b) R. Warrass, J.-M. Wieruszeski, C. Boutillon, G. Lippens, *J. Am. Chem. Soc.* **122**, 1789–1795 (2000); (c) K. Elbayed, M. Bourdonneau, J. Ferrer, T. Richert, J. Raya, J. Hirschinger, M. Piotto, *J. Magn. Reson.* **136**, 127–129 (1999); (d) J. Ferrer, M. Piotto, M. Bourdenneau, D. Limal, G. Guichard, K. Elbayed, J. Raya, J. Briand, A. Bianco, *J. Am. Chem. Soc.* **123**, 4130–4138 (2001); (e) C. H. Gotfrdsen, M. Grøtli, M. Willert, M. Meldal, J. Ø. Duus, *J. Chem. Soc. Perkin Trans.* **1**, 1167–1171 (2000).

23. (a) Y. R. de Miguel, N. Bampos, K. M. Nalin de Silva, S. A. Richards, J. K. M. Sanders, *Chem. Commun.*, 2267–2268 (1998); (b) Y. Ng, J. Meillon, T. Ryan, A. P. Dominey, A. P. Davis, J. K. M. Sanders, *Angew. Chem. Int. Ed.* **40** (9), 1757–1760 (2001).

24. P. A. Keifer, *Curr. Opin. Biotech.* **10**, 34–41 (1999).

CHAPTER

5

MULTIVARIATE TOOLS FOR REAL-TIME MONITORING AND OPTIMIZATION OF COMBINATORIAL MATERIALS AND PROCESS CONDITIONS

RADISLAV A. POTYRAILO, RONALD J. WROCZYNSKI,
JOHN P. LEMMON, WILLIAM P. FLANAGAN, and
OLTEA P. SICLOVAN

5.1. INTRODUCTION

Combinatorial and high-throughput methods are opening up the exploration of multidimensional chemical composition and process parameter space at a previously unavailable level of detail. This is achieved by combining the chemical intuition and courage with recently available means for reliable, rapid, and automated synthesis and evaluation of materials libraries. At present, combinatorial and high-throughput methods have been applied not only to discovery of materials in chemistry and materials science[1-3] but also for optimization of materials compositions. Recent examples include optimization of homogeneous and heterogeneous catalysts,[4,5] multicomponent inorganic films,[6] nanoscale materials,[7,8] phosphors,[9] and others.

High-throughput approaches can provide important time savings in optimization of process parameters of materials fabrication similar to optimization of pharmaceutically relevant reactions[10] through parallel reactions and automation.[11,12] However, until now, high-throughput optimization of process parameters has not been explored in much detail. Limited studies of variation of process conditions include annealing temperature and atmosphere content in processing of tricolor phosphors,[9] polymer blend phase behavior at different temperatures,[13] temperature-modulated dewetting effects,[14] and effects of different process conditions on end-use performance of combinatorial arrays of organic coatings.[15] Selection of optimum process conditions in combinatorial microreactors is essential if the combi-

Analysis and Purification Methods in Combinatorial Chemistry, Edited by Bing Yan.
ISBN 0-471-26929-8   Copyright © 2004 by John Wiley & Sons, Inc.

natorial synthesis process is to be correlated with the synthesis process on a more conventional scale and the materials are to have desired chemical properties. The reaction optimization process is a trade-off between the desire for best performance and least experimental investment.[12] Thus it is highly desirable to increase the optimization throughput by performing parallel reactions at different process conditions and by rapid nondestructive measurement of chemical properties of materials.

This chapter analyzes the needs for reaction monitoring and optimization in combinatorial materials science. It is demonstrated that these needs can be achieved by the combination of the advanced analytical instrumentation, the proper design of the measurement setups, and analysis of the multivariate data collected during these experiments.

5.2. REACTION MONITORING AND OPTIMIZATION NEEDS

A significant part of current research efforts in the area of combinatorial materials science is dedicated to the development of methods for in situ monitoring of combinatorial reactions in both discovery and optimization phases. Measurement tools that would enable better characterization and would support rapid analysis are highly desired. As indicated in Vision 2020,[16] an integration of analytical equipment with the combinatorial reactors is one of the top priorities for characterization of combinatorial libraries. As the first step, new analytical tools are urgently needed for in situ measurements within individual combinatorial reactor cells. When developed, such integrated systems would enable more rapid, accurate characterization of combinatorial samples. Ultimately a top priority goal is to achieve integration of the system from sample preparation to end-use. This will require longer term research to develop a work station approach that will integrate reactor sample preparation, reaction analysis and characterization, management of the data produced (storage, interpretation, and informatics), and end-use across industries that apply combinatorial methods.[16]

In situ monitoring of combinatorial reaction provides several attractive options for high-throughput screening. Real-time observation of the reaction progress in combinatorial reactors can tremendously speed up the materials discovery process by providing previously unavailable information about the starting reaction components evolving into the reaction and the dynamics of progress of multiple reactions at once at each reaction phase. Monitoring of reaction components can provide valuable feedback information to control and rapidly optimize reaction parameters. Overall, in situ monitoring of combinatorial reactions includes all of the attractive features of in-line detection methods, such as automation, no sample removal, and/or preparation steps, and thus reduced number of contamination sources. It has been statistically demonstrated that in situ measurement systems, in principle, are capable of making quality determinations to a substantially higher order of precision than the traditional off-line laboratory systems.[17–19]

In situ monitoring coupled with the adequate measurement equipment can be further adapted for applications and reaction scales beyond combinatorial and high-throughput screening.[20,21] For example, in situ measurements can be done on a 10^{-12} to 10^{-6} L scale for combinatorial materials science screening, whereas a reaction volume as large as 10^4 to 10^5 L should be monitored at a manufacturing plant. While the diversity of goals for these in situ measurements is governed by the reaction scale, spectroscopic monitoring techniques can remain the same. Thus a measurement technique initially developed for combinatorial and high-throughput screening can be of use for process monitoring where measurements are performed in the pilot and manufacturing plants.

Optical, chromatographic, electrochemical, and mass-spectrometric techniques are evolving as the most widely used analytical methodologies for direct in situ monitoring and optimization of combinatorial reactions.[4,21–24] A representative list of applications of analytical techniques recently reported for the in situ monitoring of combinatorial reactions and processes is compiled in Table 5.1.[4,25–45]

Optical detection methods have a suite of attractive features for many in situ monitoring and reaction optimization applications.[29,46] These features include nondestructive measurement capabilities, fast response, low detection limit, means to analyze materials with no sample preparation, ease of multiplexing for parallel measurements of multiple compounds in the library, and spatial-mapping capabilities.[47,48] One or several lightwave parameters (amplitude, wavelength, polarization, phase, and temporal property) can be measured and related to a material property of interest. Complementing "direct" measurements, in which the spectroscopic features of the material are measured, an "indirect" material determination can be used in which an auxiliary reagent is employed.[49]

In addition to the analytical tools adapted from the laboratory use for the high-throughput screening applications, sensors capable of combinatorial screening are being developed for the measurement of a diverse range of chemical and physical parameters.[49–54] Examples of sensor applications for monitoring of combinatorial reactions are presented in Table 5.1. For combinatorial screening, sensors have a wide variety of attractive capabilities. Indeed, already miniaturized sensors are used to detect quantitatively small amounts of materials. Sensors are also often multiplexed into arrays

Table 5.1. Selected Examples of Real-Time Monitoring of Combinatorial Reactions: Instrumentation and New Knowledge

Materials System	Instrumentation	Knowledge	Reference
Solid-phase synthesis of trisubstituted amines	Five-step reaction sequence on resin is followed the reaction step-by-step using the sequence of five analytical tools as a function of experiment time: single-bead IR, ^{1}H MAS NMR, 2D MAS COSY, MAS HMQC, ^{13}C MAS NMR	Generation of a protocol of analytical tools that allows a chemist to decisively evaluate synthetic steps, verify new building blocks, and detect possible side reactions prior to or during actual library construction	25
Metal alloys catalyst candidates	Serial MS screening of a $15 \times 15 \times 15$ libraries with 120 different compositions	Kinetics of catalytic reactions	26
96-Capillary array for palladium-catalyzed annulation	Nonaqueous multiplexed capillary electrophoresis	Rapid determination of catalytic activity, selectivity, and kinetics of the various combinations	4
Solid-phase organic synthesis products	Single bead FTIR	Reaction kinetics, conversion yield	27
Resin-bead supported combinatorial libraries	Hyperspectral IR imaging for monitoring of catalytic reactions with the screening time independent on the number of elements in the library	Kinetics of catalytic reactions	28
Solid-phase peptide synthesis products	Near-IR multispectral imaging based on scanning acousto-optic tunable filter	Simultaneous determination of kinetics of multiple reactions	29

Catalytic dehydrogenation of cyclohexane to benzene	Resonance-enhanced multiphoton ionization for low parts per billion and high parts per trillion detection capability	Determination of activities of catalyst sites by monitoring of a single-reaction product; possibility for monitoring of multiple reaction products to determine catalyst selectivity	30
Amorphous microporous mixed oxide catalysts	IR thermography for gas-phase screening of catalyst candidates (sample size 200 µg)	Kinetics of catalytic reactions	31
Bead-bound catalysts	IR thermography for solution-phase screening of 3150 potential catalysts bound to 300 to 500µm diameter polymer beads	Kinetics of catalytic reactions	32
Styrene-polymerization catalysis	IR thermographic imaging of organometallic catalysts	Reaction kinetics from the time dependence of the heat generation	33
Catalytic activity of V_2O_5 in oxidation of naphthalene to naphthoquinone by O_2	Fluorescence and thermographic imaging for monitoring of catalytic reactions	Determination of nonspecific temperature increase by thermography and species-specific concentration maps by fluorescence	34
48 element array of epoxy formulations	Automated large sample array differential scanning calorimeter for process optimization for screening studies of multiple variable arrays	Cure kinetics	35

Table 5.1. *Continued*

Materials System	Instrumentation	Knowledge	Reference
Discrete array of inorganic oxide films	Pulsed laser deposition with in situ monitoring of growth surface with reflection high-energy electron diffraction (RHEED)	One-lot optimization of epitaxial growth process by using a carrousel type masking plate; variable growth conditions include pressure, temperature, laser energy, and laser repetition rate	36
Electrochemical catalysts	Fluorescence imaging of catalysts for oxidation of methanol using a pH indicator in discovery and focus libraries	Kinetics of catalytic reactions	37
Catalytic hydrogen-producing materials	Near-IR reflection sensor array for 2D mapping of H_2 from catalytic hydrogen-producing materials	Kinetics of catalytic reactions	38
Organic coating formulations	Optimization of processing conditions (curing parameters) in fabrication of UV-cured automotive organic protective coatings; fluorescence of a viscosity-sensitive molecular probe monitored during curing of coatings	Rapid decoupling of temperature and radiation effects in curing of UV curable coating formulations by using multiple coatings and process conditions at once	39
Chemical and biochemical catalysts	96-Thermistor array for detection of temperature changes with high resolution	Correlation of catalysts concentration and time-dependent recorded maximum temperature	40

Libraries of polymer/pigment compositions	Fluorescence spectroscopy and imaging for the evaluation of oxidative stability (weathering) of polymer/pigment compositions under conventional test conditions	Multiple levels of end-use testing conditions provide more reliable ranking of performance of materials; similar kinetic rates of weathering of polymers with quite different pigments were found.	41
One-dimensional coiled libraries of polymer/UV absorber compositions	Fluorescence spectroscopy and imaging for the evaluation of oxidative stability (weathering) of polymer/UV absorber compositions under accelerated test conditions; environmental stress is applied to only local regions, followed by high-sensitivity spatially resolved characterization	Ranking of polymer/UV absorber compositions equivalent to traditional weathering data while achieved 20 times faster.	42
Polyphasic fluid reactions	A microreactor for liquid/liquid isomerization and gas/liquid asymmetric hydrogenation based on dynamic sequential operation	Reaction rate is proportional to the catalyst concentration; the rate decreases with increasing surfactant concentration, no change in the enantiomeric excess was observed	43
Polymer synthesis	In-line GPC for reaction optimization	Determination of activation energy of polymerization	44
Siloxanerubber/carbon black nanocomposites	Automated scanning probe microscope	Study of curing rate of siloxane rubber matrix on roughness and conductivity of composites	45

for improvement of selectivity and overall performance. The data analysis from sensor arrays is well understood, as are different types of sensor transduction mechanisms, and these are available to fit particular HT applications. Thus it is relatively straightforward to adapt the advances in the sensor development for combinatorial screening applications.

In materials and catalyst research, combinatorial methods and associated instrumentation are increasingly being used also for optimizing process conditions.[55] The availability of analytical tools that can handle multiple reactors in real time and in parallel make possible the efficient studies of multiple reactions conducted under different process conditions of temperature, pressure, and other variables simultaneously. The new tools have extended that capability dramatically affording a much higher degree of flexibility in the number of reactors and the parameters that can be independently varied.[55]

Common goals of optimization of combinatorial reactions include maximizing reaction yield, reduction of reaction by-products, and improvement of reaction reproducibility. The varied and optimized parameters depend on the nature of the process and can include reaction conditions (temperature, pressure, solvent or carrier gas), processing conditions (addition sequence and timing), and reaction interferences (concentration of various species).[10] The set of parameters that describes how a reaction is carried out defines a reaction space, and reaction optimization explores as wide a variety of reaction spaces as possible to achieve the desired goal.

5.3. MULTIVARIATE MEASUREMENT TOOLS

When an analytical instrument collects quantitative data from a combinatorial experiment, the accuracy of determinations often depends on the ability to provide an interference-free response. The interferences can arise from a variety of sources and can include chemical and environmental interferences. The ability to provide accurate data improves with the increase of the information content or dimensionality of the collected data per combinatorial sample. Analytical instruments can be classified according to the type of data they provide as zero-, first-, second-, and higher order instruments. Such classification of analytical instruments is well accepted[56] and will be applied here for description of capabilities of instruments for real time monitoring and optimization of combinatorial reactions. We will first discuss this classification principles with the schematic of such classification illustrated in Figure 5.1. This general analysis then will be followed by a discussion of representative examples from combinatorial screening (Figures 5.2–5.5). The discussion will provide the needed understanding, in

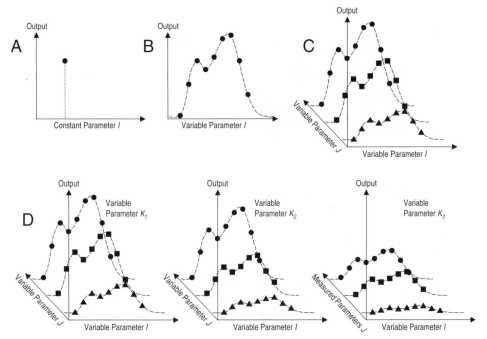

Figure 5.1. Schematic of different orders of analytical measurements. Orders: (*A*) zero; (*B*) first; (*C*) second; (*D*) third.

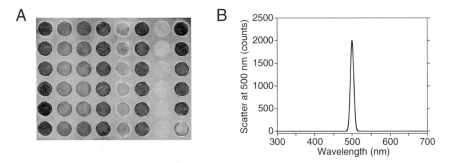

Figure 5.2. Example of a zero-order measurement of combinatorial materials. Abrasion resistance of coatings from measurements of scattered light from each organic coating in a 48-element array at a single wavelength upon abrasion test. (*A*) Reflected-light image of the coatings array; (*B*) representative data from a single material in the array.

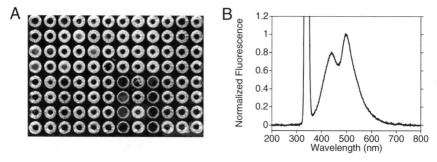

Figure 5.3. Example of a first-order measurement of combinatorial materials. Polymer branching from measurements of fluorescence spectra from each polymerized material in a 96-element microreactor array at a single excitation wavelength. (*A*) Reflected-light image of the microreactor array; (*B*) representative fluorescence spectrum from a single microreactor in the array.

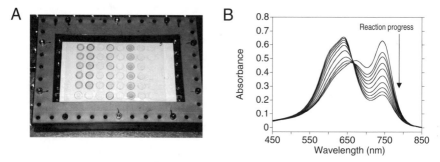

Figure 5.4. Example of a second-order measurement of combinatorial materials. Sensor materials as a 48-element sensor materials array. (*A*) general view of the sensor materials array in a gas flow-through cell; (*B*) representative absorption spectra from a single material in the array collected over a period of time of reaction of this sensor material with a vapor of interest.

capabilities and applicability, of each type of the instrument for the analysis of combinatorial samples.

A measurement system that generates a single data point for each combinatorial sample is a zero-order instrument, as shown in Figure 5.1*A*. A single number is a zero-order tensor, the same as is known in mathematics.[56] First-order measurement systems generate a string of multiple measurements for each combinatorial sample (see Figure 5.1*B*). For example, optical measurements can be done at a single wavelength over the course of a reaction in monitoring the reaction's progress. The variable parameter

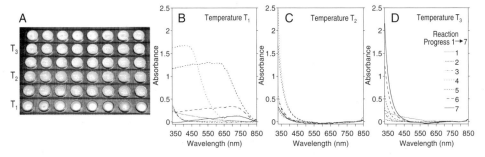

Figure 5.5. Example of a third-order measurement of combinatorial materials. Oxidative stability of polymers from measurements of UV-VIS reflection spectra from each polymeric composition in a materials array as a function of reaction temperature and time. (A) General view of the materials array on a gradient temperature heater; (B) representative UV-VIS spectra from a single material in the array as a function of reaction time and temperature. Reaction temperatures: $T_1 > T_2 > T_3$. Reaction progress is shown as spectra changes from spectrum 1 to spectrum 7.

is the time to complete the reaction. Alternatively, an optical spectrum of the final combinatorial material can be produced by the first-order measurement system. The variable parameter in this case is the wavelength. As may be seen, measurements provided by the first-order measurement system are of the same nature, whether, for example, temporal or spectral in response. Second-order measurement systems generate a second-order tensor of data for each sample. This is a matrix of the instrument's response upon a change of the two independent variables (see Figure 5.1C). Depending on the particular need in the combinatorial screening, higher-order measurement systems are also possible. For example, Figure 5.1D shows a response of the third-order system, where a matrix of an instrument response upon the change of the two independent variables is further modulated by a third independent variable.

Clearly, the complexity of measurement systems increases dramatically with an increase in the measurement order. However, despite instrument complexity, the higher order instruments have obvious advantages for reaction monitoring and optimization. These advantages are illustrated below with real examples in the combinatorial screening of materials in our laboratories.

An example of combinatorial screening using a zero-order measurement approach is illustrated in Figure 5.2. Abrasion resistance of organic protective coatings was determined from measurements of scattered light from each coating in a 48-element array.[57] A simple zero-order measurement approach was useful because it provided the required information about the end-use performance of the protective coatings *after* an abrasion test (see Figure 5.2*A*). Measurements of the abrasion-induced increase in the amount of light scatter were performed at a single wavelength as shown in Figure 5.2*B*. A single wavelength measurement was adequate for the correlation between the high-throughput and traditional evaluation of abrasion resistance of coatings.[57]

An example of the first-order measurement approach of combinatorial materials is illustrated in Figure 5.3. Measurements of fluorescence spectra of solid polymerized materials were performed directly in individual microreactors. A view of the 96-microreactor array is shown in Figure 5.3*A*. Several chemical parameters in the combinatorial samples were identified from these measurements. The spectral shape of the fluorescence emission with an excitation at 340 nm provided the information about the concentration of the branched product in the polymer and the selectivity of a catalyst used for the melt polymerization.[21,58–61] A representative fluorescence spectrum (along with an excitation line at 340 nm) from a single microreactor in the array is illustrated in Figure 5.3*B*. The first-order measurements were used for the optimization of melt-polymerization reaction conditions as described in Section 5.1.

When measurements are done with a first-order instrument, and there is another independent variable involved, this constitutes a second-order measurement approach. This type of screening was used for the in situ monitoring of melt-polymerization reactions (see Section 5.4) and for the evaluation of the sensor materials. Figure 5.4*A* shows a 48-element array of sensor materials positioned in a gas flow-cell for the monitoring of the materials response upon exposure to vapors of interest. For the evaluation of sensor materials, absorption spectra were collected over a period of time of reaction of these sensor materials with a vapor of interest. Results of these measurements are illustrated in Figure 5.4*B*.

Adding two more independent variable parameters to the response of a first-order system obviously makes it a third-order measurement approach. An example of such system is spectroscopic monitoring (wavelength is the variable parameter *I*) of the reaction progress of combinatorial materials (time is the variable parameter *J*) at different process temperatures (temperature is the variable parameter *K*). Alternatively, a second-order system can be implemented for the measurements as a function of one or more independent parameters. Examples of second-order systems are excitation-emission luminescence measurement systems, GC-MS, and HPLC-diode array UV systems, among others. An example of one of our third-order measurement approaches for combinatorial screening is illustrated in Figure 5.5. It was implemented for the determination of oxidative stability of polymers under different process conditions (temperature *I* and time *J*).

Materials were positioned onto a gradient temperature stage as shown in Figure 5.5*A*. Measurements of reflection spectra of each material sample were performed with a scanning fiber-optic probe during the oxidation reaction in order to monitor the oxidation-induced color change. An example of a typical evolution of the reflection spectra as a function of oxidation time and temperature is presented in Figure 5.5*B*. Another example of the application of third-order measurements is in the monitoring of curing of coating; in this case a scanning fluorescence system was used to monitor an array of organic UV-cured automotive organic protective coating formulations. The third-order measurements provided the means for the high-throughput optimization of curing parameters. Fluorescence spectral changes in a viscosity-sensitive molecular probe embedded in the coatings formulations were monitored during curing of coatings. Rapid decoupling of temperature and radiation effects during the curing of these coating formulations was achieved by using multiple coatings and process conditions at once.[39]

As shown in these examples, the increase in the measurement dimensionality (i.e., the order of analytical instrumentation) improves the analytical capabilities of the screening systems and makes possible their use for reaction monitoring and optimization. These capabilities include increased analyte selectivity, more simple approach to reject contributions from interferences, multicomponent analysis, and an outlier detection. Importantly, second- and higher order measurement approaches benefit from the improved performance, even in the presence of uncalibrated interferences.[62] Since, in many cases, knowing all potential interferences and including these variations into the calibration model is very difficult, this so-called second-order advantage is significant.[63] Analysis in the presence of unknown interferences is not possible with zero- or first-order data unless full selectivity is somehow guaranteed.[64] This advantage applies to situations where measurements are done as a function of variables of different nature that can include instrumental variables (i.e., wavelength, mass-to-charge ratio) and process variables (reaction temperature, pressure, etc.). Thus second- and higher order measurement approaches are attractive for monitoring several components in combinatorial reactions.

5.4. MULTIVARIATE DATA ANALYSIS TOOLS

Multivariate data analysis in combinatorial screening involves three categories of multivariate data sources. The first category includes all types of measurement systems that generate multivariate data for a combinatorial sample. The second category includes univariate data from samples in combinatorial libraries that are analyzed with respect to multiple parameters (starting components, process conditions, etc.). The third category is the combination of the first two categories, as the multivariate measurement data from each combinatorial sample is further analyzed with respect to multiple parameters.

Massive data generated in the reaction monitoring and optimization leads to the need for effective data analysis and interpretation to identify the trends and relationships in the data.[48] Advanced mathematical and statistical chemometric techniques are used in combinatorial and high-throughput screening to determine, often by indirect means, the properties of substances that otherwise would be very difficult to measure directly.[65,66] In combinatorial and high-throughput screening, chemometrics has been successfully applied for visualization and pattern recognition,[20,67] lead identification and optimization,[68,69] library design,[70,71] process optimization,[72,73] and development of predictive models and quantitative structure-activity relationships (QSARs) and quantitative structure-property relationships (QSPRs).[67,70] The implementation of multivariate calibration improves analysis selectivity when spectroscopic or any other types of measurements are performed with relatively nonselective or low-resolution instruments.[60,74]

In combinatorial materials science, pattern recognition techniques are used to find similarities and differences between samples of combinatorial libraries, and they serve as a powerful visualization tool and can warn of the occurrence of abnormalities in the measured data.[20] The correlated patterns revealed in large data sets by these techniques can determine the structural relationships among screening hits and can significantly reduce data dimensionality to make the database more manageable. Methods of pattern recognition include principal components analysis (PCA), hierarchical cluster analysis (HCA), soft independent modeling of class analogies (SIMCA), and neural networks, among some others.[75,76]

Multivariate calibration methods offer several advantages over univariate calibration methods.[77,78] Signal averaging is achieved, since more than one measurement channel is employed in the analysis. Also concentrations of multiple species may be measured if they are present in the calibration samples. A calibration model is built by using responses from calibration standards. The analysis of unknown samples will suffer if a species is present in the sample that is not accounted for in the calibration model. This is mitigated somewhat by the ability to detect whether a sample is an outlier from the calibration set. Multivariate calibration approaches permit selective quantitation of several analytes of interest in complex combinatorial libraries using low-resolution instruments when overlapping responses from different species preclude the use of univariate analysis. Quantitative analysis is often performed using linear and nonlinear multivariate calibration methods such as partial least-squares (PLS),[27] locally weighted regression (LWR),[74] and others.

Multivariate measurement and data analysis tools has been shown to be an essential analytical tool for the monitoring of solid phase organic synthesis and other types of reactions. In particular, the attractiveness of PLS method has been demonstrated for in situ monitoring of combinatorial reactions using single-bead FTIR spectroscopy.[27] Using these tools, quantitative and qualitative analyses of combinatorial samples were performed with IR spectra containing severely overlapped bands. This analysis was possible because the primary loading factor in PLS analysis displays only spectral features that have undergone changes during a reaction. Disappearing and emerging organic functional groups generate negative and positive signals in the primary loading factor, allowing enhanced qualitative analysis of the reaction. The scores of the primary factors of spectra taken at various times during a reaction provide quantitative information and allowed the study of the reaction kinetics directly on solid support. A limitation of PLS method is that while it does reveal the changes due to reaction, it cannot provide spectra for pure component.[27] This limitation can be overcome by using other multivariate analysis tools such as evolving factor analysis[79,80] and multivariate curve resolution methods.[81,82] Section 5.4 demonstrates the use of these multivariate tools.

5.5. APPLICATION EXAMPLES

In the examples provided in this section, combinatorial methods were used to improve the properties of an industrial aromatic polymer, such as melt-polymerized bisphenol-A polycarbonate.[83–85] The reactions were performed in 96-well microtiter glass plates that served as 96-microreactor arrays[84] in a sequence of steps of increasing temperature[86,87] with a maximum temperature of 280°C. An example of one of the 96-microreactor arrays after melt-polymerization is shown in Figure 5.3A. For melt-polymerization of bisphenol-A polycarbonate, the starting reaction components included diphenyl carbonate and bisphenol-A monomers and a catalyst (e.g., NaOH).[84,86,87] The materials codes used in the examples are presented in Table 5.2. Intermediate species include polycarbonate oligomers and phenol.[88] The bisphenol-A polycarbonate polymer often contains a branched side product that produces a detectable fluorescence signal and other species that can include nonbranched end-groups and cyclics.[59]

We used fluorescence spectroscopy for nondestructive chemical analysis of melt-polymerized bisphenol-A polycarbonate.[58,59] The key attractive

**Table 5.2. Materials Used in Combinatorial
Optimization of Reaction Parameters**

Material Code	Material Name
A	Diphenyl carbonate monomer
B	Bisphenol-A monomer
C	Melt-polymerization catalyst

feature of fluorescence analysis is in quantitation of low concentrations of species. In addition a high selectivity in fluorescence measurements can be achieved by optimizing the excitation wavelength to collect fluorescence from certain species and reduce the fluorescence signal from interferences. Fluorescence analysis provided selective chemical determinations in a variety of polymers.[89–97] Importantly, the spectral features of bisphenol-A polycarbonate were found to be well correlated with the chemical properties of interest.[21,59–61]

5.5.1. High-Throughput Optimization of Process Parameters

To develop the melt-polymerization polymers using a combinatorial chemistry methodology,[84,98] it is critical to find the optimal process parameters. Unlike the traditional melt-polymerization scheme which involves stirring the reaction components in a vacuum,[86,87,99,100] the combinatorial approach is based on thin-film melt-polymerization at atmospheric pressure and without stirring.[84,98]

Our high-throughput multiparameter reaction optimization methodology included fabrication of materials arrays over a wide range of process conditions, application of nondestructive measurement techniques to collect optical spectra from the fabricated materials, multivariate data analysis to extract the desired spectral descriptors from the spectra, correlation of the variation in these spectral descriptors with the variation in process conditions, and identification of the levels of process conditions that satisfy two predetermined reaction requirements. The first requirement includes identification of process conditions that provide the largest material differentiation at a constant ratio of two reaction components A and B (ratio A/B) and increasing concentration of the third reaction component C. The second requirement includes minimum reaction variability, as reactions are performed in different microreactors under identical process conditions.

The process conditions include intra- and inter-array variable parameters. Intra-array parameters include reaction volume, ratio of reaction com-

Table 5.3. Variable Input Parameters (Process Conditions) for High-Throughput Multiparameter Optimization of Polymerization Conditions of Combinatorial 96-Microreactor Arrays

Parameter Type	Parameter Name	Studied Levels of Parameters
Intra-array	Reaction volume (μL)	150, 200, and 250
	Ratio of starting components A/B (mol/mol)	1, 1.2, and 1.4
	Amount of catalyst C (10^{-5} mole C per mole B)	0.5, 1, 2, and 4
Inter-array	Flow rate of inert gas (L/min)	4, 6, and 8
	Dwell time (min)	10, 20, and 30

ponent A to reaction component B (ratio A/B), and concentration of component C. These parameters were varied at different levels across multiple elements of the 96-microreactor array within each experiment. Inter-array parameters include flow rate of inert gas and dwell time. These parameters are varied at different levels in different experiments. The process conditions and their levels are listed in Table 5.3. These process conditions were selected based on previous work on this type of polymerization reaction.[84]

Fluorescence spectra are collected under excitation conditions that are optimized to correlate the emission spectral features with parameters of interest. Principal components analysis (PCA) is further used to extract the desired spectral descriptors from the spectra.[75,101] The PCA method is used to provide a pattern recognition model that correlates the features of fluorescence spectra with chemical properties, such as polymer molecular weight and the concentration of the formed branched side product, also known as Fries's product,[60] that are in turn related to process conditions.[61] The correlation of variation in these spectral descriptors with variation in the process conditions is obtained by analyzing the PCA scores. The scores are analyzed for their Euclidean distances between different process conditions as a function of catalyst concentration. Reaction variability is similarly assessed by analyzing the variability between groups of scores under identical process conditions. As a result the most appropriate process conditions are those that provide the largest differentiation between materials as a function of catalyst concentration and the smallest variability in materials between replicate polymerization reactions.

Multiple melt-polymerization reactions are performed using mixtures of A and B at different ratios with different amounts of catalyst C. Catalyst amounts are expressed as certain fractions of 10^{-5} mol of catalyst per mole of component B. In process optimization experiments, polymeric materials are fabricated in five microreactor arrays under different intra- and inter-array reaction parameters as illustrated in Table 5.3. The combinations of these parameters were provided from design of experiments. Ranges of these parameters were selected to preserve the rank order of a variety of control catalysts from a lab-scale to combinatorial reactors.[86,98] All parameters described in Table 5.3 are considered process conditions because once chosen they are not altered throughout screening of new catalysts.[98] Replicates ($n = 8$) of similar conditions in each 96-microreactor array are randomized in groups of four to reduce possible effects of any position-induced variability. In addition one set of eight microreactors in each microreactor array is always left blank and one of the microreactors contains a thermocouple (see Figure 5.3A).

The experimental setup for the automated acquisition of fluorescence spectra from materials in 96-microreactor arrays is shown in Figure 5.6. The fluorescence measurements of polymers in each microreactor were performed through the glass bottom of the microreactors using a white light source, a monochromator, a portable spectrofluorometer, and a translation stage. The white light source was coupled to the monochromator for selection of the excitation wavelength and further focused into one of the arms of a "six-around-one" bifurcated fiber-optic reflection probe. Emission light

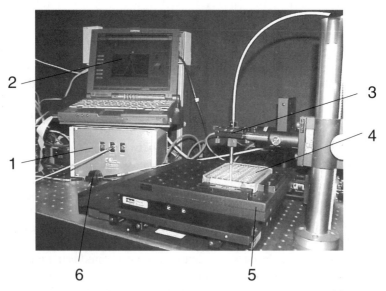

Figure 5.6. Automated multifunctional high-throughput screening system: (1) spectrofluorometer; (2) data acquisition computer; (3) fiber-optic probe; (4) combinatorial array; (5) translation stage; (6) in-line optical filter.

from the polymeric material in each microreactor was collected when the common end of the fiber-optic probe was positioned near the bottom of the reactor at a 45° angle to the normal to the surface. The second arm of the probe was coupled to the portable spectrofluorometer through an in-line optical filter holder. The holder contained a long-pass optical filter to block excitation light from entering the spectrofluorometer. The spectro-fluorometer was equipped with a 200-μm slit, 600-groove/mm grating blazed at 400 nm and covering the spectral range from 250 to 800 nm, and a linear CCD-array detector. The fluorescence spectra reported here were not corrected by the spectral response of the optical system.

For fluorescence analysis of solid polymers, each microreactor array was arranged on an X-Y translation stage, and the common end of the fiber-optic probe was held stationary to measure emission spectra. The size of the excitation beam of the fiber-optic probe was in the range from 1 to 4 mm, depending on the probe-microreactor distance. Data acquisition and automatic control of the X-Y translation stage were achieved with a computer using a program that also provided adequate control of the data acquisition parameters and real-time communication with the translation stage.

Prior to PCA, fluorescence spectra from five 96-microreactor arrays were appropriately preprocessed. The preprocessing included a baseline sub-traction, normalization of spectra by the fluorescence intensity at 500 nm, selection of an appropriate spectral range for the PCA, and mean-centering the spectra. The spectra from the empty and thermocouple-containing microreactors were excluded from the spectral descriptor analysis. Principal components analysis was used to extract the desired spectral descriptors from the spectra. PCA is a multivariate data analysis tool that projects the data set onto a subspace of lower dimensionality with removed collinearity. PCA achieves this objective by explaining the vari-ance of the data matrix $\mathbf{X}$ in terms of the weighted sums of the original vari-ables with no significant loss of information. These weighted sums of the original variables are called principal components (PCs). In applying PCA, we express the data matrix $\mathbf{X}$ as a linear combination of orthogonal vectors along the directions of the principal components:

$$\mathbf{X} = \mathbf{t}_1 \mathbf{p}_1^T + \mathbf{t}_2 \mathbf{p}_2^T + \cdots + \mathbf{t}_A \mathbf{p}_K^T + \mathbf{E}, \tag{5.1}$$

where $\mathbf{t}_i$ and $\mathbf{p}_i$ are, respectively, the score and loading vectors, K is the number of principal components, $\mathbf{E}$ is a residual matrix that represents random error, and T is the transpose of the matrix.

PCA determines a spectral descriptor for each fluorescence spectrum. The spectral descriptor is represented as a vector in PCA space. This vector is described by a unique combination of the respective scores of principal components. The spectral descriptors from materials produced under the same processing conditions are considered as a cluster S in the PCA space without the use of automated clustering algorithms.[75,101] Each cluster S is represented by its mean and standard deviation with respect to kth principal component. The Euclidean distance E between two clusters of spectral descriptors is calculated as

$$E_{ij} = \left\{ \sum_{1}^{n} (W_k(S_{ki} - S_{kj}))^2 \right\}^{1/2},$$ (5.2)

where i and j are indexes of clusters S_i and S_j, respectively, E_{ij} is the Euclidean distance between these clusters, W_k is the weighting factor (equal to captured percent variance) of the kth principal component, and n is the number of principal components used for multivariate analysis.

Calculations of means and standard deviations of E_{ij} according to Eq. (5.2) were performed by a Monte Carlo simulation program. For these simulations, the means and standard deviations of S_{ki} and S_{kj} were initially calculated from the multiple spectral descriptors of each cluster. Further, both the mean of each cluster and its standard deviation (assuming the normally distributed error) were entered into the program. Finally, 10,000 iterations were performed to calculate the mean and standard deviation for a given E_{ij}. From a variety of available approaches for cluster analysis,[76] we selected analysis of Euclidean distances because it provides the information about both the distance between clusters and the spread of each cluster. Further, although it is possible to perform calculations of Euclidean distances on raw spectra, we performed the PCA first, to reduce the noise in the data.

For the determination of the key process parameters and their respective values, fluorescence spectra from five 96-microreactor arrays were collected and processed. The normalized spectra are presented in Figure 5.7. The spectral features of the polymeric materials in the microreactors contain a wealth of information about the chemical properties of the materials that were extracted using PCA. According to the PCA results, the first two principal components (PCs) accounted for more than 95% of the spectral variation among all spectra. Thus the first two PCs were used for an adequate description of the fluorescence spectra. Results of the principal components analysis of the spectra from all 96-microreactor arrays as a function of catalyst concentration C are presented in Figure 5.8. The plot demonstrates the existence of the major general trend in the spectral descriptors where the variation in scores of both PCs strongly depends on concentration of component C for all screened process parameters.

To determine other key process variables, besides the concentration of component C, that affect performance of the combinatorial thin-film

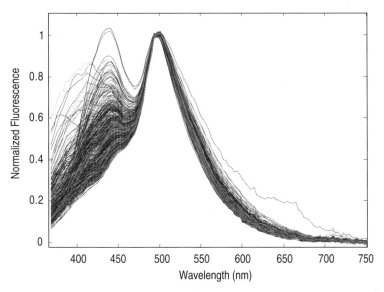

Figure 5.7. Normalized fluorescence spectra of the polymeric materials made in the 96-microreactor arrays under all experimental conditions. From ref. 73.

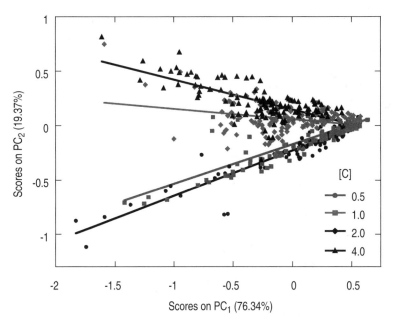

Figure 5.8. Results of the principal components analysis of the spectra of the polymeric materials made in the 96-microreactor arrays under all experimental conditions as a function of concentration of catalyst. From ref. 73.

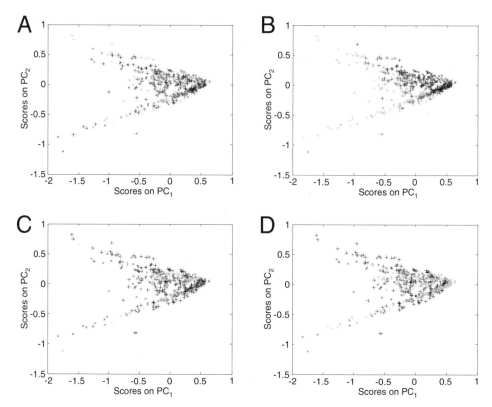

Figure 5.9. Results of the principal components analysis of the spectra of the polymeric materials made in the 96-microreactor arrays with variable intra-array (*A* and *B*) and inter-array (*C* and *D*) parameters: (*A*) volume; (*B*) ratio *A/B*; (*C*) flow rate; (*D*) dwell time. Levels: asterisks—smallest; plus—medium; X—largest. For values, see Table 5.3. From ref. 73.

polymerizations, PCA results can be analyzed as a function of individual intra- and inter-array variables. Figure 5.9 demonstrates PCA results as a function of variable volume, ratio *A/B*, flow rate, and dwell time. These data illustrate that for two intra-array parameters, such as the sample volume and ratio *A/B*, significant clustering of spectral descriptors was observed to be a function of the ratio *A/B*. For two inter-array parameters, such as flow rate and dwell time, the clustering of spectral descriptors was insignificant.

We next performed a more extensive analysis to consider the spectral descriptors in *individual* 96-microreactor arrays. The data were analyzed as a function of catalyst concentration, reaction volume, and ratio *A/B* at different flow rates of inert gas and dwell times. In our analysis are evaluated

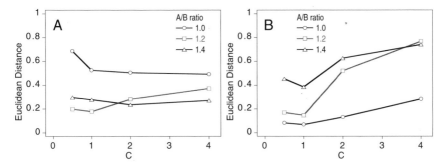

Figure 5.10. Results of calculation of Euclidean distances of spectral descriptors in polymer arrays produced under inter-array conditions: (*A*) 8-L/min flow rate of inert gas and 10-minute dwell time and (*B*) 6-L/min flow rate of inert gas and 20-minute dwell time. Maximum Euclidean distance indicates best conditions for material differentiation as a function of catalyst concentration *C*. From ref. 73.

the effects of these process parameters on the properties of polymerized materials and the reproducibility of replicate polymerizations in different 96-microreactor arrays. Minimum reaction variability was characterized by the smallest spread in clustering of spectral descriptors. We found that the variation of flow rate of inert gas and dwell time affected the reproducibility of replicate polymerizations. Changes of reaction volume did not significantly affect the clustering of spectra for any of inter-array conditions. The most pronounced dependence of spectral features was provided by the variation in the ratio *A/B*.

To determine the optimal levels of the identified process conditions, a more detailed evaluation was performed. Euclidean distances between different clusters of spectral descriptors and the uncertainty in these distances were computed using Eq. (5.2). Calculations were performed between spectral descriptors associated with materials produced with different amounts of catalyst and ratios *A/B* in all 96-microreactor arrays.

Figure 5.10 provides plots of the calculations of Euclidean distance between the spectral descriptors of two representative 96-microreactor arrays. The largest Euclidean distances indicate the best conditions for material differentiation. The best inter-array conditions were found to be a 6-L/min flow rate of inert gas and 20-min dwell time. The best intra-array conditions were a combination of the catalyst concentration of 2 to 4 equivalents and ratio *A/B* of 1.2 to 1.4 (Figure 5.10*B*). Results for the reaction variability for these representative microreactor arrays are presented in Figure 5.11. The smallest relative standard deviation (RSD) of spectral features indicates the best reaction reproducibility. This figure illustrates that

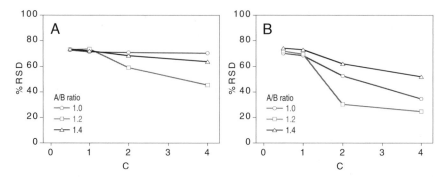

Figure 5.11. Results of calculation of reaction variability as %RSD of PC$_1$ and PC$_2$ scores of spectral descriptors in polymer arrays produced under inter-array conditions: (*A*) 8-L/min flow rate of inert gas and 10-minute dwell time and (*B*) 6-L/min flow rate of inert gas and 20-minute dwell time. Smallest relative standard deviation (RSD) of spectral features indicates best reaction reproducibility. From ref. 73.

the smallest RSD was achieved with the conditions of microreactor array processed under a 6-L/min flow rate of inert gas and 20-min dwell time, ratio *A/B* of 1.2 over the concentration range of catalyst from 2 to 4.

As a result the combinatorial polymerization system was optimized for the best processing parameters using a set of input variables that included reactant parameters (relative amounts of starting components and catalyst loading) and processing variables (reaction time, reaction temperature, and inert gas flow rate). The measured output parameters were the chemical properties of materials and variability of the material formation within each of the microreactors as measured noninvasively using optical spectroscopy.

5.5.2. Outliers in Multivariate Measurements

Several statistical tools are available that ensure the quality of the data analyzed using multivariate tools, such as PCA. These tools are multivariate control charts and multivariate contributions plots. In the multivariate control charts two statistical indicators of the PCA model such as Hotelling's T^2 and Q values are plotted as a function of combinatorial sample or time. The significant principal components of the PCA model are used to develop the T^2 chart and the remaining PCs contribute to the Q chart. A graphic representation of T^2 and Q statistics for multiple samples in the PCA model is given in Figure 5.12. The sum of normalized squared scores, the T^2 statistic, gives the measure of variation within the PCA model and determines statistically anomalous samples:[102]

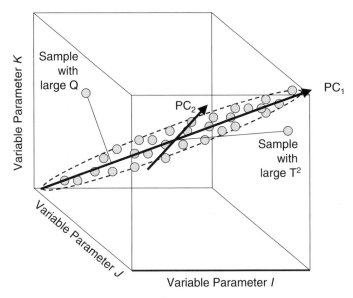

Figure 5.12. Graphical representation of T^2 and Q statistics for multiple samples in the PCA model.

$$T_i^2 = \mathbf{t}_i \lambda^{-1} \mathbf{t}_i^T = \mathbf{x}_i \mathbf{P} \lambda^{-1} \mathbf{P}^T \mathbf{x}_i^T , \qquad (5.3)$$

where $\mathbf{t}_i$ is the ith row of $\mathbf{T}_k$, the matrix of k scores vectors from the PCA model, λ^{-1} is the diagonal matrix containing the inverse of the eigenvalues associated with the K eigenvectors (principal components) retained in the model, $\mathbf{x}_i$ is the ith sample in $\mathbf{X}$, and $\mathbf{P}$ is the matrix of K loadings vectors retained in the PCA model (where each vector is a column of $\mathbf{P}$). The Q residual is the squared prediction error and describes how well the PCA model fits each sample. It is a measure of the amount of variation in each sample not captured by K principal components retained in the model:[102]

$$Q_i = \mathbf{e}_i \mathbf{e}_i^T = \mathbf{x}_i (\mathbf{I} - \mathbf{P}_k \mathbf{P}_k^T) \mathbf{x}_i^T , \qquad (5.4)$$

where $\mathbf{e}_i$ is the ith row of $\mathbf{E}$, and $\mathbf{I}$ is the identity matrix of appropriate size $(n \times n)$.

The multivariate T^2 and Q statistics control charts for the fluorescence spectra of the one of the 96-microreactor arrays discussed in Section 5.1 are presented in Figure 5.13. These control charts illustrate that several samples exceed the 95% confidence limits for the T^2 and Q statistics described by

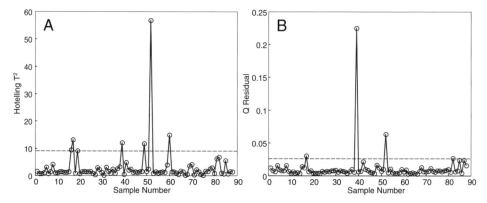

Figure 5.13. Multivariate statistical indicators for combinatorial production of melt polycar-bonate in one of the 96-microreactor arrays: (A) T^2 control chart; (B) Q residual control chart; (dotted lines in A and B) 95% confidence limits.

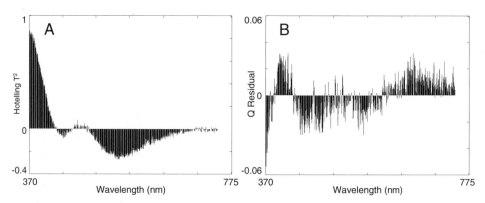

Figure 5.14. Multivariate statistical indicators for combinatorial production of melt polycar-bonate in one of the 96-microreactor arrays: (A) T^2 contributions plot to sample 52; (B) Q residual contributions plot to sample 39.

the PCA model. The largest T^2 score is associated with sample 52 (see Figure 5.13A). The largest Q residual score is associated with sample 39 (see Figure 5.13B). The contributions plots of these statistical parameters can be used to track the origin of the largest contributors to these alarms in samples 52 and 39 (see Figure 5.14).

5.5.3. Demonstration of Optimized Process Conditions

Optimization of process parameters of melt-polymerization conditions described in Section 5.5.1 resulted not only in the improved reproducibility of replicate polymerization reactions and improved discrimination between materials with different levels of catalyst C, but also in the improved homogeneity of polymers within individual microreactors. Typical examples of fluorescence spectra from polymer materials produced in replicate microreactors under nonoptimized and optimized conditions are presented in Figure 5.15. This comparison is performed for materials with increasing levels of catalyst C and constant levels of ratio A/B and volume. These plots illustrate that upon finding the optimal reaction conditions, the reproducibility of replicate polymerization reactions is improved. In addition the discrimination among materials of increasing concentration of component C is also improved, as evidenced by the more reproducible fluorescence spectra under identical process conditions.

Under the optimized reaction conditions, the relative importance of variable reaction volume and reaction ratio A/B was confirmed. No significant difference in materials properties was found upon changing of the reaction volume under optimized reaction conditions as illustrated in Figure 5.16A. In contrast, the variation in ratio A/B had a pronounced effect (see Figure 5.16B) as predicted by the descriptor analysis.

The high spatial polymer homogeneity in individual microreactors is another powerful indicator of the reproducible polymerization process on

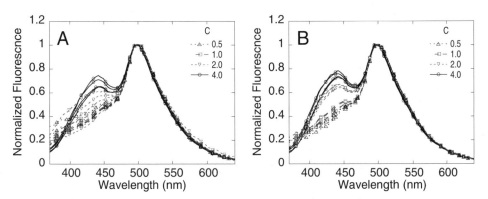

Figure 5.15. Typical material variability under (A) nonoptimized and (B) optimized reaction conditions: (A) 8-L/min flow rate of inert gas and 10-minute dwell time and (B) 6-L/min flow rate of inert gas and 20-minute dwell time. Intra-array materials' parameters, ratio $A/B = 1.2$; reaction volume = 200 μL. From ref. 73.

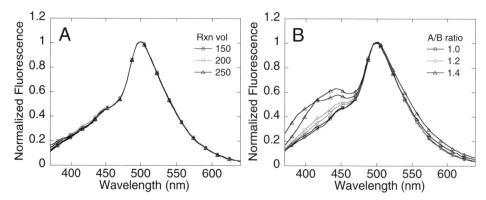

Figure 5.16. Effects of (*A*) reaction volume and (*B*) ratio *A/B* under optimized reaction conditions of 6-L/min flow rate of inert gas and 20-minute dwell time. Intra-array materials' parameters: (*A*) Ratio *A/B* = 1.0, catalyst concentration C = 1.0; (*B*) reaction volume = 200 μL, catalyst concentration C = 1.0. From ref. 73.

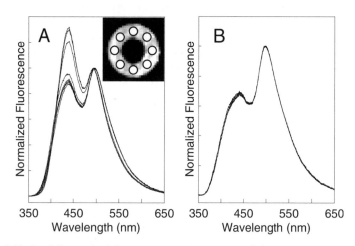

Figure 5.17. Spatially resolved fluorescence emission spectra of solid bisphenol-A polycarbonate in a melt-polymerization microreactor before (*A*) and after (*B*) optimization of processing conditions. (*Inset*) Microreactor cross section and locations of spectral measurements. From ref. 73.

the combinatorial scale. This variability is determined as differences in measured fluorescence spectra which are caused by the spatial variation in the chemical composition in the formed polymer within each microreactor. Figure 5.17 illustrates the fluorescence spectra of polymers within individual microreactors under nonoptimized and optimized polymerization conditions. The evaluations were performed by measuring multiple regions of polymer in each microreactor. The small beam size (ca. 1 mm in diameter) permitted the detailed evaluation of the spatial distribution of polymer formation in each of the microreactors in the array. These measurements of variability can be provided only on a solid material directly in each microreactor because this information is lost after dissolving the polymer for traditional GPC analysis.

5.5.4. Determination of Contributing Factors to Combinatorial-Scale Chemical Reactions

Monitoring of melt-polymerization reactions of bisphenol-A polycarbonate was performed in individual microreactors intended for combinatorial screening. Fluorescence spectroscopy coupled with the multivariate analysis of kinetic data provided previously unavailable combinatorial and high-throughput screening capabilities for tracking the progress of the reaction in real time and without sample preparation steps and determination of critical contributing factors to combinatorial reactions. Spectral data were generated using a UV light source coupled via a fiber-optic probe to microreactors and to a portable spectrofluorometer based on a linear CCD-array detector described in Section 5.5.1. The data collected during the reaction was analyzed using two chemometric tools to learn about the nature and concentrations of chemical species participating in the reaction. First, the contributing chemical factors in the reaction were determined using evolving factor analysis.[79,80] Second, pure spectra and concentration profiles of these contributing factors were estimated using the multivariate curve resolution method.[81,82]

The evolving factor analysis is useful for processing of multivariate data produced by dynamic systems. In the monitored chemical reactions the measured spectra evolved with time because of the consumption of starting products during the reactions and because of a generation of intermediate and final products. The evolving factor analysis was used to estimate the number of independent factors in the spectral data set at each time step that contributed to the spectral variation observed during a polymerization reaction. For this analysis fluorescence data were baseline-corrected and spectral range was limited to 340 to 600 nm to reduce noise in the data (Figure 5.18A). Results of the forward analysis are presented in Figure 5.18B. As the first step in the forward analysis, the eigenvalue of the first measured spectrum was calculated. Then eigenvalues were calculated of matrices as new measured spectra were appended. The forward analysis suggests that several independent factors have evolved into the data set. In the reverse analysis (see Figure 5.18C), eigenvalues were calculated for the last spectra first, and subsequent eigenvalues were calculated as spectra

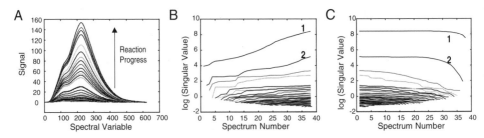

Figure 5.18. Determination of contributing factors from fluorescence spectra collected in real time during a combinatorial melt polymerization reaction of bisphenol-A polycarbonate in one of the microreactors. (A) Temporal evolution of fluorescence spectra; (B) forward evolving factor analysis results; (C) reverse evolving factor analysis results. Numbers 1 and 2 in (B) and (C) are the first and second reaction components, respectively. From ref. 47.

were appended in a reverse fashion. The reverse analysis gives results of when factors evolved out of the data set.[102] The results of our reverse analysis suggest that only one factor remained at the end of the melt-polymerization process. Other factors disappeared at the end of the reaction. The times for factors appearing and disappearing identified in the evolving factor analysis indicate when different chemical species are present in the process. This is helpful when that information is not available at the onset. Results from the evolving factor analysis clearly identified the existence of at least two independent factors.

After the number of contributing factors was determined using evolving factor analysis, their pure spectra and concentration profiles were estimated using the multivariate curve resolution method. To be effective, this method requires one to apply a correct set of constraints to obtain a good solution, even though often multiple solutions can fit the data equally well.[102] Several constraints were applied for this data set. First, weak components, which were estimated with evolving factor analysis, were assigned as a single component because of their strong correlation. Second, spectra of pure components and relevant concentrations were set to be nonnegative at any time. Third, initial concentrations C_i of the components were set as $C_1 = 0$ for the first (strongest) component and $C_2 > 0$ for the second component. The calculated concentration profiles of the first and second components are presented in Figure 5.19A and 5.19B, respectively. The concentration profile of the first component was found to correspond to the experimentally observed increase in fluorescence due to formation of the branched polycarbonate side product. The calculated concentration profile of the second component was assigned to a combination of initial and intermediate

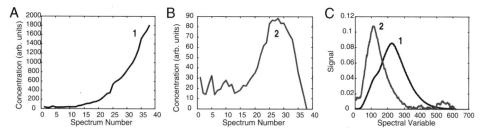

Figure 5.19. Determination of concentration profiles and spectral characteristics of pure components from fluorescence spectra collected in real time during a combinatorial melt-polymerization reaction of bisphenol-A polycarbonate in one of the microreactors. (*A*) Concentration profile of the first component calculated using multivariate curve resolution; (*B*) concentration profile of the second component calculated using multivariate curve resolution; (*C*) spectra of two pure components calculated using multivariate curve resolution. Numbers 1 and 2 are the first and second reaction components, respectively. From ref. 47.

species generated during the reaction. The fluorescence spectra of two pure components were also calculated (Figure 5.19*C*).

5.6. CONCLUSIONS AND OUTLOOK

Combinatorial and high-throughput methodologies are now recognized as the breakthrough technologies in materials science. Their success greatly depends on the development of analytical tools with new and improved capabilities. As indicated at several recent symposia on analytical instrumentation, these new developments should focus on both new instrumentation for analysis of diverse materials properties and new mathematical approaches for processing of large amounts of data.[103–105] An important aspect of the instrumental advances is the integration of new in situ measurement capabilities with microreactors. Similar to advantages provided by in situ methods in catalysis on traditional scale,[106] recent catalyst discoveries have already demonstrated[107] that the high-throughput experimentation coupled with in situ reaction monitoring complement each other and significantly contribute to success of the combinatorial and high-throughput screening programs.

New combinatorial schemes are needed where various reaction variables can be altered simultaneously and where important cooperative effects can be carefully probed to provide better understanding of variation sources. New performance models must be developed that can relate intrinsic and performance properties and can provide transfer functions of scalability of performance of chemical reactors from nano- to macroscale. Advances in data analysis and chemometrics should lead to better means of real-time analysis of complex multidimensional data arrays. Advances in computational chemistry should lead to more capabilities for virtual combinatorial and high-throughput screening, that is, selection of materials by evaluating their desirability in a computational multivariate model. Overall, gathered and analyzed combinatorial and high-throughput screening data promise to provide new opportunities for structure activity relationships analysis and for the more cost- and time-effective materials design.

ACKNOWLEDGMENTS

The author thanks co-workers, whose names are in the references, for their contributions. This work has been supported by US Department of Energy (Grant DEFC07-01ID14093) and by the National Institute of Standards and Technology Advanced Technology Program (Grant 70NANB9H3038).

REFERENCES

1. B. Jandeleit, D. J. Schaefer, T. S. Powers, H. W. Turner, W. H. Weinberg, *Angew. Chem. Int. Ed.* **38**, 2494–2532 (1999).

2. R. Malhotra, ed., *Combinatorial Approaches to Materials Development*, American Chemical Society, Washington, DC, Vol. 814 (2002).

3. I. Takeuchi, J. M. Newsam, L. T. Wille, H. Koinuma, E. J. Amis, eds., *Combinatorial and Artificial Intelligence Methods in Materials Science*, Materials Research Society, Warrendale, PA, Vol. 700 (2002).

4. Y. Zhang, X. Gong, H. Zhang, R. C. Larock, E. S. Yeung, *J. Comb. Chem.* **2**, 450–452 (2000).

5. S. Thomson, C. Hoffmann, S. Ruthe, H.-W. Schmidt, F. Schuth, *Appl. Catal. A* **220**, 253–264 (2001).

6. A. Ohtomo, T. Makino, K. Tamura, Y. Matsumoto, Y. Segawa, Z. Tang, G. K. L. Wong, H. Koinuma, M. Kawasaki, *Proc. SPIE-Int. Soc. Opt. Eng.*, **3941**, 70–81 (2000).

7. A. W. Bosman, A. Heumann, G. Klaerner, D. Benoit, J. M. J. Frechet, C. J. Hawker, *J. Am. Chem. Soc.* **123**, 6461–6462 (2001).

8. A. M. Cassell, S. Verma, L. Delzeit, M. Meyyappan, J. Han, *Langmuir* **17**, 260–264 (2001).

9. X.-D. Sun, C. Gao, J. Wang, X.-D. Xiang, *Appl. Phys. Lett.* **70**, 3353–3355 (1997).

10. A. W. Czarnik, S. H. DeWitt, eds., *A Practical Guide to Combinatorial Chemistry*, American Chemical Society, Washington, DC, 1997.

11. J. M. Newsam, F. Schuth, *Biotechnol. Bioeng. (Comb. Chem.)* **61**, 203–216 (1998/1999).

12. V. W. Rosso, J. L. Pazdan, J. J. Venit, *Org. Process Res. Dev.* **5**, 294–298 (2001).

13. J. C. Meredith, A. Karim, E. J. Amis, *Macromolecules* **33**, 5760–5762 (2000).

14. J. C. Meredith, A. P. Smith, A. Karim, E. J. Amis, *Macromolecules* **33**, 9747–9756 (2000).

15. R. A. Potyrailo, W. G. Morris, B. J. Chisholm, J. N. Cawse, M. J. Brennan, C. A. Molaison, L. Hassib, W. P. Flanagan, H. Reitz, G. F. Medford, *Pittsburgh Conference on Analytical Chemistry and Applied Spectroscopy,* March 17–22 (2002). New Orleans, Paper 751.

16. Vision 2020. *Technology Roadmap for Combinatorial Methods*, 2001.

17. F. Lombard, *Chemom. Intell. Lab. Syst.* **37**, 281–289 (1997).

18. J. A. Calvin, *Chemom. Intell. Lab. Syst.* **37**, 291–294 (1997).

19. G. J. Lyman, *Chemom. Intell. Lab. Syst.* **37**, 295–297 (1997).

20. R. A. Potyrailo, in K. H. J. Buschow, R. W. Cahn, M. C. Flemings, B. Ilschner, E. J. Kramer, S. Mahajan, eds., *Encyclopedia of Materials: Science and Technology*, Elsevier, Amsterdam, Vol. **2**, 1329–1343 (2001).

21. R. A. Potyrailo, *Proc. SPIE-Int. Soc. Opt. Eng.*, **4578**, 366–377 (2002).

22. H.-U. Gremlich, *Biotechnol. Bioeng. (Comb. Chem.)* **61**, 179–187 (1998/1999).

23. M. E. Swartz, ed., *Analytical Techniques in Combinatorial Chemistry*, Dekker, New York (2000).

24. S. Schmatloch, M. A. R. Meier, U. S. Schubert, *Macromol. Rapid Comm.* **24**, 33–46 (2003).

25. Y. Luo, X. Ouyang, R. W. Armstrong, M. M. Murphy, *J. Org. Chem.* **63**, 8719–8722 (1998).

26. P. Cong, R. D. Doolen, Q. Fan, D. M. Giaquinta, S. Guan, E. W. McFarland, D. M. Poojary, K. W. Self, H. W. Turner, W. H. Weinberg, *Angew. Chem. Int. Ed.* **38**, 484–488 (1999).

27. B. Yan, H. Yan, *J. Comb. Chem.* **3**, 78–84 (2000).

28. J. Lauterbach, C. M. Snively, G. Oskarsdottir, in R. Malhotra, ed., *Combinatorial Approaches to Materials Development*, American Chemical Society, Washington, DC, Vol. **814**, 65–5 (2002).

29. M. Fischer, C. D. Tran, *Anal. Chem.* **71**, 2255–2261 (1999).

30. S. M. Senkan, *Nature* **394**, 350–353 (1998).

31. A. Holzwarth, H.-W. Schmidt, W. Maier, *Angew. Chem. Int. Ed.* **37**, 2644–2647 (1998).

32. S. J. Taylor, J. P. Morken, *Science.* **280**, 267–270 (1998).

33. E. W. McFarland, W. H. Weinberg, *Trends Biotechnol.* **17**, 107–115 (1999).

34. H. Su, E. S. Yeung, *J. Am. Chem. Soc.* **122**, 7422–7423 (2000).

35. P. J. Stirn, T. C. Hofelich, R. R. Tamilarasan, in I. Takeuchi, J. M. Newsam, L. T. Wille, H. Koinuma, E. J. Amis, eds., *MRS Symposium Proceedings. Combina-*

torial and Arficial Intelligence Methods in Materials Science; Materials Research Society, Warrendale, PA, Vol. **700**, 145–150 (2002).

36. R. Takahashi, Y. Matsumoto, H. Koinuma, M. Lippmaa, M. Kawasaki, in I. Takeuchi, J. M. Newsam, L. T. Wille, H. Koinuma, E. J. Amis, eds., *MRS Symposium Proceedings. Combinatorial and Arficial Intelligence Methods in Materials Science*; Materials Research Society, Warrendale, PA, Vol. **700**, 13–18 (2002).

37. E. Reddington, A. Sapienza, B. Gurau, R. Viswanathan, S. Sarangapani, E. S. Smotkin, T. E. Mallouk, *Science* **280**, 1735–1737 (1998).

38. T. F. Jaramillo, A. Ivanovskaya, E. W. McFarland, *J. Comb. Chem.* **4**, 17–22 (2002).

39. R. A. Potyrailo, *Pittsburgh Conference on Analytical Chemistry and Applied Spectroscopy,* March 17–22, New Orleans, (2002), Paper 755.

40. A. R. Connolly, J. D. Sutherland, *Angew. Chem. Int.. Ed.* **39**, 4268–4271 (2000).

41. R. A. Potyrailo, J. E. Pickett, *Angew. Chem. Int. Ed.* **41**, 4230–4233 (2002).

42. R. A. Potyrailo, R. J. Wroczynski, J. E. Pickett, M. Rubinsztajn, *Macromol. Rapid Comm.* **24**, 123–130 (2003).

43. C. de Bellefon, N. Tanchoux, S. Caravieilhes, H., V. Grenouillet, *Angew. Chem. Int. Ed.* **39**, 3442–3445 (2000).

44. R. Hoogenboom, M. W. M. Fijten, C. Brändli, J. Schroer, U. S. Schubert, *Macromol. Rapid Commun.* **24**, 98–103 (2003).

45. R. Neffati, A. Alexeev, S. Saunin, J. C. M. Brokken-Zijp, D., Wouters, S. Schmatloch, U. S. Schubert, J. Loos, *Macromol. Rapid Commun.* **24**, 113–117 (2003).

46. B. Yan, H.-U. Gremlich, S. Moss, G. M. Coppola, Q. Sun, L. Liu, *J. Comb. Chem.* **1**, 46–54 (1999).

47. R. A. Potyrailo, *Trends Anal. Chem.* **22**, 374–384 (2003).

48. R. Hoogenboom, M. A. R. Meier, U. S. Schubert, *Macromol. Rapid Commun.* **24**, 15–32 (2003).

49. R. A. Potyrailo, S. E. Hobbs, G. M. Hieftje, *Fresenius' J. Anal. Chem.* **362**, 349–373 (1998).

50. A. Mandelis, C. Christofides, *Physics, Chemistry and Technology of Solid State Gas Sensor Devices*, Wiley, New York, 1993.

51. H. Baltes, W. Göpel, J. Hesse, eds., *Sensors Update*, VCH, Weinheim, 1996.

52. J. Janata, M. Josowicz, P. Vanysek, D. M. DeVaney, *Anal. Chem.* **70**, 179R–208R (1998).

53. K. J. Albert, N. S. Lewis, C. L. Schauer, G. A. Sotzing, S. E. Stitzel, T. P. Vaid, D. R. Walt, *Chem. Rev.* **100**, 2595–2626 (2000).

54. S. Soloman, *Sensors Handbook*, McGraw-Hill, New York, 1999.

55. R. Malhotra, Epilogue, in R. Malhotra, ed., *Combinatorial Approaches to Materials Development*, American Chemical Society: Washington, DC, Vol. **814**, 165–172 (2002).

56. K. S. Booksh, B. R. Kowalski, *Anal. Chem.* **66**, 782A–791A (1994).

57. R. A. Potyrailo, B. J. Chisholm, D. R. Olson, M. J. Brennan, C. A. Molaison, *Anal. Chem.* **74**, 5105–5111 (2002).

58. R. A. Potyrailo, J. P. Lemmon, *Method and apparatus for obtaining fluorescence data*, US Patent 6166804 (2000).

59. R. A. Potyrailo, J. P. Lemmon, *Method for direct measurement of polycarbonate compositions by fluorescence*, US Patent 6193850 (2001).

60. R. A. Potyrailo, R. J. May, R. E. Shaffer, J. P. Lemmon, R. J. Wroczynski, *Method for high-throughput fluorescence screening of polymerization reactions*, PCT Int. Appl. WO 02/33384 A1 (2002).

61. R. A. Potyrailo, J. P. Lemmon, T. K. Leib, *Aromatic polycarbonate characterization*, PCT Int. Appl. WO 02/33386 A1: 2002.

62. E. Sanchez, B. R. Kowalski, *J. Chemom.* **2**, 265 (1988).

63. Z. Lin, K. S. Booksh, L. W. Burgess, B. R. Kowalski, *Anal. Chem.* **66**, 2552–2560 (1994).

64. K. S. Booksh, Z. Lin, Z. Wang, B. R., Kowalski, *Anal. Chem.* **66**, 2561–2569 (1994).

65. B. K. Lavine, Chemometrics, *Anal. Chem.* **72**, 91R–97R (2000).

66. B. K. Lavine, J. Workman, Jr. *Anal. Chem.* **74**, 2763–2770 (2002).

67. K. Rajan, C. Suh, A. Rajagopalan, X. Li, in I. Takeuchi, J. M. Newsam, L. T. Wille, H. Koinuma, E. J. Amis, eds., *MRS Symposium Proceedings. Combinatorial and Arficial Intelligence Methods in Materials Science*, Materials Research Society, Warrendale, PA, Vol. **700**, 223–232 (2002).

68. J. Singh, M. A. Ator, E. P. Jaeger, M. P. Allen, D. A. Whipple, J. E. Soloweij, S. Chowdhary, A. M. Treasurywala, *J. Am. Chem. Soc.* **118**, 1169–1679 (1996).

69. D. T. Stanton, T. W. Morris, S. Roychoudhury, C. N. Parker, *J. Chem. Inf. Comput. Sci.* **39**, 21–27 (1999).

70. C. H. Reynolds, *J. Comb. Chem.* **1**, 297–306 (1999).

71. A. Linusson, J. Gottfries, F. Lindgren, S. Wold, *J. Med. Chem.* **43**, 1320–1328 (2000).

72. S. Guessasma, G. Montavon, C. Coddet, in I. Takeuchi, J. M. Newsam, L. T. Wille, H. Koinuma, E. J. Amis, eds., *MRS Symposium Proceedings. Combinatorial and Arficial Intelligence Methods in Materials Science*, Materials Research Society, Warrendale, PA, Vol. **700**, 253–258 (2002).

73. R. A. Potyrailo, R. J. Wroczynski, J. P., Lemmon, W. P. Flanagan, O. P. Siclovan, *J. Comb. Chem.* **5**, 8–17 (2003).

74. R. A. Potyrailo, R. J. May, *Rev. Sci. Instrum.* **73**, 1277–1283 (2002).

75. K. R. Beebe, R. J. Pell, M. B. Seasholtz, *Chemometrics: A Practical Guide*, Wiley, New York (1998).

76. M. Otto, *Chemometrics: Statistics and Computer Application in Analytical Chemistry*, Wiley-VCH, Weinheim, Germany (1999).

77. H. Martens, T. Nœs, *Multivariate Calibration*, Wiley, New York (1989).

78. J. M. Henshaw, L. W. Burgess, K. S. Booksh, B. R. Kowalski, *Anal. Chem.* **66**, 3328–3336 (1994).

79. M. Meader, *Anal. Chem.* **59**, 527–530 (1987).

80. C. S. P. C. O. Silva, J. C. G. Esteves da Silva, A. A. S. C. Machado, *Appl. Spectrosc.* **48**, 363–372 (1994).

81. S. Nigam, A. de Juan, V. Cui, S. C. Rutan, *Anal. Chem.* **71**, 5225–5234 (1999).

82. R. Tauler, A. K. Smilde, J. W. Henshaw, L. W. Burgess, B. R. Kowalski, *Anal. Chem.* **66**, 3337–3344 (1994).

83. R. A. Potyrailo, J. P. Lemmon, J. C. Carnahan, R. J. May, T. K. Leib, in *Pittsburgh Conference on Analytical Chemistry and Applied Spectroscopy*, March 12–17, New Orleans, (2000), Paper 1173.

84. J. C. Carnahan, J. P. Lemmon, R. A. Potyrailo, T. K. Leib, G. L. Warner, *Method for parallel melt-polymerization*, US Patent 6307004 B1 (2001).

85. J. P. Lemmon, R. J. Wroczynski, D. W. Whisenhunt Jr., W. P. Flanagan, *222nd ACS National Meeting* August 26–30, Chicago, 2001.

86. J. A. King Jr., in T. E. Long and M. O. Hunt, eds., *Solvent-Free Polymerizations and Processes. Minimization of Conventional Organic Solvents*; American Chemical Society, Washington, DC, Vol. **713**, 49–77 (1998).

87. P. J. McCloskey, T. B. Burnell, P. M. Smigelski, *Alkali metal salts of oxoacids of sulfur as polymerization catalysts*, US Patent 6184334 B1 (2001).

88. R. A. Potyrailo, R. E. Shaffer, P. J. McCloskey, *In situ determination of DPC and BPA in polycarbonate by Raman spectroscopy*, PCT Int. Appl. WO 02, 33388 A1 (2002).

89. V. F. Gachkovskii, *Vysokomolekul. Soedin.* **7**, 2199–2205 (1965).

90. Y. Nishijima, *J. Polym. Sci. Part C* **31**, 353–373 (1970).

91. N. S. Allen, J. Homer, J. F. McKellar, *Analyst* **101**, 260–264 (1976).

92. H. Morawetz, *Science* **203**, 405–410 (1979).

93. S. W. Beavan, J. S. Hargreaves, D. Phillips, *Adv. Photochem.* **11**, 207–303 (1979).

94. N. S. Allen, in L. S. Bark, N. S. Allen, eds., *Analysis of Polymer Systems*, Applied Science Publ., Barking, UK, 79–102 (1982).

95. M. H. Chipalkatti, J. J. Laski, in M. W. Urban, C. D. Craver, eds., *Structure-Property Relations in Polymers. Spectroscopy and Performance*; American Chemical Society, Washington, DC, Vol. **236**, 623–642 (1993).

96. I. Soutar, L. Swanson, in N. S. Allen, M. Edge, I. R. Bellobono, E. Selli, eds., *Current Trends in Polymer Photochemistry*, Ellis Horwood, New York, 1–22 (1995).

97. H. Itagaki, Fluorescence spectroscopy, in T. Tanaka, ed, *Experimental Methods in Polymer Science*, Academic Press, San Diego, CA, 155–260 (2000).

98. J. P. Lemmon, R. J. Wroczynski, D. W. Whisenhunt Jr., W. P. Flanagan, *Polymer Preprints* **42**, 630–631 (2001).

99. S. N. Hersh, K. Y. Choi, *J. Appl. Polym. Sci.* **41**, 1033–1046 (1990).

100. Y. Kim, K. Y. Choi, *J. Appl. Polym. Sci.* **49**, 747–764 (1993).

101. H. Martens, M. Martens, *Multivariate Analysis of Quality: An Introduction*, Wiley, New York (2001).

102. B. M. Wise, N. B. Gallagher, *PLS_Toolbox Version 2.1 for Use with MATLAB*; Eigenvector Research, Inc.: Manson, WA (2000).

103. R. A. Potyrailo, organizer, *Invited Symposium "Analytical Challenges and Opportunities in Combinatorial Chemistry"*, Pittsburgh Conference on Analytical Chemistry and Applied Spectroscopy, March 12–17, New Orleans (2000).

104. R. A. Potyrailo, organizer, *Invited Symposium "Analytical Tools For High Throughput Chemical Analysis And Combinatorial Materials Science"*, Pittsburgh Conference on Analytical Chemistry and Applied Spectroscopy, March 4–9, New Orleans (2001).

105. S. Borman, *C&EN* **80**, 49–52 (2002).

106. J. F. Haw, ed., *In Situ Spectroscopy in Heterogeneous Catalysis*, Wiley, New York (2002).

107. M. Hunger, J. Weitkamp, *Angew. Chem. Int. Ed.* **40**, 2954–2971 (2001).

CHAPTER
6

MASS SPECTROMETRY AND SOLUBLE POLYMERIC SUPPORT

CHRISTINE ENJALBAL, FREDERIC LAMATY, JEAN MARTINEZ, and JEAN-LOUIS AUBAGNAC

6.1. INTRODUCTION

Polymeric supports of variable solubility have been investigated as an alternative to insoluble supports used in solid-phase synthesis.[1,2] Reactions are performed in homogeneous media by choosing an appropriate solvent that solubilizes the polymer, and purification is performed by precipitation. This methodology benefits both solution-phase and solid-phase syntheses.[3,4] Moreover compound characterization can be easily undertaken at any stage of the synthesis, since the support is soluble in standard spectroscopic solvents.[5,6] A direct real-time control is possible, whereas a solid-phase protocol relies on a "cleave and analyze" strategy that consumes compound, imparts delay, and thus can only be accomplished at the end of synthesis. For these reasons soluble polymeric supports are preferred to conventional insoluble supports (resins, plastic pins), and they are compatible with analytical techniques such as NMR and mass spectrometry.

Mass spectrometry allows performing compound characterization in a sensitive, specific, and rapid manner (high-throughput) that is particularly important in CombiChem applications. The development of ionization techniques such as ESI[7] and MALDI[8] allows the analysis of high molecular weight compounds broadening the field of polymer analysis by mass spectrometry.[9]

Among all reported polymers used as soluble support, poly(ethyleneglycol) (PEG) is the most popular.[10] Such material is relatively polydisperse with average molecular weights around 2000, 3400, or 5000. Water-soluble PEG of lower molecular weights (<1000 mass units) are widely used as biocompatible polymers.[11] Indeed, the solubility in water increases with decreasing molecular weight. In the frame of liquid-phase syntheses, two

Analysis and Purification Methods in Combinatorial Chemistry, Edited by Bing Yan.
ISBN 0-471-26929-8 Copyright © 2004 by John Wiley & Sons, Inc.

options are reported in the literature concerning the number of anchoring sites on each oligomer: monofunctional PEG (MeO–PEG–OH) and bifunctional PEG (HO–PEG–OH) were employed. The former provides only one hydroxyl group for chemical modifications, whereas the latter allows to double the loading of the polymer. Both molecular weight and number of reactive sites govern the precipitation properties of the polymer and are chosen from a synthetic point of view.

Mass spectrometry was already used to monitor reactions carried out on PEG without releasing the growing compound from the support. Both MALDI and ESI ionization techniques were reported. Relevant examples will be chosen from the literature. The pros and cons of the two analytical strategies will be discussed.

6.2. RESULTS AND DISCUSSION

6.2.1. Analysis of PEG by MALDI Mass Spectrometry

MALDI was the first ionization method used to monitor peptide synthesis carried out on the monofunctional MeO–PEG–OH possessing an average molecular weight of 5000.[5] The recorded positive ion mass spectra showed one gaussian ion distribution corresponding to monoprotonated species. Thus all ions in the cluster were separated by 44 Da, the mass of the ethylene oxide unit. Characterization of the product was performed step by step along the synthetic scheme by choosing a reference ion in the positive mass spectrum of the native PEG and adding to that ion the theoretical mass increment corresponding to the hydroxyl substitution. The observed ions corresponded to the calculated one within 1 Da error. Such iterative method provided a convenient way to rapidly monitor syntheses on soluble support. The same methodology was applied to the monitoring of more complex chemistry performed on PEG.[12,13] For instance, characterization of a ruthenium catalyst anchored to bifunctional PEG_{3400} was performed with the α-cyano-4-hydroxycinnamic acid matrix. As expected for the analyses of polymers by desorption methods, ionization occurred by cationization,[9,14] with both sodium and potassium cations. In our case the most abundant distribution corresponded to the potassium adducts, thus verifying that high molecular weight PEG accommodate preferentially alkali metal ions of large ionic radii.[14] The two overlapping ion distributions displayed in Figure 6.1 were sufficiently well defined to confirm the substituent structure. The most abundant ion at m/z 3772.8 corresponds to the potassium adduct of the oligomer possessing 60 ethylene oxide units. These ruthenium-based substituents exhibit different ionization properties compared to peptides.

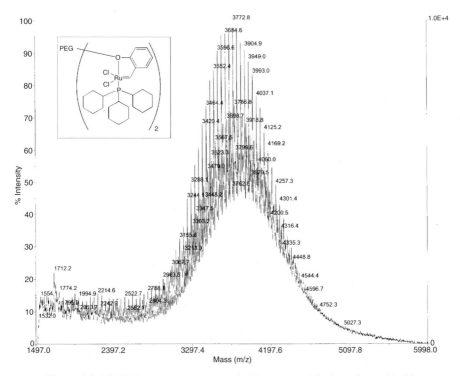

Figure 6.1. MALDI mass spectrum recorded in α-cyano-4-hydroxycinnamic acid.

Whatever the chemistry, MALDI mass spectrometry has proved to be very effective in validating results of synthesis performed on PEG. Nevertheless some drawbacks were spotted. First, a chemical reactivity directed by the acidic properties of the matrix co-crystallized with the sample under investigation was observed with acid-sensitive Schiff bases.[15] The side reaction was avoided when a nonacidic matrix (2,4,6-trihydroxyacetophenone) was used. Second, mixture analysis was quite difficult to conduct due to the presence of large overlapping distributions. For these two reasons ESI mass spectrometry was also investigated to monitor PEG-mediated syntheses[6,16] as it requires no specific sample treatment prior to analysis and is easily coupled to a liquid chromatography to profile mixtures.

6.2.2. Analysis of PEG by ESI Mass Spectrometry

ESI mass spectra of PEG was far more difficult to analyze than the spectra recorded by MALDI because multicharged species were produced.[6,16] Several ion clusters possessing a charge state varying from +2 to +6, accord-

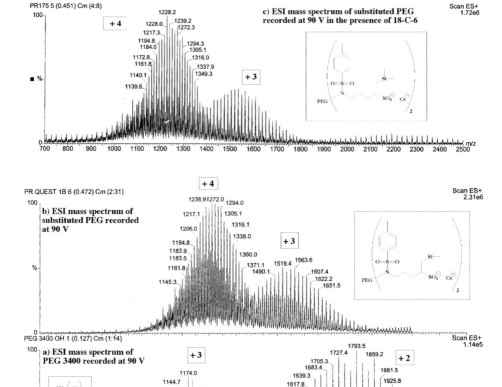

Figure 6.2. ESI positive mass spectra recorded at 90 V: (*a*) PEG$_{3400}$, (*b*) substituted PEG$_{3400}$, (*c*) substituted PEG$_{3400}$ with 18-C-6.

ing to the sample and the experimental conditions, were present in the positive ion mass spectra of heavy PEG (PEG$_{2000}$ and above). The charge state *z* of each distribution must be defined by searching the mass difference between two consecutive ions (44/*z*). As illustrated in Figure 6.2*a*, the most abundant ion cluster of native PEG$_{3400}$ contained three consecutive ions at *m/z*: 1683.4, 1705.3, and 1727.4. The constant mass difference of 22 Th[17] (44/2) indicated a doubly charged distribution. The mass increment Δ*m*

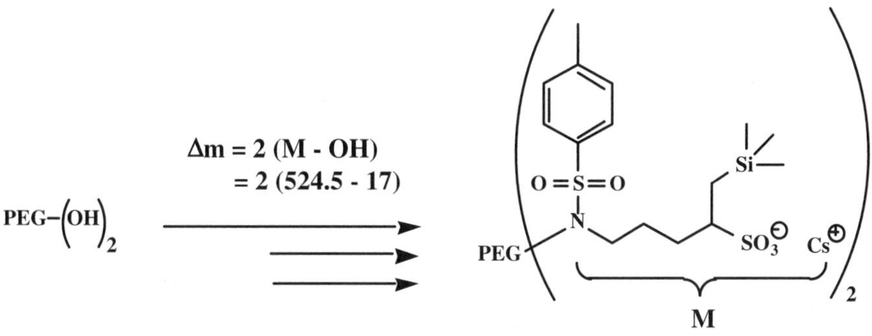

$$\Delta m = 2 \, (M - OH)$$
$$= 2 \, (524.5 - 17)$$

PEG$-$(OH)$_2$

Scheme 6.1. Studied substituted PEG$_{3400}$ and corresponding calculation of mass increment.

between the starting material and the studied product must be calculated for one chosen distribution ($\Delta m/z$). For the studied compound (shown in Scheme 6.1), the mass increment equaled to 507.5 mass units for each anchoring site (524.5 for the mass of the amino substituent, 17 for the mass of the hydroxyl goup that was functionalized). The expected mass shift between the distribution of the two quadruply charged ions of native PEG$_{3400}$ and the product should be $(2 \times 507.5)/4 = 253.75$ Th. The mass spectrometry analysis of this compound could not be performed in the positive mode because of its salt character. The analysis in the positive mode required either protonation of the sulfate group or retention of the cesium cation on the sulfate anion to yield a neutral molecule that could then undergo ionization. Unfortunately, the two situations cannot be differentiated from the recorded data as replacement of the cesium cation by a proton led to a mass difference of 132 Da, which equals to 3 ethylene oxide units (44 Da).

One could simply assume that the most abundant ion of any distribution in the starting material mass spectrum would lead as well to the most intense signal in the corresponding distribution of the product mass spectrum, but this was never the case due to some spectral discrimination. Measuring the mass shift between one distribution in the native PEG$_{3400}$ mass spectrum and the corresponding distribution in the product mass spectrum could be troublesome. Such simple protocol worked pretty well as long as two similarly charged ions belonging to the starting and final product mass spectra exhibited the expected mass difference. As an illustration (Figures 6.2a and 6.2b), the ion at m/z 941.7 in the quadruply charged cluster of native PEG$_{3400}$ led to the ion at m/z 1194.8 (calculated 1195.4) in the corresponding distribution of the product. One has to keep in mind that the described calculation only takes into account the substituent mass. The calculated mass increment ($\Delta m/z$) will be correct if the two considered distributions exhibit not only the same charge state but also the same type of charge (proton or alkali metal). Knowledge of the ionization type (protonation or cationization) is thus required prior to any spectral assignment. The situation became even more complicated since the ionization mechanism was found to occur by both protonation and cationization. The difficulty resided in the fact that the response was not homogeneous across the mass spectrum, some distributions were composed of protonated species and others were exhibiting cationized ions. For example, one of the most abundant doubly charged ion of PEG_{3400} at m/z 1705.3 (n = 77) is doubly protonated, whereas the corresponding ion (n = 77) in the triply charged distribution at m/z 1144.7 is both protonated and cationized ($2H^+/Na^+$). Just applying the calculated mass shift on that distribution [($2 \times$ 507.5)/3 = 338.3 Th] would have led to an incorrect ion attribution: for example, the ion at m/z 1232.6 (n = 83, $2H^+ + Na^+$) in the triply charged cluster of native PEG_{3400} (Figure 6.2a) led to the ion at m/z 1563.6 (n = 83, $3H^+$) and not 1232.6 + 338.3 = 1570.9 Th in the corresponding distribution of the product (Figure 6.2b). The extent of cationization versus protonation, which is clearly seen in Figure 6.2b as overlapping ion distributions, was found to vary according to experimental conditions, in particular, the sample concentration influenced greatly the ionization mechanism.[18]

Although mass spectra of polymers are most commonly described in the literature in the positive mode, the negative ion mass spectrum of the studied sample was recorded (Figure 6.3). This mass spectrum exhibited only doubly charged species of the type $(M-2H)^{2-}$ due to the two negative charges of the two sulfonate end groups. In this case the pre-formed doubly charged anions were directly detected without requiring the ionization step.[19] A unique ion distribution was thus obtained, providing an ESI mass spectrum similar to the MALDI mass spectrum. The ion attribution was far simpler than that for the positive mass spectrum, confirming the structure of the end groups (mass increment of 391 mass units corresponding to the free sulfonate ion). Acquisition of such negative ion mass spectrum was only possible if a negatively charged moiety was present on the polyethylene oxide skeleton.

Taking into account that the most abundant ion in one distribution of the native PEG was not necessarily the same in the corresponding distribution of the product and that the ionization mechanism was fluctuating, automatic simulation of PEG product spectra by simply shifting the ion distributions recorded in native PEG mass spectrum was not possible.

The aim was to produce in ESI mass spectrometry the simplest mass spectra as possible to automate both data acquisition and interpretation

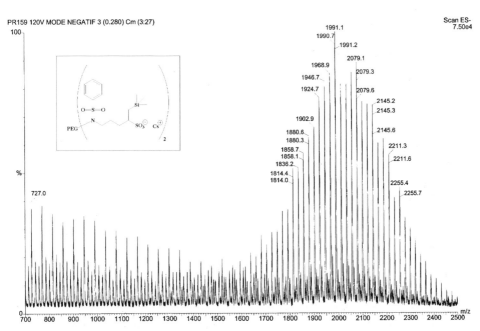

Figure 6.3. ESI negative mass spectrum of substituted PEG$_{3400}$ recorded at 120 V.

opening the field of high-throughput analyses required in the frame of combinatorial strategies. Attempts to simplify the ESI mass spectra to favor low-charge state distributions and a single mechanism of ionization were undertaken to facilitate ion assignments. Although PEG were known to exhibit high affinity for alkali metal,[14,20] extensive cationization was never reached even with the addition of highly concentrated NaI solutions. Three phenomena—polydispersity—multicharging and composite ionization— were cumulated and provided relatively complex mass spectra that necessitated skilled analysts to unravel them.

The breakthrough came from the addition of an alkali cation scavenger (crown ether 18-C-6) that could suppress all cationized ions provided that its concentration was at least 10^{-4} M[21], as shown in Figure 6.2c. All studied samples behaved similarly, indicating that this protocol could be easily implemented without any specific sample preparation or instrument tuning. Automated analyses were performed and no cross-contamination between samples were observed. An algorithm[21] developed in-house, and based on the calculation previously described (mass increment, charge state, type of ionization), enabled the automated PEG mass spectra to make interpretations according to the pass/fail criterion used in the profiling of libraries:[22] the ions of the expected products were detected and the reaction was declared successful; otherwise, the synthesis was rejected.

In the case of nonquantitative PEG-supported syntheses, purification by ion-exchange chromatography has been reported in the literature prior to fraction characterization.[23] Direct analysis of mixture was attempted by LC/MS[21] to avoid time-consuming preparative sample fractionation. The behavior of PEG_{3400} samples on reversed-phase LC columns was our first concern, since relatively sharp peaks are required to ensure a correct component separation in mixtures. PEG of lower molecular weights have been analyzed by liquid chromatography[24] and their constitutive oligomers were separated under specific conditions. On the other hand, fast chromatography (2.0×2.1 mm, 3.5μm) has shown single signals for each type of PEG samples possessing molecular weights under 1000.[25] Detection was performed in SIR (one monoprotonated ion per sample) to follow easily the chromatographic behavior. In our case discrimination among oligomers belonging to the same compound had to be avoided. Baseline separation was only expected among differently substituted PEG_{3400}. Since this polymer is very polydisperse ($60 \leq n \leq 100$), tailing on the stationary phase and peak width were our main concerns, corresponding to the literature data on the low molecular weight of PEG.[25] It appears that PEG oligomers have been partitioned onto the stationary phase but within a single tough enlarged chromatographic signal under gradient elution. We used differently substituted PEG that were separated in large column dimensions (250×4.6 mm, 5μm, 30 min gradient). An example is given in Figure 6.4.

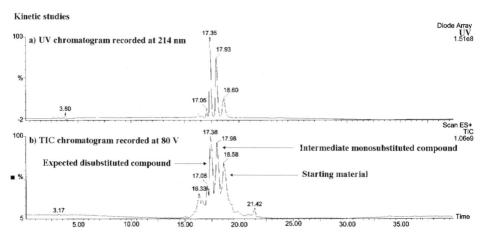

Figure 6.4. LC/MS experiment: (*a*) UV chromatogram, (*b*) TIC chromatogram. (ref. 21: Enjalbal, C., Anal. Chem., **75**, 175–184 (2003). Copyright 2002 American Chemical Society)

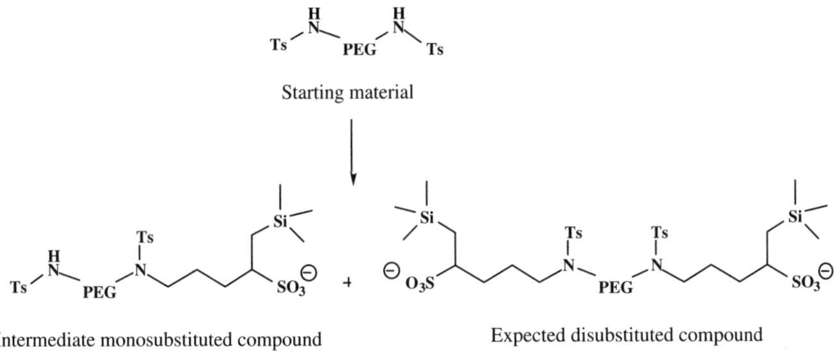

Starting material

Intermediate monosubstituted compound   Expected disubstituted compound

Scheme 6.2. Reaction on PEG$_{3400}$ monitored by LC/MS.

Our kinetic study[21] followed the formation of the product shown in Figure 6.2 according to the synthetic pathway represented in Scheme 6.2. Three compounds were evidenced that corresponded to the starting and expected products, together with the monosubstituted intermediate product. To interpret easily LC/MS data, we generated the mass spectra from a portion of the chromatographic signal to sample only few oligomers and thus provided rather narrow ion distributions. Unfortunately, any attempts to decrease the run times led to poorly resolved PEG chromatographic signals.

6.3. CONCLUSION

The use of soluble supports to generate combinatorial libraries has been reported in the literature,[26–28] but complete automation for both synthesis and analysis must be effected to probe such methodology as a viable alternative to solid-phase synthesis protocols. The discussed results indicate that poly(ethylene glycol) samples could be handled under certain operation conditions as standard chemicals and subjected to high-throughput combinatorial analytical process.

REFERENCES

1. E. Bayer, M. Mutter, *Nature* **237**, 512 (1972).
2. H. Hans, M. M. Wolfe, S. Brenner, K. D. Janda, *Proc. Nat. Acad. Sci. USA.* **92**, 6419–6423 (1995).

3. D. J. Gravert, K. D. Janda, *Curr. Opin. Chem. Biol.* *1*, 107–113 (1997).

4. D. J. Gravert, K. D. Janda, *Chem. Rev.* *33*, 489–509 (1997).

5. R. Thürmer, M. Meisenbach, H. Echner, A. Weiler, R. A. Al-Qawasmeh, W. Voelter, U. Korff, W. Schmitt-Sody, *Rapid Commun. Mass Spectrom.* *12*, 398–402 (1998).

6. B. Sauvagnat, C. Enjalbal, F. Lamaty, R. Lazaro, J. Martinez, J.-L. Aubagnac *Rapid Commun. Mass Spectrom.* *12*, 1034–1037 (1998).

7. J. B. Fenn, M. Mann, C. K. Meng, S. F. Wong, C. M. Whitehouse, *Science 246*, 64–71 (1989).

8. M. Karas, D. Bachmann, U. Bahr, F. Hillemkamp, *Int. J. Mass Spectrom. Ion Proc.* *78*, 53–68 (1987).

9. G. Montaudo, R. P. Lattimer, *Mass Spectrometry of Polymers*, CBC Press, Boca Raton, FL (2002).

10. P. Wentworth, *Trends Biotechnol.* *17*, 448–452 (1999).

11. K. J. Wu, R. W. Odom, *Anal. Chem. News Features*, 456A–461A (1998).

12. B. Sauvagnat, F. Lamaty, R. Lazaro, J. Martinez, *Tetrahedron Lett.* *41*, 6371–6375 (2000).

13. S. Varray, C. Gauzy, F. Lamaty, R. Lazaro, J. Martinez, *J. Org. Chem.* *65*, 6787–6790 (2000).

14. S. Keki, L. S. Szilagyi, G. Deak, M. Zsuga, *J. Mass Spectrom.* *37*, 1074–1080 (2002).

15. C. Enjalbal, B. Sauvagnat, F. Lamaty, R. Lazaro, J. Martinez, P. Mouchet, F. Roux, J.-L. Aubagnac, *Rapid Commun. Mass Spectrom.* *13*, 1775–1781 (1999).

16. F. Nativel, C. Enjalbal, F. Lamaty, R. Lazaro, J. Martinez, J.-L. Aubagnac, *Eur. Mass Spectrom.* *4*, 233–237 (1998).

17. R. G. Cooks, A. L. Rockwood, *Rapid Commun. Mass. Spectrom.* *5*, 93 (1991).

18. S. Varray, J.-L. Aubagnac, F. Lamaty, R. Lazaro, J. Martinez, C. Enjalbal, *Eur. J. Anal. Chem.* *28*, 263–268 (2000).

19. K. L. Busch, S. E. Unger, A. Vincze, R. G. Cooks, T. Keough, *J. Am. Chem. Soc.* *104*, 1507–1511 (1982).

20. M. J. Bogan, G. R. Agnes, *J. Am. Soc. Mass Spectrom.* *13*, 177–186 (2002).

21. C. Enjalbal, F. Lamaty, P. Sanchez, E. Suberchicot, P. Ribière, S. Varray, R. Lazaro, N. Yadav-Bhatnagar, J. Martinez, J.-L. Aubagnac, *Anal. Chem.*, *75*, 175–184 (2003).

22. N. Shah, M. Gao, K. Tsutsui, A. Lu, J. Davis, R. Scheuerman, W. L. Fitch, R. L. Wilgus, *J. Comb. Chem.* *2*, 453–460 (2000).

23. S. Drioli, F. Benedetti, G. M. Bonora, *React. Funct. Poly.* *48*, 119–128 (2001).

24. Y. Wang, H. Rashidzadeh, B. Guo, *J. Am. Soc. Mass Spectrom.* *11*, 639–643 (2000).

25. J. J. Palmgrén, E. Toropainen, S. Auriola, A. Urtti, *J. Chromatog. A976*, 165–170 (2002).

26. E. Bayer, M. Mutter, G. Holzer, in R. Walter, J. Meienhofer, eds., *Peptides: Chemistry, Structure and Biology*, Ann Arbor Sci. Publ., Ann Arbor, MI, pp. 425–432 (1975). K. Rissler, N. Wyttenbach, K. O. Bornsen, *J. Chromatog. A822*, 189–206 (1998).

27. C. M. Sun, *Comb. Chem. High Through. Screen. 2*, 299–318 (1999).

28. B. Sauvagnat, K. Kulig, F. Lamaty, R. Lazaro, J. Martinez, *J. Comb. Chem. 2*, 134–142 (2000).

PART
II

HIGH-THROUGHPUT ANALYSIS FOR LIBRARY QUALITY CONTROL

CHAPTER

7

HIGH-THROUGHPUT NMR TECHNIQUES FOR COMBINATORIAL CHEMICAL LIBRARY ANALYSIS

TING HOU and DANIEL RAFTERY

7.1. INTRODUCTION

Recently, due to its capabilities for providing efficient synthesis and product screening strategies, combinatorial chemistry methods have been rapidly and widely adopted with great impact in areas such as pharmaceutical research, organic synthesis, and catalysis discovery.[1,2,3,4] New chemical synthetic strategies such as combinatorial chemistry[5,6] and parallel organic synthesis methods[7,8] have revolutionized the way new compounds are synthesized, screened, and characterized. For example, the application of combinatorial chemistry and high-throughput screening in drug discovery has changed the traditional serial process of drug candidate identification and optimization. It is now possible to generate chemical libraries with large numbers of compounds in a relatively short time and to screen them against a variety of targets to find the "lead" compounds of interest for further study and optimization. As a result of the accelerated pace of drug discovery and the large numbers of new compounds that are being synthesized in combinatorial chemical processes, high-throughput analysis methods are highly desired. Efforts have been made to meet this requirement in a number of analytical areas, particularly high-throughput mass spectrometry.[9,10]

Due to its high resolution and unique capabilities in structure determination, nuclear magnetic resonance (NMR) spectroscopy is extremely useful for a variety of chemical analyses. In particular, the need for high-throughput NMR approaches for drug discovery has been expressed and such methods are being developed.[11] There are several areas in which this development is being focused, including NMR drug screening, complex mixture analysis, and combinatorial library characterization.

Analysis and Purification Methods in Combinatorial Chemistry, Edited by Bing Yan.
ISBN 0-471-26929-8   Copyright © 2004 by John Wiley & Sons, Inc.

In the area of drug screening, several new NMR techniques provide methods to detect weak protein-ligand interactions that are difficult to observe using fluorescent screening approaches, but provide an alternative starting point for drug development. For example, Fesik and co-workers described methods to investigate structure-activity relations (SAR) by NMR[12] to determine protein-drug interactions and to aid in the discovery and refinement of potential drug molecules. Two-dimensional (2D) protein NMR experiments were performed using 10 to 100 potential drug compounds at the same time at the rate of 1 every 10 minutes, allowing 1000 to 10,000 samples to be screened over a 24-hour period by an automatic sample changer.[13,14] Shapiro and co-workers developed a method based on ligand diffusion, which they called affinity NMR, to analyze small ligands binding to protein targets.[15] A relatively large numbers of ligands can be analyzed simultaneously in the same NMR tube using these methods. Another related technique, called NOE pumping, has also been reported by Shapiro and co-workers.[16,17] An alternative technique, which is based on the transfer NOE, allows identification and structural characterization of biologically active molecules from a mixture.[18] This technique could aid in screening mixtures of compounds for biological activity. For lead generation, the SHAPES strategy was developed, which uses NMR to detect binding of a limited but diverse library of small molecules to a potential drug target.[19] The information obtained is used to refine the library in an iterative fashion. More recently Prestegard and co-workers have described NMR experiments to use competitive binding as a method for determining potential ligands of interest.[20] To further improve the throughput, flow-injection NMR spectroscopy has been adopted in compound library screening. Research in this area has recently been reviewed by Stockman et al.[21]

To obtain improved structural information in complex mixtures, there are also significant and increasing efforts to improve the coupling of NMR with chromatographic separation methods such as LC, HPLC, and CE.[22,23] The requirement for higher sensitivity and small working volumes imposed by chromatographic methods has led to the development, refinement, and application of a number of techniques, including microbore HPLC, flow-through NMR probes, micro- or nanoliter volume NMR probes, and solvent suppression pulse sequences.[23,24,25,26,27] In addition to the traditional automatic sample changer, flow probes[23] have been adopted to achieve rapid NMR analysis of up to 500 samples per day.[28] This technique has also been interfaced to standard 96-well microtiter plates, which are the popular formats for combinatorial libraries.

Despite these advances NMR measurements are usually made on a single sample at a time, and throughput is limited by this serial approach.

To overcome this limitation, parallel detection methods, which have been implemented for a number of different analytical techniques for increasing throughput,[29,30,31,32] are highly desirable for NMR detection. To date, several parallel NMR methods have been developed,[33,34,35,36,37,38,39,40] including a new methodology called multiplex NMR[37,38,39] developed in our laboratory. Recent developments in high-throughput NMR indicate that, with further development, parallel NMR techniques will play an important role in high-throughput chemical analysis and drug screening.

7.1.1. High-Throughput Analytical Techniques for Combinatorial Library Analysis

The concept of combinatorial chemistry was first introduced by Merrifield[41] in 1986, and involved solid-phase synthesis methods to generate small mixtures of immobilized peptides. This idea was further developed in the early 1990s by Geysen, Dibo, Houghten, and their co-workers.[42,43,44] Their technique represented a completely different synthesis method capable of generating vast number of compounds in a relatively short period of time compared to classical synthetic chemistry. Combinatorial chemistry was initially developed for solid-phase peptide and oligonucleotide synthesis, but because of the poor bioavailability of these biopolymers, the interests of synthesis have shifted to peptidomimetics (compounds that can mimic the critical features of the molecular recognition process of the parent peptide) and small organic molecules ($MW < 700\,\mathrm{Da}$).[45] In its original implementation the split-couple-recombine (or split-and-mix) methodology was adopted to produce one to several mixtures or libraries of products through a randomization process.[41] Although the split-and-mix synthesis technique is still used by many chemists, it has been largely replaced by automated parallel solid-phase and solution-phase synthesis, which are of high speed and spatially addressable.[6,7,8] Compared to combinatorial chemistry, parallel synthesis approaches can produce large collections of individual compounds (as opposed to mixtures) in larger quantities that can be purified and more rigorously analyzed, which makes it easier to characterize, screen, and maintain the combinatorial libraries. However, unless specifically stated otherwise, we will include both parallel and split-and-pool syntheses under the definition of combinatorial chemistry in our discussions below.

Combinatorial approaches make it relatively straightforward to generate chemical libraries with vast number of new compounds; thus synthesis is often no longer the rate-limiting step in drug discovery. Since almost all analytical characterization tools are serial techniques, the purification and analysis of the chemical libraries has instead become a new bottleneck. This, of course, imposes new analytical challenges in the fields associated with combinatorial chemistry, and more rapid and robust analytical approaches are in great need to keep pace with the new paradigm.

Analytical techniques such as IR, NMR, fluorescence spectroscopy, MS, GC, and HPLC have historically been used in drug discovery and other areas to characterize new compounds. These techniques are now being adapted for use in combinatorial chemistry and high-throughput screening applications to facilitate the analysis of new compounds and the identification of promising leads. However, each of the analytical techniques has its own limitations in the analysis of combinatorial libraries. For example, techniques such as GC and fluorescence spectroscopy are not applicable to the identification of most new chemical entities, while IR and MS are not specific enough to determine chemical structures unambiguously, and other techniques, such as HPLC and NMR spectroscopy, suffer from low throughput or sensitivity. Due to its high speed and high sensitivity, mass spectrometry is one of the most widely used techniques in combinatorial library analysis,[46] and high-throughput MS methods have been developed to accommodate the needs of analyzing large numbers of compounds in combinatorial libraries. Chemiluminescence nitrogen detection (CLND),[47] a method that provides linear response to the absolute amount of nitrogen from most nitrogen-containing analytes, can be used as a tool to determine the concentrations of pure compounds accurately and rapidly down to low-picomole levels. CLND is easy to couple with other techniques, such as GC, LC, and recently has been utilized as a quantitative tool in combinatorial library analysis.[48,49]

Mass spectrometry provides the means to characterize new compounds based on their molecular weight and to a lesser extent structure. At the same time, MS methods are often high-throughput and can be coupled with chromatographic separation techniques. There are two widely adopted approaches for high-throughput mass spectrometric analyses of combinatorial libraries: flow injection analysis mass spectrometry (FIA-MS) and liquid chromatography mass spectrometry (LC-MS). Due to its simplicity in design and operation, FIA-MS has become the most widely utilized mass spectrometric method for characterizing combinatorial chemical libraries, and it is also the mass spectrometric based technique with the highest throughput.[46] In addition to coupling an FIA-MS system with an auto-sampler, parallel sampling methodologies have been implemented into the FIA-MS system,[50,51] and therefore the throughput of FIA-MS has been improved significantly.

LC-MS is usually used for rapid purity assessment of the compound libraries. Because LC-MS has advantages over FIA-MS in its ability to analyze mixtures and to provide more complete characterization of compound libraries, it has replaced FIA-MS as the standard analysis tool in many pharmaceutical laboratories.[52] Two approaches have been adopted to increase the throughput of LC-MS analysis: fast LC-MS,[53,54] which utilizes a fast (2–3 min) chromatographic separation step, and parallel LC-MS,[32,55,56] in which multiple HPLC columns are coupled to a single mass spectrometer. With these developments, LC-MS has become more suitable for combinatorial library analysis. In addition to FIA-MS and LC-MS, supercritical fluid chromatography (SFC)-MS[57,58] and CE-MS[59,60,61] have also been adopted for combinatorial library analysis. SFC has a higher chromatographic resolution and faster separation speed than HPLC, and CE is a particularly useful in separating for ionic compound and chiral mixtures.

However, mass spectrometry encounters problems when compounds with the same molecular weight are present. These compounds could be stereo-isomers, positional isomers, or simply molecules that happen to have the same molecular weights. This situation often occurs in tightly focused or closely related combinatorial libraries. To solve this ambiguity, NMR spectroscopy is usually employed to determine the structures of the compounds involved. NMR spectroscopy is a very powerful and widely adopted analytical tool for structural and quantitative determinations in chemistry, biology, material science, and many other areas. It is also a nondestructive technique and capable of deriving information on molecular dynamics. But traditional NMR spectroscopy suffers from low sensitivity and slow analysis speed; thus it is not an ideal tool to analyze the large number of compounds present in the combinatorial libraries. To meet the new challenges emerging from combinatorial chemistry and parallel synthesis methods, efforts are being made to make NMR more suitable for analyzing vast number of samples quickly, and a number of these developments are described below.

7.1.2. High-Throughput Serial NMR Analysis

A number of factors contribute to the slow speed of NMR analysis, including sample preparation, sample loading/unloading, shimming to ensure high-resolution spectra, frequency locking, as well as data acquisition and analysis time. To increase the speed of NMR analysis, the most straightforward way is to reduce the time needed for preparing, loading, and retrieving the NMR samples. Automatic sample-changing mechanisms have been integrated with NMR spectrometers for at least 10 years.[62] Such systems include a sample holder tray in which a large numbers of samples are placed; a mechanism to track each sample; and a robotic device to load the samples into standard 3 mm, 5 mm, or 10 mm NMR tubes, and to move sample tubes pneumatically into and out of the magnet. Many NMR labs, such as those in University of Minnesota, University of California at

Los Angeles, Conoco Research, and IBM, started to use automatic sample changers to speed up NMR analysis in late 1980s, with throughput rates of up to 100 samples per day on a single NMR spectrometer.[62] In 1988 Abbott Laboratories developed their first automation systems to achieve totally automated NMR analysis, and both the turnaround time and the cost per sample were lowered significantly.[63] For example, the number of NMR samples run per year at Abbott had grown to nearly 60,000 in 1995 upon the implementation of three totally automated NMR systems. Currently automatic sample changers compatible with different NMR systems are provided by the major NMR system manufacturers, including Bruker, Varian, and JEOL. These commercially available automatic sample changers are capable of carrying out tasks such as recording the sample identification number, preparing sample solution, pipetting solution into a clean NMR tube, inserting an NMR tube into the magnet, and then removing the NMR tube and disposing of the sample when the NMR experiment is finished. Currently the use of NMR systems equipped with automatic sample changers has become routine for many NMR laboratories with high-volume needs.

Conventional high-resolution NMR spectroscopy requires the use of precision glass NMR tubes, which are delicate and expensive. Sample preparation involves filling NMR tubes, loading tubes into the magnet one at a time, and removing each tube from magnet after the NMR experiments are finished. Although the use of robotic sample changers has made the automation of the sample preparation and changing possible, the use of NMR tubes limits its speed and efficiency for high-throughput applications. To avoid the steps of transferring samples from a reservoir into NMR tubes and thus further increasing the throughput, an alternative solution is to use a flow NMR probe.[28,64] Such flow probes, which are also known as "tubeless" NMR probes, have a permanently mounted sample chamber or flow cell, on which an RF detection coil is usually wrapped. The sample is introduced into the flow cell, and a rinsing step is necessary to eliminate the carryover from the previous sample. Because the RF coil is mounted closer to the sample, an increased filling factor is obtained. Therefore flow NMR probes usually have higher sensitivity than conventional NMR probes. However, because of the permanent mounting of the RF coil on the flow cell, it is not possible to spin the sample as is done with normal NMR tubes. Therefore the resolution of flow NMR probes is typically poorer. Standard flow NMR probes have detection volumes that range from 60 to $250\,\mu L$,[65,66,67,68] which are smaller volumes than those used in conventional 5-mm NMR probes. Recent developments in microcoil probe technologies have pushed the sample volume requirements in flow NMR to the nanoliter to few microliter range.[25]

To introduce samples directly from a sample reservoir into a flow NMR probe, flow injection analysis (FIA) NMR or direct injection (DI) NMR can be used.[69,70] FIA-NMR and DI-NMR can be viewed as "columnless" LC-NMR. Since the slow separation process is not present, these two techniques are capable of carrying out high-throughput NMR analysis. FIA-NMR benefits from the speed and simplicity of FIA,[71] which is an analytical technique based on the injection of a liquid sample into a nonsegmented carrier stream and the transportation of the injected sample to a detector or sensor. FIA-NMR requires smaller sample volumes and has higher throughput rates. DI-NMR system has been further simplified from the FIA-NMR system by removing the mobile phase and the pump, but leaving the injector valve and connection tubing. DI-NMR has higher sensitivity because there is no dilution step, and the sample is easier to recover by returning the sample in the flow cell to its original sample plate after the NMR experiment.[72] Both FIA-NMR and DI-NMR can be carried out under continuous-flow or stopped-flow conditions, and are easily interfaced with 96- or 384-well microtiter plates that are standard sample formats for combinatorial chemistry and parallel synthesis. With the use of an automatic sample handler, a flow NMR probe is capable of running as many as 600 to 700 samples every 24 hours.[11] Current flow NMR capabilities are sure to increase the application of NMR spectroscopy to the characterization of combinatorial chemical libraries.

7.1.3. Microcoil NMR Techniques

As the NMR signal depends on the very small population difference between the upper and lower energy states of the nuclei of interest, NMR suffers from intrinsically poor sensitivity compared to other analytical techniques. For example, Fourier-transform infrared spectroscopy (FTIR) has limits of detection (LODs) in the range of 10^{-12} to 10^{-15} mol,[73] mass spectrometry has LODs around 10^{-19} mol,[74] but the LODs for conventional NMR are typically 5×10^{-9} mol.[75] The low-sensitivity problem is even more serious when ^{13}C or ^{15}N is detected. Nevertheless, for a typical 0.5 mL, 0.1 mM sample, a sufficient signal-to-noise (S/N) ratio of over 100:1 can be obtained within a few scans. In the past decade or so, many technical advances have been developed to increase the NMR sensitivity. One approach has been to increase the static magnetic field strength, which increases both S/N and the dispersion of NMR chemical shifts. Recently high-resolution NMR spectrometers with the superconducting magnetic field strength up to 21.1 T (operating at a frequency of 900 MHz for protons) have been developed. A second approach is to improve the radio frequency (RF) coil design in NMR probes by decreasing the coil noise and increasing the coil quality factor. Two examples of such improvements are the use of high-temperature superconducting materials in probe construction[76,77] and the application of cryogenic NMR probes.[13,78] The first cryogenic probe demonstrated with a practical implementation was constructed by Styles et al.,[79] and the schematic diagram of this probe is shown in Figure 7.1.

Although expensive, cryogenic probes are now available commercially, and are capable of delivering three to four times better S/N than conventional NMR probes, which represents a substantial benefit for many NMR experiments. A third approach is to use extremely small RF coils, namely microcoils (see Fig. 7.2), in NMR probes as the magnetic resonance detector. The definition of "microcoil" is somewhat arbitrary, but in general, microcoils refer to coils with the active volume in the range of 1 μL or less.[25] Although it has been known for quite a long time that decreasing the diameter of the coil increases its sensitivity,[80,81,82,83] it is only recently that microcoils were adopted successfully in acquiring high-resolution NMR spectra.

The first implication of small RF coils in NMR probe was reported by Odeblad[82] in 1966, who used probes with solenoid microcoils ranging in diameter from 200 to 1000 μm (with a fixed length of just over 1 mm) and continuous wave (CW) techniques to study the physical chemistry of mucus secreted from cells in the human cervix. The best resolution reported by Odeblad was a line width of 13 Hz, and the limit of detection (LOD) was 0.4 μg of pure water. In 1979 Shoolery[83] reported the first application of reduced size RF coils in high-resolution NMR spectroscopy. He showed the acquisition time could be reduced by a factor of 40 for a given S/N by decreasing the sample vial diameter (and the closely fitted concentric RF coils) from 10 to 1.7 mm. [13]C spectra of a 1 mg sample or ^{1}H spectra of a 1 μg sample could be obtained in 16 hours. Recently Wu et al.[84,85,86] have successfully decreased the microcoil detecting volume down to the nanoliter range, with the LODs of less than 1 nanomole, a significant achievement.

A problem exists with these very small detection coil, namely that the magnetic susceptibility of the coil material does not match that of the sample, resulting in large spectral linewidths. Significant efforts have been made to deal with this susceptibility mismatching problem, and an often used solution is to make a coil with zero magnetic susceptibility by combining metals such as Cu and Rh, which have opposite susceptibilities. An alternative approach is to use "zero-susceptibility" materials,[87,88,89] which are specially designed composite materials consisting of matched diamagnetic and paramagnetic components that result in very low magnetic susceptibility, such that there is no perturbation of the main magnetic field in the sample region that is produced by the RF coil. Varian has recently introduced a new probe called the "Nanoprobe," in which zero-susceptibility materials were used to construct the 5-mm i.d. solenoid coil. With the

A

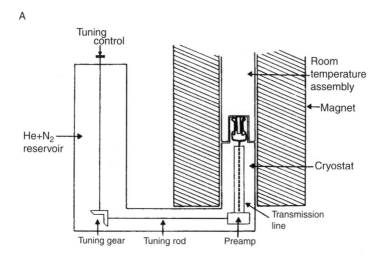

Tuning control

Room temperature assembly

Magnet

He+N$_2$ reservoir

Cryostat

Transmission line

Tuning gear   Tuning rod   Preamp

B

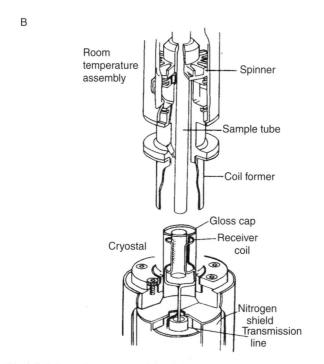

Room temperature assembly

Spinner

Sample tube

Coil former

Gloss cap

Receiver coil

Cryostal

Nitrogen shield
Transmission line

Figure 7.1. (*A*) Schematic representation of the arrangement of a cryogenic NMR probe, superconducting magnet, and room-temperature assembly. (*B*) Detail of the cryogenic probe and room-temperature assembly in the vicinity of the coil. (Adapted with permission from *J. Magn. Reson.*, 60: 397. Copyright Academic Press, 1984).

Figure 7.2. Picture of a typical solenuidal microcoil. This particular microcoil was wrapped using a 130-μm diameter copper wire on borosilicate glass capillary. The coil has an i.d. of 860 μm and a length of 1.6 mm.

capability of spinning the sample at the magic angle, mass limited solid or gel-like samples can be detected with good spectral resolution. The application of the Nanoprobe in combinatorial chemistry has been demonstrated by Fitch et al.[90]

One new approach is to match the magnetic susceptibility of the materials used in probe construction to a non-zero value. Olson et al.[24] have obtained high-resolution NMR spectra using this susceptibility matching approach. A microcoil with an active volume of 5 nL was immersed into a fluid with a magnetic susceptibility similar to that of the copper coil, and thus compensated for the susceptibility mismatch at the air/copper interface. The spectral line width achieved was under 1 Hz, and the LODs were in the 10 to 100 pmol range.

Typically the sensitivity of a NMR probe is represented by the S/N for a given analyte concentration. The general equation for the S/N of an NMR probe after a 90° pulse is given by[81]

$$\frac{S}{N} = K\eta M_0 \left(\frac{\mu_0 Q \omega_0 V_c}{4 F k T_c \Delta f} \right)^{1/2}, \tag{7.1}$$

where K is a numerical factor (~1) dependent on the geometry of the receiving coil, η is the filling factor (the ratio of the sample and RF coil volumes), M_0 is the nuclear magnetization which is proportional to the magnetic field strength B_0 and the sample concentration, μ_0 is the permeability of free space, Q is the quality factor of the coil, ω_0 is the Larmor angular frequency, V_c is the volume of the coil, F is the noise figure of the preamplifier, k is Boltzmann's constant, T_c is the probe (as opposed to sample) temperature, and Δf is the bandwidth (in Hz) of the receiver. It can be seen that the concentration sensitivity S_c (S/N per μM concentration of analyte) is poor for microcoils. This is due to the fact that microcoil probes have very small observation volumes and therefore contain a very small amount of analyte. However, if the sample can be concentrated into a small volume, then the microcoil can more easily detect the signal. This high mass sensitivity S_m (S/N per μmol of analyte) is characteristic of microcoil NMR probes. In essence, the use of microcoil probes enhances the mass sensitivity S_m at the expense of the concentration sensitivity S_c.[26] To better understand the relationship between sensitivity and coil diameter, a detailed analysis was reported by Peck et al.[91] Their results showed that mass sensitivity increases monotonically with decreasing coil diameter within the 1 mm to 50 μm range they studied. However, the concentration sensitivity decreases, and therefore there is a trade-off between S_c and S_m that depends on coil diameter.

Another issue that affects the microcoil NMR probe is the trade-off between sensitivity and resolution. As can be seen in equation (7.1), an increase in the fill factor η increases the S/N of the microcoil; however, a larger filling factor also results in a closer proximity of the sample to the coil. As a result any magnetic susceptibility mismatch plays a more significant role in distorting the spectral line shape, and therefore poor resolution usually results.[86] Webb and Grant[92] reported studies to quantify this trade-off, and found that with a filling factor of up to 28% there was little loss in spectral resolution. However, if the filling factor was increased further to 50%, the line width significantly increased to above 3 Hz.

Currently the most widely adopted coil type for microcoil NMR probe construction is of solenoid geometry, due to the relative ease in manufacturing and increased sensitivity compared to other geometries. However, the miniaturization of other coil geometries has also been investigated. For example, a 2-mm diameter double-saddle Helmholz coil has been used by Albert et al.[93] to acquire capillary LC-NMR spectra. This approach has been further optimized for the detection of 5-μL spinning samples contained in 1-mm diameter capillary tubes.[94] Miniaturization of a birdcage resonator has also been carried out by Roffman et al., who constructed a high-pass birdcage coil with an i.d. of 3 mm and eight 0.5 mm wide rings made of zero-susceptibility materials for microscopic imaging studies.[95] Coils geometries such as the Helmholtz and birdcage coils have their major axis parallel to the static magnetic field (B_0) direction; therefore magnetic susceptibility mismatch from such coils are minimized, and sample spinning is easy to realize. Helmholtz or birdcage coils can more easily be constructed to operate at very high resonance frequencies due to their lower intrinsic inductance. Finally, with the recent development of microfabrication techniques, it is now possible to make precision detection coils with diameters less than 100 µm with mechanical stability and a variety of desired geometries. One interesting example is the planar microcoils designed by Peck et al.,[96] and a similar approach made by Magnin et al.[97] In these cases the problem of matching the magnetic susceptibility of the microfabricated planar coils to the substrate still needs to be overcome. Due to the desire to work with smaller samples volumes and high sensitivity, it is clear that with further development microcoils are certain to play an increasingly important role in NMR analysis and will eventually become one of the many routine NMR techniques for general NMR users.

7.1.4. NMR Coupled with Chromatographic Separation Methods

Over the past two decades most analytical techniques have been coupled to separation methods, and therefore on-line determination of individual compounds in mixtures can often be carried out rapidly. Since NMR is not an ideal technique for the structure elucidation of mixtures of unknown compounds, coupling NMR chromatographic separation techniques such as LC or CE can overcome this challenge. But this does not immediately alleviate the major drawback of NMR, namely its relatively low sensitivity compared to other analytical techniques. However, with the recent development of higher magnetic fields, microcoil, and cryogenic NMR probe techniques that have significantly improved sensitivity, this problem is considerably reduced compared to what it was 10 years ago. In order to save time, a direct coupling of the separation technique with NMR is normally adopted, and a so-called flow NMR probe that incorporates a static sample volume is used as the detector.

LC-NMR is the most popular and well-established technique for direct, on-line coupling of chromatography with NMR. LC-NMR instruments are commercially available, and LC-NMR has become a routine analysis method in many NMR laboratories. LC-NMR was first introduced by Watanabe et al.[98] in 1978, in which a stopped-flow approach was adopted to detect the LC effluent by NMR. In 1979 Bayer et al.[99] reported the first on-line coupling of NMR with an LC. LC-NMR only gained popularity in the mid-1990s, after significant efforts had been made to develop suitable flow NMR probes, optimize the NMR acquisition conditions, and enhance NMR sensitivity by increasing the magnetic field and/or reducing the NMR coil size.[22,23,25,26] Currently LC-NMR is performed under both stopped-flow and continuous-flow conditions. Under stopped-flow conditions, longer NMR acquisition times are available to carry out more complicated and time-consuming experiments. Under continuous-flow conditions, the NMR line width, signal intensity, and the effective longitudinal (T_1) and transverse (T_2) relaxation times can all be affected by the flow rate.[23] Therefore careful consideration must be taken to choose the optimal flow rate in order to achieve best results.

In standard HPLC-NMR, the relatively large volume and flow rates make it too expensive to use most deuterated solvents. In order to eliminate the problems caused by large, residual protonated solvent peaks (such as HOD in D_2O), solvent suppression techniques are usually employed in LC-NMR. But the use of multiple solvents with changing concentrations under flowing conditions in LC-NMR makes solvent suppression more complicated than their use in other areas. Suppression techniques such as Multigate (a modified version of WATERGATE)[100] and WET (water suppression enhanced through T_1 effects)[27] have been developed to suppress multiple solvent peaks, whose frequencies change under gradient elution conditions. For continuous-flow experiments, solvent suppression techniques such as presaturation and diffusion-based filters often fail because a portion of the affected spins is constantly replaced by fresh spins.[22,101] However, the WET technique can be used under flow because it is fast (<100 ms) and can achieve a large suppression factor (>10^4) in a single scan.[27] Despite all these developments some problems associated with solvent suppression still exist, such as the difficulty in observing spectral regions near the solvent resonance.

In addition to LC-NMR, the coupling of other important separation techniques, such as supercritical fluid chromatography (SFC) and supercritical fluid extraction (SFE), with NMR has been investigated. The use of supercritical solvents such as CO_2 eliminates the need for solvent suppression, since no ^1H solvent signals are present in the spectra obtained. SFC-NMR can give results comparable or even exceeding HPLC-NMR; however, it has its own disadvantage of longer T_1 relaxation times of compounds in supercritical CO_2 compared to traditional solvents. SFE-NMR also offers the advantage of directly monitoring the extraction process. Studies in these areas have recently been reviewed by Albert.[102]

With the development of microcoil NMR techniques, it is now possible to couple NMR with some microseparation techniques, such as microbore LC,[86,93,103] capillary electrophoresis (CE),[84,85] and capillary electrochromatogrphy (CEC),[104,105,106] where the inner diameters of the separation channels are in the range of 20 to 75 μm. After coupling these small separation channels with NMR microcoils whose diameters are in the same range, it is possible to achieve maximum mass sensitivity for NMR detection. While the concentration sensitivity of this approach is poor compared to that of a standard 5-mm NMR probe, it is interesting to note that the concentration LODs of microcoils, which are in the mM range, are approximately the same as those for the UV detectors in microcapillary separation columns.[26] Since most NMR microprobes use a flow design, it is fairly easy to carry out NMR detection under either on-line or stopped-flow conditions. Besides the advantages of improved chromatographic resolution, capillary separation has the added benefit of reduced solvent consumption. Therefore it is possible to use a variety of deuterated solvents in the separation process, and solvent suppression is simplified. One interesting recent development is the use of capillary isotachophoresis (cITP), which makes it possible to use CE techniques to both separate and concentrate samples using capillary isotachophoresis (cITP).[107] Concentration enhancements as large as a factor of 100 were reported. Recently Sweedler, Larive, and co-workers have extended this approach to the analysis of a combinatorial chemistry compound attached on a single polymer resin bead.[108]

7.2. PARALLEL NMR TECHNIQUES

Most NMR measurements are currently made on a single sample at a time, and the throughput is therefore limited by this serial detection fashion. To overcome this limitation and make NMR suitable for analyzing large numbers of samples at a reasonable rate, the throughput of NMR detection methods needs to be improved. One approach that has been applied to a number of analytical techniques is to use parallel methods to improve throughput, that is, to measure multiple samples simultaneously (see Figure 7.3). There are several different approaches that can be taken to realize parallel detection in NMR spectroscopy. A straightforward method would be to simply wrap the excitation and detection coil around a number of smaller samples, as indicated in Figure 7.4A. In this method, however, it is not easy to separate signals from different samples, and the sensitivity suffers due to the low filling factor η (see Eq. 7.1). An alternative method that restores the filling factor is to wrap individual coils around different samples, as shown in Figures 7.4B and 7.4C. The individual detection coils can be connected to separate RF circuits with complete electrical and magnetic isolation between the detection coils, as indicated in Figure 7.4B. An alternative approach that simplifies the probe and spectrometer hardware requirements is to connect parallel detection coils to the same RF circuit, as shown in Figure 7.4C. In this approach a magnetic field gradient is necessary to differentiate signals from different sample coils.

A Serial analysis

detector

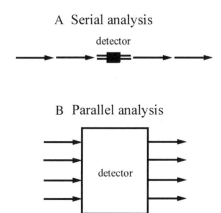

B Parallel analysis

Figure 7.3. Schematic diagrams illustrating (*A*) serial and (*B*) parallel approaches applied to analytical techniques to increase throughput.

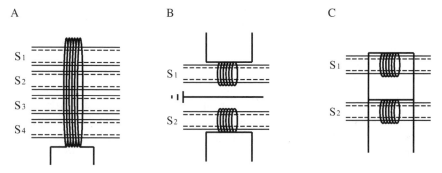

Figure 7.4. Several parallel approaches for NMR detection of multiple samples. Multiple samples detected in (*A*) the same RF coil, (B) different RF coils using separate RF circuits, and (*C*) different RF coils in the same RF circuit. *S* indicates separate sample volumes.

7.2.1. Multiple-Sample Analysis Using a Single NMR Coil

The simplest approach of using a single coil to analyze multiple samples was first demonstrated by Banas in 1969,[33] who used small capillary tubes that could be fit inside a standard 5-mm NMR tube, as shown in Figure 7.5. An additional advantage of this method is that it provided internal standardization of the frequency, so that very small differences in chemical shift could be measured precisely. Unfortunately, there was no easy way to separate the spectra from different samples at that time. Additionally the

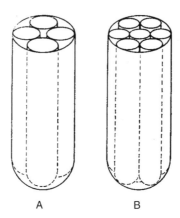

A   B

Figure 7.5. An illustration of two 5-mm NMR sample tubes with (*A*) four and (*B*) seven cap-illary tubes inside used by Banas to carry out multiple-sample detection. (Adapted with per-mission from *Appl. Spectrosc.*, 23: 281. Copyright Society for Applied Spectroscopy, 1969).

sensitivity suffered, due to the low filling factor of this method as compared to using a sample that better fills the NMR probe volume.

An improvement on this approach, in which the spectra could be sepa-rated, was demonstrated recently by Ross et al.[40] Nine samples were placed inside a single 5-mm NMR tube (as shown in Figure 7.6), which was then loaded into a triple-axis pulsed field gradient probe. The spectra from different samples were separated using a multidimensional NMR pulse sequence called chemical-shift imaging (CSI), which is capable of separat-ing chemical shift from overlapping spatial positions.[109,110] Two-dimensional Fourier transformation of the signal with respect to incremented x and y magnetic field gradient values gave the spatial distribution of individual spectra, and allowed the separation of the different chemical shift signals corresponding to each of the samples. However, a large number of phase encoding steps are necessary to differentiate the samples, and therefore the experiment can be time-consuming. An additional problem is that the signal-to-noise (S/N) ratio for each sample is intrinsically low due to the small effective filling factor. In Ross's experiments it required five minutes to acquire spectra on the nine samples, with additional time for loading and unloading the samples. Magnetic susceptibility mismatching between the capillaries and the samples inside them caused the spectral line width to be approximately 5 Hz.

7.2.2. Multiple Electrically Decoupled Coils

A second approach for parallel NMR analysis uses a separate RF coil and impedance-matching circuitry for each sample. When the individual detec-

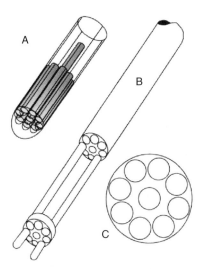

Figure 7.6. Schematic drawing of the apparatus (*A* and *B*) and its cross section (*C*) used to fit nine capillaries into a standard 5-mm NMR tube. The capillaries are fixed at the bottom and the top in Plexiglas carriers (*B*). (Adapted with permission from *Angew. Chem. Int. Edit.*, 40: 3243. Copyright John Wiley and Sons, Inc., 2001).

tion coils are connected to separate RF circuits, as indicated in Figure 7.4*B*, the circuits will normally couple, and the resonant frequencies of the coils will split by as much as several MHz. This mutual coupling effect makes it difficult to excite and detect the two samples. However, by completely isolating the two resonant circuits, both in the sample region and in rest of the RF circuit, it is possible to detect the signal emanating from all the samples simultaneously without cross talk. As indicated in Figure 7.4*B*, the addition of a ground plane separates the inductors electrically and magnetically. The advantages of this approach include the potential for running different pulse sequences for each sample, the simple differentiation of the signals from individual samples that can be routed into separate receivers or time-multiplexed into a single receiver using an appropriate RF switching network, and maintenance of maximum sensitivity. The major disadvantage of this approach is the increased complexity of the probe geometry and circuitry that results from the multiple resonance circuits, and the need for shielding to establish high electrical and magnetic isolation between the detection coils. These issues are exacerbated when more coils are incorporated.

The first NMR application of this approach (albeit observing different NMR nuclei) was in solid-state NMR spectroscopy by Oldfield.[34] For solids the line width is intrinsically large, and only ~1 ppm magnetic field homogeneity was needed. In Oldfield's design, which is shown in Figure 7.7, two

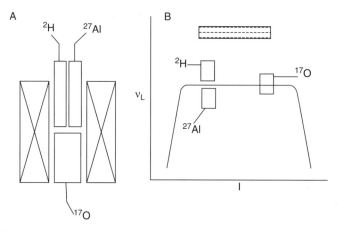

Figure 7.7. Illustration of the multiple probe designed by Oldfield. (*A*) The triple-probe arrangement showing the nuclei observed; (*B*) schematic axial field plot showing the sample location in a conventional 5-mm NMR probe (*top*) and three radial solenoid coil probes (*bottom*). (Adapted with permission from *J. Magn. Reson. A*, 107: 255. Copyright Academic Press, 1994).

side-by-side 5-mm solenoid coils occupied the upper region of the magnet, while a third solenoid (which could be spinning) occupied the lower part of the magnet, with about 3-cm vertical separation between upper and lower coils. Operating at 11.7 Tesla, the lower coil was tuned to ^{17}O at 67.77 MHz, while the top two coils were tuned to ^2H (76.74 MHz) and ^{27}Al (130.27 MHz), respectively. Tuning to very different frequencies made isolation of the different circuits much easier. Linewidth of ~1 ppm was obtained for all three coils simultaneously using samples of ^2H_2O, $H_2^{17}O$, and an aqueous solution of $^{27}Al(NO_3)_3$.

The first application of this approach using the same resonant frequency was accomplished by Fisher et al.,[35] in which two coils both tuned to the carbon frequency of 75.44 MHz at 7.05 Tesla were used. This dual-coil probe is shown in Figure 7.8*A*. The solenoid coils were arranged vertically and separated by a horizontal piece of copper-plated circuit board grounded to the probe body. An aluminum plate at the base of the probe separated the variable capacitors used for impedance-matching each of the two circuits. A commercial spectrometer was used as the transmitter, and the RF excitation pulse was channeled through a power splitter to separate duplexer circuits (consisting of crossed-diodes and quarter-wavelength cables). In addition to the receiver channel of the commercial spectrometer, a second receiver channel was constructed by the authors. Samples of ^{13}C labeled methanol and carbon tetrachloride were loaded into the two coils, respec-

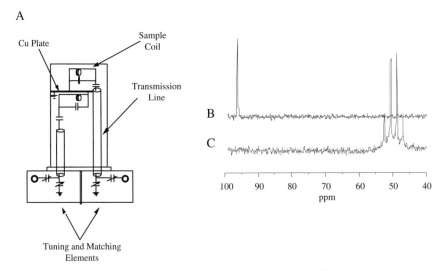

Figure 7.8. (*A*) Dual isolated-coil NMR probe, and representative ^{13}C spectra of (*B*) ^{13}CCl$_4$ and (*C*) ^{13}CH$_3$OH obtained simultaneously using this probe. (Adapted with permission from *J. Magn. Reson.*, 138: 160. Copyright Academic Press, 1999).

tively. As the NMR spectra in Figure 7.8*B* and 8.4*C* show, there was no cross-talk between the two coils.

The concept of using decoupled coils has been further developed by Li et al.[36] In this case an RF switch was used for time-domain multiplexing of the signals into a single receiver channel. A four-sample system (shown in Figure 7.9*A*) was constructed using printed circuit boards for operation at 6 Tesla in a wide-bore (89-mm) magnet that incorporated microcoils with observe volumes of 28 nL. The microcoil-based resonant circuits were mounted one above another with a vertical spacing of 5 mm between adjacent coils. Alternate coils were rotated 90° with respect to each other to reduce the electromagnetic coupling. The matching networks were also placed at 90° to each other, again to reduce coupling. The whole system was surrounded by a container filled with FC-43, a magnetic susceptibility matching fluid. Although the original implementation required four preamplifiers in the receiver chain, this was due to a relatively high-loss (2.3 dB) switch being used. In subsequent realizations, a much lower loss (0.1 dB) switch has been substituted, which results in a simplified receiver design. The time required for shimming the four coils to 2 to 4 Hz was typically less than 30 minutes. Both one-dimensional and two-dimensional correlation spectroscopy (COSY) experiments were carried out at 250 MHz for protons, and the spectra obtained from four solutions of 250 mM fruc-

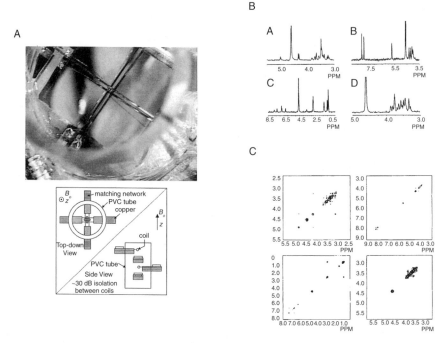

Figure 7.9. (A) Photograph of the four-coil probe developed by Li et al. showing the configuration of the microcoils and matching networks (*top*), and the schematic of coil arrangement (*bottom*): each coil is oriented at 90° with respect to adjacent coils. (B) One-dimensional spectra of (*top left*) galactose, (*bottom left*) adenosine triphosphate, (*top right*) chloroquine, and (*bottom right*) fructose (all 250 mM in D_2O) acquired using the four-coil probe. (C) COSY spectra of the same four samples as in (B) acquired using the four-coil probe. (Adapted with permission from *Anal. Chem.*, 71: 4815. Copyright American Chemical Society, 1999).

tose, chloroquine, galactose, and adenosine triphosphate, all in D_2O are shown in Figure 7.9*B* and 7.9*C*. No signal bleed-through was observed from one spectrum to another. Since the acquisition time was approximately an order of magnitude less than the recycle delay, a full factor of four improvement in throughput was achieved; in fact the number of coils could be increased to improve the temporal efficiency further, assuming these additional coils could be shimmed.

Recent developments have included the design of a two-coil probehead, with each coil having a much larger volume (15 μL), allowing the study of metabolites at concentrations below 10 mM.[111] In this instance the solenoid coils were separated horizontally, and had electrical isolations greater than 30 dB. The switching of the two coils between the transmitter and receiver was realized by a network containing two RF switches, as shown in Figure

7.10A. Both of the coils were double-tuned to proton and nitrogen frequencies to allow multiple-nuclear and multiple-sample experiments to be carried out. The COSY and HMQD spectra (Figure 7.10B and 7.10C) obtained using this two-coil probe show no evidence of cross-talk between the two coils.

7.2.3. Multiple Coils Connected in Parallel

A third approach in parallel NMR detection is to connect multiple coils together in the same RF circuit (see Figure 7.4C). The first implementation of this method, termed multiplex NMR, was achieved by MacNamara et al.,[37] and a picture of the multiplex NMR probe head is shown in Figure 7.11. The NMR coils were of solenoid geometry constructed from five turns of polyurethane-coated high-purity 42-gauge copper wire wrapped around glass capillaries. The coil had an i.d. of 0.8 mm and length of 0.5 mm, and the sample volume was 60 nL. The inter-coil (center-to-center) distance was 3.2 mm. The entire coil array was surrounded by Fluorinert FC-43, a susceptibility matching fluid that has a very similar volume magnetic susceptibility to copper and has been shown to improve magnetic field homogeneity by minimizing field distortions induced by copper NMR coils.[24] A single resonance was observed for the circuit, and the circuit had a tuning range of ~2 MHz with a quality factor (Q) value of 60. Teflon tubes were attached to the capillaries to allow flow introduction of samples. A standard NMR spectrometer was used in these experiments.

Because all the sample coils are wired into the same resonant circuit, there is no ill effect of the coupling between different coils in the multiplex NMR probe. However, since the detected signal is a combination of all the samples, it is necessary to differentiate the signals emanating from a particular sample. To carry out this task, magnetic field gradients were applied across the different coils. To illustrate how the field gradient works, a schematic diagram is shown in Figure 7.12 for the hypothetical case of a single analyte such as water in all four coils. Figure 7.12A shows the situation after the probe has been properly shimmed and no linear field gradient is applied to the sample region. The four sample regions experience the same magnetic field, B_0, and consequently the NMR signals from all four coils coincide; thus the spectrum only contains one peak. However, when a field gradient is applied to the sample region, each of the sample coils experiences a field given by $B_0 + G_z \times z_i$, where G_z is the strength of the linear field gradient and z_i is the vertical position of the ith sample coil. The resultant spectrum contains four lines, since each sample experiences a different magnetic field and each peak can be identified with a particular coil (see Figure 7.12B).

A

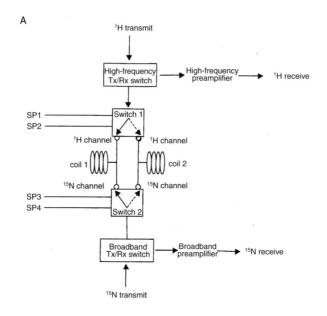

B

C

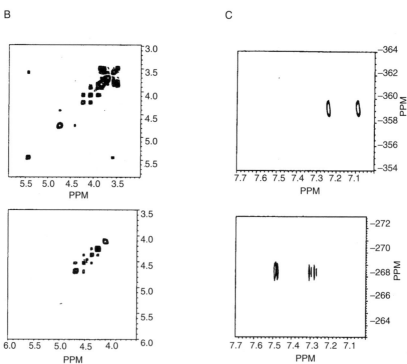

The simplest approach to generate a linear magnetic field gradient is to offset the z^1 shimming from its optimal value. This linear gradient frequency labels each of the different analytes' signals, so that they can be identified individually with their corresponding sample coil. Both 1D and 2D experiments have been carried out using this approach, in which spectra with and without a linear field gradient were acquired.[37,38] Several spectral analysis methods, including peak picking, spectral subtraction, and multiplication

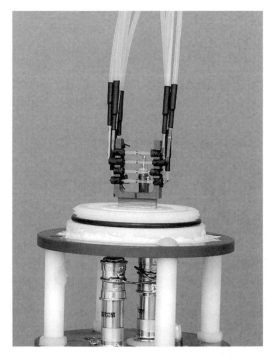

Figure 7.11. A picture of the four-coil multiplex NMR probe, with Teflon tubes attached to the sample capillaries to allow flow sample introduction. The sample volume for each coil was approximately 60 nL.

Figure 7.10. (*A*) Schematic illustration of the RF switching network used to switch one coil of the two-coil probe (developed by Zhang et al.) into the resonant circuit at a time. (*B*) COSY spectra of (*top*) sucrose and (*bottom*) adenosine triphosphate (both 20 mM in D_2O) loaded into the two coils of the two-coil probe, respectively. (*C*) HMQC spectra acquired from (*top*) ^{15}N-labeled (50%-labeled) ammonium chloride in D_2O and (*bottom*) unlabeled formamide, which were loaded in the two coils, respectively. (Adapted from *J. Magn. Reson.*, 153: 254. Copyright Academic Press, 2001).

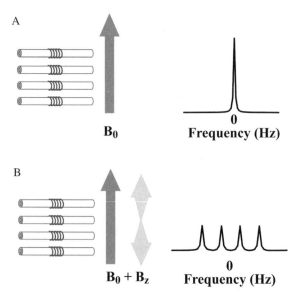

Figure 7.12. Schematic illustration of the use of a field gradient to differentiate signals from each sample in the multiplex NMR probe. The four coils are wired in parallel in a single resonant circuit. The single-arrowed line indicates the static magnetic field (7.05 T). (*A*) A simulated spectrum of four water samples in the four coils is shown at right for the case of no applied field gradient. All coils experience the same field, and therefore only one resonance is detected. (*B*) An applied field gradient (double-arrow line) changes the field experienced by each sample, and thus four resonances are observed.

methods, have been used by these authors to generate subspectra of the samples in different coils.

Two-dimensional spectroscopy can also be carried out using the multiplex NMR approach.[38] For example, correlated spectroscopy (COSY) experiments have been performed with a linear z-gradient applied during the data acquisition period. Spectral analysis methods can also be used for postprocessing the data. For the latter process the authors suggested zeroing all the peaks within a certain bandwidth from the diagonal, with this bandwidth set to be less than the smallest J-coupling observed in the spectrum. Using 0.5-M samples of ethanol, 1-propanol, dichloroacetic acid and acetaldehyde in D_2O, it was shown that the appropriate individual subspectra could be generated four at a time, and the subspectra of 1-propanol and ethanol are shown in Figure 7.13.

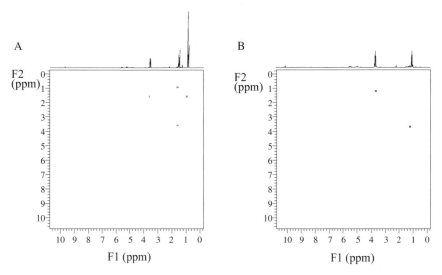

Figure 7.13. Two-dimensional subspectra: (*A*) 1-propanol and (*B*) ethanol generated from zero-gradient and gradient shifted 2D COSY spectra of four coils using the subtraction method. (Reproduced with permission from *Anal. Chim. Acta*, 400: 297. Copyright Elsevier Press, 1999).

Advanced Field Gradient Methods

The application of the small static magnetic field gradients (most easily generated by offsetting the z^1 shimming parameter) works well for spectra in which there is no significant spectral overlap between compounds, or for situations where the samples to be analyzed are quite similar. For example, parallel process monitoring could be accomplished using such an approach. In cases where spectra contain many resonances that overlap, it is much more difficult to assign resonances based on frequency shifts. To solve this problem, it is preferable to use larger, pulsed field gradients (PFGs) instead of the small static gradients to distinguish the different samples. PFGs are generated by a pair of oppositely wound, concentric gradient coils, which are situated around the RF coil in the probe. PFG coils that can generate large gradients of 10 to 60 G/cm are routinely used in commercial NMR probes, and their actuation can be easily realized during the pulse sequence. The exquisite control of RF pulses and PFGs currently available in modern NMR spectrometers allows a number of alternative, and more advanced, approaches to realize the multiple-sample analysis using the multiplex NMR probe. In particular, rapid selective excitation and chemical shift imaging (CSI)[109,110] methods with the application of large PFGs on the order

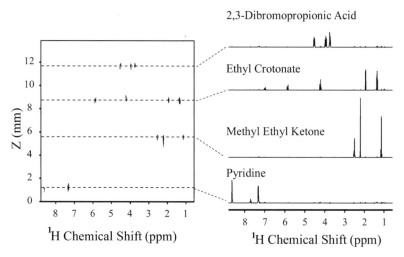

Figure 7.14. 2D contour plot and spectra of 2,3-dibromopropionic acid, ethyl crotonate, methyl ethyl ketone and pyridine (0.1 M in CDCl₃) generated using the CSI method. (Reproduced with permission from *Anal. Chem.*, 73: 2541. Copyright American Chemical Society, 2001).

of a few to $10\,G\,cm^{-1}$ have been carried out.[39] As shown in Figure 7.14, CSI acquisition using an incremented gradient value in the pulse sequence results in a frequency versus spatial correlation that separates the frequency components of the four samples based on their spatial position in the probe. Taking slices along the frequency dimension allows one to observe the individual spectrum from each sample coil. In this experiment, 12 gradient increments were required, along with 8 acquisitions per increment to cancel artifacts due to the non-ideal RF homogeneity of the short solenoid coils used in the experiment. Such long acquisition times need to be reduced before high-throughput can be achieved using CSI methods. Nevertheless, for samples at low concentration and with short T_1 relaxation times, the CSI method is efficient, particularly for samples with low concentration that benefit from signal averaging.

A second approach involves selective excitation, as has been used for slice selection in magnetic resonance imaging,[112] and uses the pulse sequence shown in Figure 7.15A. Application of the gradient causes the individual spectra from the samples in the four coils to be centered at different frequencies, as described by the following equation:

$$\omega_i = \gamma B_0(1 - \sigma_i) + \gamma G_z z_i, \tag{7.3}$$

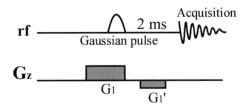

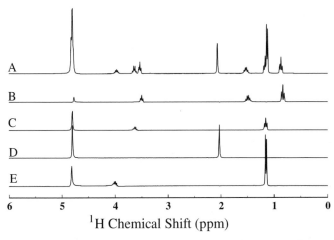

Figure 7.15. I. Pulse sequence for the rapid selective excitation experiment. (*Bottom*) Isolated ¹H spectra of two sets of samples obtained using the rapid selective excitation methodology. Normal ¹H Bloch decay spectrum of the first set of four samples (*A*) and the resulting sub-spectra of 0.5 M (*B*) 1-propanol, (*C*) 2-propanol, (*D*) acetic acid, and (*E*) ethanol in D₂O. (Adapted with permission from *Anal. Chem.*, 73: 2541. Copyright American Chemical Society, 2001).

where ω_i is the frequency of the compound in the ith coil, γ is the gyromagnetic ratio σ_i is the electronic shielding constant of the compound in the ith coil (and is responsible for the NMR chemical shift), G_z is the linear field gradient strength along the z direction, and z_i is distance of the ith coil from the center of the gradient coil along the z direction. The applied gradient G_z should be large enough such that the bandwidth of each spectrum (typically 10 ppm for proton spectroscopy) is smaller than the frequency separation between the samples when the gradient is applied. The RF pulse is frequency selective, with a bandwidth greater or equal to 10 ppm, and centered at frequency ω_1 for the first pulse, ω_2 for the second, and so on. Spectra for each sample were acquired sequentially in rapid succession, and the results are shown in Figure 7.15 for the case of 1-propanol, 2-propanol, acetic acid, and ethanol (each 500 mM in D_2O) loaded into the four coils, respectively. Selective excitation is an efficient method of data collection when the effective repetition time, TR_{eff} (given by n samples times data acquisition time), is less than three times the relaxation time T_1 value of any sample.

While the authors demonstrated this approach with four parallel coils and line widths of 2 to 3 Hz per sample, the maximum number of parallel detection coils and the ultimate achievable resolution of this approach have yet to be determined. An issue that will limit very high numbers of parallel coils is the fact that the S/N varies as $1/\sqrt{n}$ in this coil configuration, where n is the number of parallel coils.[113] One can compensate by increasing the size of the coil volume, essentially trading some mass sensitivity for improved concentration sensitivity. Alternatively, the use of multiple resonant circuits[35,36] or even switched coils can potentially overcome this limitation. The design of multicoil high-throughput NMR probes is currently an active area of research.

Automation for Higher Throughput

In the multiplex NMR experiments described above, sample loading and changing were carried out manually; that is, samples were injected into the sample regions of the multiplex NMR probe using a syringe. This slow sample-changing step is a significant bottleneck in the attempt to increase the throughput of the multiplex NMR analysis of multiple samples further. Therefore automatic sample changing methods need to be adopted for further improvements. Automated methods have been implemented for single sample NMR probes, as discussed above.[69,70,71,72] To achieve these goals, efforts are being made to make the sample-changing step faster.[113] FIA controlled multiple-sample analysis was initially investigated, in which an FIA controller was used to control the flow introduction of four samples into the multiplex NMR probe simultaneously. In this case the time required to acquire spectra for 4 samples with 8 transients each using the rapid selective excitation method is 80 s, and it took approximately 30 s to load 4 new samples. No contaminating carryover signals from prior samples were observed. These experiments show that 4 samples can be detected in less than 2 minutes, which correspond to a factor of 6 to 10 over current commercially available methods. A factor of 4 was obtained by the use of the parallel 4-coil multiplex NMR probe, and an additional factor of 1.5 to 2.5 in speed was gained due to the small volume of the samples used in these experiments, which are faster to load and remove from the probe. However, this approach still involves manual steps and thus is not as effi-

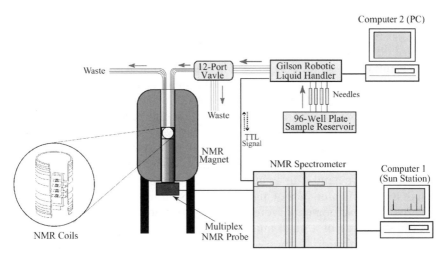

Figure 7.16. Schematic illustration of the automated flow-through setup for multiple-sample NMR analysis using a robotic liquid handler, which is controlled by a PC computer and able to communicate with the NMR spectrometer.

cient as a fully automated method. To realize the complete automatic sample changing, recent experiments using a robotic liquid handler (Gilson, Inc.) as the sample delivering system have recently been reprinted.[113] A schematic diagram for the automated multiplex NMR analysis is shown in Figure 7.16. The liquid handler is controlled by a PC computer, and the communication between the liquid handler and the NMR spectrometer is established through TTL signals. In this system the delivery of samples of small volumes (~25 µL) into the multiplex NMR probe with high precision was investigated, and total computer control is straightforward to realize. NMR experiments using the setup of Figure 7.16 show that the NMR analysis of multiple sample chemical libraries can be achieved at a rate of two samples a minute.[113]

7.3. FUTURE DIRECTIONS

Serial NMR analysis methodologies with high-throughput capabilities have been studied extensively, and are becoming routine. However, the high-throughput requirements for the analysis of chemical libraries by NMR dictate that alternative approaches need to be formulated. The development of different techniques for parallel NMR data acquisition has occurred only over the past few years, but this is clearly a rapidly evolving area of research. Currently most of the development in parallel NMR detection has been in the area of probe hardware design, and in the optimization of different pulse sequences and data processing techniques. Although one is ultimately limited by the extent of the homogeneous region of the magnet, a major thrust is to increase the number of coils. This may well involve the incorporation of microfabrication methods, as have already been prototyped for coil production.[96,114] The ultimate aim of parallel NMR probe technologies will likely be to optimize the interface with widely accepted sample formats such as the 96-well plate. However, the number of NMR sample coils that will optimize the throughput is yet to be determined. Hand in hand with an increase in the number of coils, is the requirement for the maintenance of high spectral resolution. This requirement may involve the development of zero-susceptibility materials for very small coils, as the shimming of a small number of sample coils is already challenging. Besides combinatorial library analysis and drug screening there are a number of applications for parallel high-throughput NMR. Clearly, the tremendous advances in hyphenated microseparation/NMR detection can benefit from the incorporation of parallelism. The incorporation of capillary separation with a multicoil NMR probe and the associated capillary fluidics to deliver the samples to and from the coils is the next step in probe development.[115] We anticipate further advances along these lines. Also exciting will be the interface of such "intelligent" NMR probe and fluidic systems to other integrated detection modalities such as fluorescence, absorbance, and even mass spectrometry to provide an integrated system capable of delivering unprecedented structural information from complex samples.

ACKNOWLEDGMENTS

The authors would like to thank Greg Fisher, Ernie MacNamara, Megan Macnaughtan, Chris Petucci, Jay Smith, and Jun Xu for their contribution to the work presented here. Financial support from the National Science Foundation, the Alfred P. Sloan Foundation, the Purdue Research Foundation, and the Lemelson Foundation is gratefully acknowledged.

REFERENCES

1. R. Bolger, *Drug Discovery Today* **4**, 251 (1999).
2. A. W. Czarnik, J. D. Keene, *Curr. Biol.* **8**, R705 (1998).
3. K. Lam, M. Lebl, V. Krchnak, *Chem. Rev.* **97**, 411 (1997).

4. J. Ellman, B. Stoddard, J. Wells, *J. Proc. Natl. Acad. Sci. USA* **94**, 2779 (1997).

5. E. M. Gordon, M. A. Gallop, D. V. Patel, *Acc. Chem. Res.* **29**, 144 (1996).

6. L. A. Thompson, J. A. Ellman, *Chem. Rev.* **96**, 555 (1996).

7. J. C. Hogan, *Nature* **384**, 17 (1996).

8. J. J. Parlow, D. L. Flynn, *Tetrahedron* **54**, 4013 (1998).

9. W. L. Fitch, *Mol. Divers.* **4**, 39 (1998).

10. J. A. Boutin, P. Hennig, P. Lambert, S. Bertin, L. Petit, J. P. Mahieu, B. Serkiz, J. P. Volland, J. L. Fauchere, *Anal. Biochem.* **234**, 126 (1996).

11. M. J. Shapiro, J. S. Gounarides, *Prog. Nucl. Magn. Reson. Spectr.* **35**, 153 (1999).

12. S. B. Shuker, P. J. Hajduk, R. P. Meadows, S. W. Fesik, *Science* **274**, 1531 (1996).

13. P. J. Hajduk, T. Gerfin, J. M. Böhlen, M. Häberli, D. Marek, S. W. Fesik, *J. Med. Chem.* **42**, 2315 (1999).

14. P. J. Hajduk, R. P. Meadows, S. W. Fesik, *Science* **278**, 497 (1997).

15. M. Lin, M. J. Shapiro, J. R. Wareing, *J. Org. Chem.* **62**, 8930 (1997).

16. A. Chen, M. J. Shapiro, *J. Am. Chem. Soc.* **120**, 10258 (1998).

17. A. Chen, M. J. Shapiro, *J. Am. Chem. Soc.* **122**, 414 (2000).

18. B. Meyer, T. Weimar, T. Peters, *Eur. J. Biochem.* **246**, 705 (1997).

19. J. Fejzo, C. A. Lepre, J. W. Peng, G. W. Bemis, Ajay, M. A. Murcko, J. M. Moore, *Chem. Biol.* **6**, 755 (1999).

20. A. H. Siriwardena, F. Tian, S. Noble, J. H. Prestegard, *Angew. Chem.-Int. Ed.* **41**, 3454 (2002).

21. B. J. Stockman, K. A. Farley, D. T. Angwin, *Method Enzymol.* **338**, 230 (2001).

22. S. A. Korhammer, A. Bernreuther, *Fresenius J. Anal. Chem.* **354**, 131 (1996).

23. K. Albert, *J. Chromatogr. A* **703**, 123 (1995).

24. D. L. Olson, T. L. Peck, A. G. Webb, R. L. Magin, J. V. Sweedler, *Science* **270**, 1967 (1995).

25. A. G. Webb, *Prog. Nucl. Magn. Reson. Spectr.* **31**, 1 (1997).

26. M. E. Lacey, R. Subramanian, D. L. Olson, A. G. Webb, J. V. Sweedler, *Chem. Rev.* **99**, 3133 (1999).

27. S. H. Smallcombe, S. L. Patt, P. A. Kiefer, *J. Magn. Reson.* **117**, 295 (1995).

28. P. A. Keifer, *Curr. Opin. Biotech.* **10**, 34 (1999).

29. K. Ueno, E. S. Yeung, *Anal. Chem.* **66**, 1424 (1994).

30. K. Asano, D. Goeringer, S. McLuckey, *Anal. Chem.* **67**, 2739 (1995).

31. M. W. Lutz, J. A. Menius, T. D. Choi, R. G. Laskody, P. L. Domanico, A. S. Goetz, D. L. Saussy, *Drug Discov. Today* **1**, 277 (1996).

32. T. Wang, L. Zeng, J. Cohen, D. B. Kassel, *Comb. Chem. High Through. Screen.* **2**, 327 (1999).

33. E. M. Banas, *Appl. Spec.* **23**, 281 (1969).

34. E. Oldfield, *J. Magn. Reson. A* **107**, 255 (1994).

35. G. Fisher, C. Petucci, E. MacNamara, D. Raftery, *J. Magn. Reson.* **138**, 160 (1999).

36. Y. Li, A. Wolters, P. Malaway, J. V. Sweedler, A. G. Webb, *Anal. Chem.* **71**, 4815 (1999).

37. E. MacNamara, T. Hou, G. Fisher, S. Williams, D. Raftery, *Anal. Chim. Acta* **397**, 9 (1999).

38. T. Hou, E. MacNamara, D. Raftery, *Anal. Chim. Acta* **400**, 297 (1999).

39. T. Hou, J. Smith, E. MacNamara, M. Macnaughtan, D. Raftery, *Anal. Chem.* **73**, 2541 (2001).

40. A. Ross, G. Schlotterbeck, H. Senn, M. von Kienlin, *Angew. Chem. Int. Edit.* **40**, 3243 (2001).

41. B. Merrifield, *Science* **232**, 341 (1986).

42. N. J. Maeji, A. M. Bray, H. M. Geyseen, *J. Immunol. Methods* **134**, 23 (1990).

43. A. Furka, F. Sebestyen, M. Asgedom, D. Dibo, *Intl. J. Pept. Prot. Res.* **37**, 487 (1991).

44. R. A. Houghten, C. Pinilla, S. E. Blondelle, J. R. Appel, C. T. Dooley, J. H. Cuervo, *Nature* **354**, 84 (1991).

45. R. D. Süßmuth, G. Jung, *J. Chromatog. B* **725**, 49 (1999).

46. X. Cheng, J. Hochlowski, *Anal. Chem.* **74**, 2679 (2002).

47. M. T. Combs, M. Ashraf-Khorassani, L. T. Taylor, *Anal. Chem.* **69**, 3044 (1997).

48. N. Shah, M. Gao, K. Tsutsui, A. Lu, J. Davis, R. Scheuerman, W. L. Fitch, R. L. Wilgus, *J. Comb. Chem.* **2**, 453 (2000).

49. D. A. Yurek, D. L. Branch, M.-S. Kuo, *J. Comb. Chem.* **4**, 138 (2002).

50. T. Wang, L. Zeng, T. Strader, L. Burton, D. B. Kassel, *Rapid Commun. Mass Spectrom.* **12**, 1123 (1998).

51. E. Moran, *IBC Conference on Combinatorial Chemistry*, San Diego, CA (1997).

52. G. A. Nemeth, D. B. Kassel, *An. Rep. Med. Chem.* **36**, 277 (2001).

53. L. Zeng, L. Burton, K. Yung, B. Shushan, D. B. Kassel, *J. Chrom. A* **794**, 3 (1998).

54. C. Garr, *The Use of Evaporative Light Scattering in Quality Control of Combinatorial Libraries*, Strategic Research Institute Proceedings of the Conference on Solid and Solution Phase Combinatorial Synthesis, New Orleans (1997).

55. V. de Biasi, N. Haskins, A. Organ, R. Bateman, K. Giles, S. Jarvis, *Rapid Commun. Mass Spectrom.* **13**, 1165 (1999).

56. L. Yang, N. Wu, R. P. Clement, P. J. Rudewicz, *Proc. 48th ASMS Conf. Mass Spectrom. Allied Topics*, 861 (2000).

57. M. C. Ventura, W. P. Farrell, C. M. Aurigemma, M. J. Greig, *Anal. Chem.* **71**, 4223 (1999).

58. T. A. Berger, W. H. Wilson, *J. Biochem. Biophys. Methods* **43**, 77 (2000).

59. Y. M. Dunayevskiy, P. Vouros, E. A. Wintner, G. Shipps, T. Carell, J. Rebek Jr., *Proc. Natl. Acad. Sci. USA* **93**, 6152 (1996).

60. J. A. Boutin, P. Hennig, P. H. Lambert, S. Bertin, L. Petit, J. P. Mahieu, B. Serkiz, J. P. Volland, J. L. Fauchère, *Anal. Biochem.* **234**, 126 (1996).

61. P. J. Simms, C. T. Jeffries, Y. J. Huang, L. Zhang, T. Arrhenius, A. M. Nadzan, *J. Comb. Chem.* **3**, 427 (2001).

62. C. G. Wade, R. D. Johnson, S. B. Philson, J. Strouse, F. J. McEnroe, *Anal. Chem.* **61**, A107 (1989).

63. M. Levenberg, *J. Automat. Chem.* **18**, 149 (1996).

64. B. J. Stockman, *Curr. Opin. Drug Discov. Develop.* **3**, 269 (2000).

65. P. A. Keifer, S. H. Smallcombe, E. H. Williams, K. E. Salomon, G. Mendez, J. L. Belletire, C. D. Moore, *J. Comb. Chem.* **2**, 151 (2000).

66. N. T. Nyberg, H. Baumann, L. Kenne, *Magn. Reson. Chem.* **39**, 236 (2001).

67. M. Spraul, M. Hofmann, R. Ackermann, A. W. Nicholls, J. P. Damment, J. N. Haselden, J. P. Shockcor, J. K. Nicholson, J. C. Lindon, *Anal. Commun.* **34**, 339 (1997).

68. See the Bruker and Varian websites: *http://www.bruker.de/*, *http://www.varianinc.com/* to obtain current specifications of their commercially available flow probes.

69. P. A. Keifer, *Drug Discov. Today* **2**, 468 (1997).

70. P. A. Keifer, *Drugs Future* **23**, 301 (1998).

71. J. Ruzicka, E. H. Hansen, *Anal. Chim. Acta* **78**, 145 (1975).

72. R. L. Haner, W. Llanos, L. Mueller, *J. Magn. Reson.* **143**, 69 (2000).

73. C. L. Putzig, M. A. Leugers, M. L. McKelvy, G. E. Mitchell, R. A. Nyquist, R. R. Papenfuss, L. Yurga, *Anal. Chem.* **66**, 26R (1994).

74. O. Vorm, P. Roepstorff, M. Mann, *Anal. Chem.* **66**, 3281 (1994).

75. A. E. Derome, *Modern NMR Techniques for Chemistry Research*, Pergamon Press, New York (1987).

76. R. D. Black, T. A. Early, P. B. Roemer, O. M. Mueller, A. Morgocampero, L. G. Turner, G. A. Johnson, *Science* **259**, 793 (1993).

77. H. D. Hill, *Appl. Supercond.* **7**, 3750 (1997).

78. R. Triebe, R. Nast, D. Marek, R. Withers, L. Baselgia, M. Häberli, T. Gerfin, P. Calderon, *40th Experimental Nuclear Magnetic Resonance Conference* Orlando, FL, p. 198 (1999).

79. P. Styles, N. F. Soffe, C. A. Scott, D. A. Cragg, F. Row, D. J. White, P. C. J. White, *J. Magn. Reson.* **60**, 397 (1984).

80. A. Abragam, *The Principles of Nuclear Magnetism*, Clarendon Press, Oxford (1961).

81. D. I. Hoult, R. E. Richards, *J. Magn. Reson.* **24**, 71 (1976).

82. E. Odeblad, *Micro-NMR in High Permanent Magnetic Fields*, Nordisk Forening for Obsterik och Gynekologi, Lund, Sweden (1966).

83. J. N. Shoolery, *Topics in Carbon-13 NMR Spectroscopy* **2**, 28 (1979).

84. N. Wu, T. L. Peck, A. G. Webb, R. L. Magin, J. V. Sweedler, *J. Am. Chem. Soc.* **116**, 7929 (1994).

85. N. Wu, T. L. Peck, A. G. Webb, R. L. Magin, J. V. Sweedler, *Anal. Chem.* **66**, 3849 (1994).

86. N. Wu, A. Webb, T. L. Peck, J. V. Sweedler, *Anal. Chem.* **67**, 3101 (1995).

87. N. Soffe, J. Boyd, M. Leonard, *J. Magn. Reson. A* **116**, 117 (1995).

88. L. F. Fuks, F. S. C. Huang, C. M. Carter, W. A. Edelstein, P. B. Roemer, *J. Magn. Reson.* **100**, 229 (1992).

89. F. O. Zelaya, S. Crozier, S. Dodd, R. McKenna, D. M. Doddrell, *J. Magn. Reson. B* **115**, 131 (1995).

90. W. L. Fitch, G. Detre, C. P. Holmes, J. N. Shoolery, P. A. Keifer, *J. Org. Chem.* **59**, 7955 (1994).

91. T. L. Peck, R. L. Magin, P. C. Lauterbur, *J. Magn. Reson. B* **108**, 114 (1995).

92. A. G. Webb, S. C. Grant, *J. Magn. Reson. B* **113**, 83 (1996).

93. K. Albert, G. Schlotterbeck, L.-H. Tseng, U. Braumann, *J. Chromatogr. A* **750**, 303 (1996).

94. G. Schlotterbeck, A. Ross, R. Hochstrasser, H. Senn, T. Kuhn, D. Marek, O. Schett, *Anal. Chem.* **74**, 4464 (2002).

95. W. U. Roffman, S. Crozier, K. Leuscher, D. M. Doddrell, *J. Magn. Reson. B* **111**, 174 (1996).

96. T. L. Peck, R. L. Magin, J. Kruse, M. Feng, *IEEE Trans. Biomed. Eng.* **41**, 706 (1994).

97. J. D. Trumbull, I. K. Glasgow, D. J. Beebe, R. L. Magin, *IEEE Trans. Biomed. Eng.* **47**, 3 (2000).

98. N. Watanabe, E. Niki, *P. Jpn. Acad. B-Phys.* **54**, 194 (1978).

99. E. Bayer, K. Albert, M. Nieder, E. Grom, T. Keller, *J. Chromatogr. A* **186**, 497 (1979).

100. C. Dalvit, S. Y. Ko, J. M. Böhlen, *J. Magn. Reson. B* **110**, 124 (1996).

101. S. X. Peng, *Biomed. Chromatogr.* **14**, 430 (2000).

102. K. Albert, *J. Chromatogr. A* **785**, 65 (1997).

103. B. Behnke, G. Schlotterbeck, U. Tallarek, S. Strohschein, L.-H. Tseng, T. Keller, K. Albert, E. Bayer, *Anal. Chem.* **68**, 1110 (1996).

104. K. Pusecker, J. Schewitz, P. Gfrörer, L.-H. Tseng, K. Albert, E. Bayer, *Anal. Chem.* **70**, 3280 (1998).

105. K. Pusecker, J. Schewitz, P. Gfrörer, L.-H. Tseng, K. Albert, E. Bayer, I. D. Wilson, N. J. Bailey, G. B. Scarfe, J. K. Nicholson, J. C. Lindon, *Anal. Commun.* **35**, 213 (1998).

106. J. Schewitz, P. Gfrörer, K. Pusecker, L.-H. Tseng, K. Albert, E. Bayer, I. D. Wilson, N. J. Bailey, G. B. Scarfe, J. K. Nicholson, J. C. Lindon, *Analyst* **123**, 2835 (1998).

107. R. A. Kautz, M. E. Lacey, A. M. Wolters, F. Foret, A. G. Webb, B. L. Karger, J. V. Sweedler, *J. Am. Chem. Soc.* **123**, 3159 (2001).

108. M. E. Lacey, J. V. Sweedler, C. K. Larive, A. J. Pipe, R. D. Farrant, *J. Magn. Reson.* **153**, 215 (2001).

109. T. R. Brown, B. M. Kincaid, K. Ugurbil, *Poc. Nat. Acad. Sci.-Biol.* **79**, 3523 (1982).

110. P. Mansfield, *Magnet. Reson. Med.* **1**, 370 (1984).

111. X. Zhang, J. V. Sweedler, A. G. Webb, *J. Magn. Reson.* **153**, 254 (2001).

112. M. A. Brown, R. C. Semelka, *MRI: Basic Principles and Applications*, Wiley-Liss Press, New York (1999).

113. M. Macnaughtan, T. Hou, J. Xu, D. Raftery, *Anal. Chem.* in press.

114. J. E. Stocker, T. L. Peck, A. G. Webb, M. Feng, R. L. Magin, *IEEE Trans. Biomed. Eng.* **44**, 1122 (1997). J. Koivuniemi, M. Kiviranta, H. Seppa, M. Krusius, *J. Low. Temp. Phys.* **110**, 255 (1998). J. A. Rogers, R. J. Jackman, G. M. Whitesides, D. L. Olson, J. V. Sweedler, *App. Phys. Lett.* **70**, 2464 (1997).

115. A. M. Wolters, D. A. Jayawickrama, A. G. Webb, J. V. Sweedler, *Anal. Chem.* **74**, 5550 (2002).

CHAPTER

8

MICELLAR ELECTROKINETIC CHROMATOGRAPHY AS A TOOL FOR COMBINATORIAL CHEMISTRY ANALYSIS: THEORY AND APPLICATIONS

PETER J. SIMMS

8.1. INTRODUCTION

Capillary electrophoresis is a separation technique that has gained popularity over the years. This has lead to several other electrokinetic chromatography modes, which include electrophoresis, capillary electrokinetic chromatography, micellar electrokinetic chromatography (MEKC), capillary zone electrophoresis, and isoelectric focusing. Capillary electrophoresis (CE) has been used successfully to separate several types of biomolecules,[1-4] separating molecules primarily by size and charge. Of all the techniques developed, the one used most predominantly is polyacrylamide gel electrophoresis, which can be used to separate proteins and peptides. Sodium dodecyl sulfate (SDS) is added to the run buffer to denature the protein, as well as to impart a uniform charge to the molecules. Once all the proteins in the mixture have the same charge, they separate by a mass-to-charge ratio as they move down the gel.[5-8] The molecular weight cutoff is dependent on the extent of cross-linking in the gel. The more cross-linking in the gel, the better is the separation of the low molecular weight proteins. SDS-Page, however, is not useful for the separation of small molecules, not only because of their size but also because of their inherent incompatibility with the run buffer.[9-14]

Capillary electrophoresis is primarily limited to small molecules that are water soluble because of their compatibility with the run buffer. Other similar techniques such as isoelectric focusing[15-17] and capillary zone electrophoresis[18-20] have aided in the separation of proteins by allowing for the separation of larger proteins. In addition these techniques can separate isoforms of proteins and peptides by using an extraordinarily low pH range. However, capillary electrophoresis cannot separate neutral compounds, and

Analysis and Purification Methods in Combinatorial Chemistry, Edited by Bing Yan.
ISBN 0-471-26929-8   Copyright © 2004 by John Wiley & Sons, Inc.

highly hydrophobic compounds can be difficult to separate due to their incompatibility with the run buffer. Charged hydrophobic compounds have been successfully separated by CE using nonaqueous run buffers[21-23] or by adding cyclodextrins to the run buffer to aid in the solubility of these compounds.[24-26] Nonetheless, these systems are incapable of separating neutral compounds from each other. The low current can additionally create long analysis times.

Micellar electrokinetic chromatography is a branch of capillary electrophoresis that allows for the separation of neutral analytes using capillary electrophoresis equipment, resulting in a separation that is similar to reverse-phase HPLC.[27-29] Another related technique is capillary electrokinetic chromatography (CEC), which separates compounds in a manner analogous to chromatography. CEC performs separations by using a capillary that is packed with a stationary phase such as C18 resin. The run buffer, which serves as the mobile phase, is typically composed of an organic modifier and an ionic buffer as the aqueous portion. When a voltage is applied across the capillary, the ionic portion of the run buffer creates an electroosmotic flow (EOF). The EOF acts like a HPLC pump, pulling the run buffer and analyte through the capillary. The analyte interacts with the stationary phase and the mobile phase in the same way as it would with HPLC. This results in a very efficient chromatographic separation. The selectivity can be manipulated, like HPLC, by adjusting the ratio of organic modifier to aqueous portion in the run buffer.[14,15] The disadvantage to this technique is that all the chromatographic methods developed thus far are isocratic separations. The run buffer used must have some ionic strength to generate the EOF. Therefore the organic modifier chosen must not precipitate the buffer, which currently makes reverse-phase or ion-exchange chromatography the only practical separations for this technique. In addition the use of buffers makes it difficult to do normal phase separations by CEC.

Micellar electrokinetic chromatography uses ionic surfactants at a concentration above the critical micelle concentration (CMC) as a component of the run buffer chosen to separate compounds. This generates a pseudostationary phase that performs the separation. This technique is therefore optimal for separating neutral and charged compounds from each other. In addition compounds that are very hydrophobic, and those typically insoluble in traditional capillary electrophoresis run separate buffers under these conditions. Neutral compounds elute in the order of their hydrophobicity.

Chiral surfactants allow for the separation of enantiomers by MEKC.[30-33] The separation of neutral compounds is dependent on the partitioning coefficient of the analyte into the micelle. The elution order for neutral compounds is often similar to reverse-phase chromatography, again, due to the fact that in the run buffer, the micelle acts like a pseudostationary phase.

Elution of charged compounds, by contrast, will depend on the charge of the micelle, which causes it to act like an ion-pairing reagent if the analyte has an opposite charge. The separation can be manipulated for optimal separation by changing the surfactant in the run buffer and the composition of the run buffer, which affects the selectivity of the run buffer system. For example, molecules that are extremely hydrophobic and do not separate with SDS in the run buffer can be separated using bile salt surfactants[34-40] or by adding an organic modifier to the run buffer.[41-45] MEKC allows for the resolution of closely related compounds because of the high efficiency generated from using the capillary tube. Since only a small amount of run buffer is used for the analysis of each sample (nanoliter volumes), a few milliters of run buffer can be used to perform several analyses. In addition only a few nanoliters of sample are required for analysis. Many separations can be performed in a short period of time, therefore making this technique useful for high-throughput analysis. Choosing this technique can be a good alternative to HPLC for combinatorial chemistry analyses. The parameters that can be changed to design an assay, and how this technique can be used to complement HPLC, will be discussed in the following chapter.

8.2. THEORY OF MICELLAR ELECTROKINETIC CHROMATOGRAPHY

Micellar electrokinetic chromatography uses surfactants at concentrations above the CMC in the run buffer during a capillary electrophoresis separation to separate neutral and hydrophobic compounds. The separation mechanism is based on the partitioning of the analyte between the micelle and the bulk aqueous solution. When a high voltage is applied across the capillary, a micelle, for example, SDS, will migrate toward the cathode because of its negative charge. However, when the applied voltage acts on the aqueous phase, the ionic particles flow toward the anode. This creates a phenomenon, known as electroosmotic flow (EOF). EOF acts like a pump for the mobile phase in HPLC and pulls the solvent in one direction. Therefore, when an analyte is partitioned into the micelle, its overall migration velocity is slowed relative to the bulk EOF. The difference in velocity of each analyte in the system leads to the separation of the compound mixture.[43-45] The following principle is demonstrated in Figure 8.1.

The surfactants move in the opposite direction of the EO flow because migration is in the direction opposite the charge on the polar portion of the molecule. However, the flow of the EOF is stronger than the migration of the micelles. Eventually the micelle migration is overcome by the EOF, and

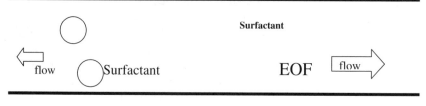

Cathode +   **Anode -**

Figure 8.1. A schematic drawing of the micellar electrokinetic chromatography process. The micelles flow in the direction opposite of their charge. The EOF will flow in the direction of the anode. The EOF is stronger than the attraction of the micelles to the opposite charge, which results in the micelles actually being pulled to the anode. This results in the elution range of the system.

they are pulled toward the anode. This results in something called the "elution range parameter." When a neutral analyte is injected into the micellar solution, a portion of the analyte gets incorporated into the micelle and migrates with the micelle. The remaining amount is free and is allowed to migrate with the bulk EO flow. Therefore the migration velocity is dependent on the distribution coefficient between the micelle and the aqueous run buffer. To calculate, one must determine the net velocity of the micelle, which can be done by using Eq. (8.1).

$$v_{mc} = v_{eo} + v_e = -\frac{\varepsilon E}{\eta}\left[\zeta_c - \frac{2\zeta_{mc}}{3}f(KA)\right],\qquad(8.1)$$

where ζ_c and ζ_{mc} are the zeta potentials of the capillary and the micelle, respectively. ε and η equals the dielectric constant and the viscosity of the running electrolyte, respectively. $f(KA)$ depends on the shape of the micelle. A equals the radius of the micelle, K is the Debye-Hückel constant, and E is the electric field strength. Once the net velocity is determined, one needs to determine the retention time of the micelle, which can be done using Eq. (8.2).

$$t_{mc} = \frac{1}{v_{eo} + v_e}.\qquad(8.2)$$

With this information one can determine the elution range parameter by using Eq. (8.3). This will determine the time range in which the analytes can be expected to elute:

$$\frac{t_0}{t_{mc}} = \frac{v_{mc}}{v_{eo}} = 1 - \frac{2\zeta_{mc}}{3\zeta_c} f(KA).$$ (8.3)

This information is helpful for designing a rapid assay. The elution range parameter will determine if the compounds will migrate within the time frame required for the analysis. The t_{mc} can also be determined experimentally using Sudan III as a tracer. Under MEKC conditions, Sudan III is completely incorporated into the micelle and moves at the same velocity. The capacity factor of MEKC depends on the compounds being analyzed. The capacity factor for neutral compounds can be determine by plugging the values into Eq. (8.4).

$$k' = \frac{t_r - t_0}{t_0 [1 - (t_r/t_{mc})]}.$$ (8.4)

Charged compounds that are charged interact with the micelle, and the calculation for the capacity factor is not straightforward. t_r is the retention time of the solute, t_0 is the retention time of an unretained molecule (usually a tracer that migrates with the EOF), and t_{mc} is the migration time of the micelle. The resolution in MEKC is often superior to that in reverse-phase HPLC, and is largely influenced by efficiency (N), selectivity (α), capacity (k') and elution range (t_0/t_{mc}). The selectivity can be determined as follows: $\alpha = k'_2/k'_1$. Just as in HPLC, the resolution increases with the square root of N. In MEKC a typical plate count can be around 200,000. The efficiency increases with the applied voltage, but usually reaches a maximum at approximately 25 kV. For two adjacent peaks the resolution can be calculated using this equation

$$R_s = \frac{\sqrt{N}}{4} \frac{\alpha - 1}{\alpha} f(k'),$$

where

$$(k') = \frac{k'}{k' + 1} \frac{1 - (t_0/t_{mc})}{1 + (t_0/t_{mc})k'}.$$

It is possible to adjust the capacity factor to the desired value by altering the surfactant concentration, since k' is directly proportional to the CMC. The capacity factor increases with increasing surfactant concentration. Adjusting these parameters will allow one to develop a set of conditions appropriate for use in analyzing sample mixtures.

8.3. SURFACTANTS

Surfactants are a critical component of the run buffer, determining how and what type of separation will be achieved with the system. Surfactants are typically composed of a hydrophobic portion, usually a lipid or a long-chain hydrocarbon, and a polar portion, which is typically charged. The charge can be either positive or negative.[46–49] One of the surfactants selected must contain a charge to promote separation. When the surfactant is added to the aqueous buffer the lipid chains orient themselves away from the water, and the polar portion orients with the water, forming the micelle. To get this to happen, the concentration of the surfactant must be above the critical micelle concentration. The choice of surfactant is important because this will ultimately determine how well the component mixture will be separated.

The most popular surfactant used is SDS. This surfactant is preferred for several reasons. It can give an elution profile that is similar to reversed-phase chromatography, which can allow for direct comparison to reverse-phase HPLC. In addition it forms micelles easily and can be used over a wide range of concentrations. Finally, it is transparent to *UV* light, which allows detection of compounds that absorb *UV* light at 200 nm. MEKC systems containing SDS in the run buffer are best used to separate compounds that are neutral,[50–52] positive,[53–55] and chiral.[24–26] This surfactant is good for compounds that have an intermediate hydrophobicity. In the case of positively charged compounds, the negatively charged portion of SDS will act as an ion-pairing reagent, which can affect the selectivity of the MEKC system. Another popular long-chain surfactant is cetyltetramethylammonium bromide. This positively charged surfactant has 14 carbons on it, and it is often used to separate compounds that are negatively charged species.[56–58] When using this surfactant, one must remember to change the polarity of the instrument in order to achieve separation. Compared to SDS run buffer systems, compounds separated using CTAB will have a shorter migration time.

In some cases ionic surfactants create a strong interaction with the analyte and cause the analyte to migrate with the micelle, increasing the migration time. Addition of a non-ionic surfactant to the run buffer can help the analysis by decreasing the surface charge. This will in turn decrease the migration time of the analyte and change the selectivity of the system. In addition, changing the charge of the analyte can lead to successful separations. Positively charged surfactants have been used to separate a variety of basic drugs.[60–61] Using ion-pairing reagents with the surfactants affects the separation. The surfactants increase the migration time of the charged species having the same charge as the surfactant and decrease the migration time of compounds that have a charge opposite of the micelle.

Other alkyl surfactants that have gained popularity are the alkyl glycosides. Micelles composed of alkyl glycosides have been used in conjunction with borate buffers, and under these alkaline conditions in situ charged micelles are formed. Varying the ionic strength and the pH of the borate in the run buffer can change the density of these micelles. El Rassi and coworkers have been successful with these types of separations.[62-64] A list of commonly used alkyl surfactants and the CMC of each is given in Table 8.1.

Bile salts are anionic surfactants that can be found in biological sources. They have steroidal structures and can form reverse micelles.[65] Bile salt surfactants are typically used to separate extremely hydrophobic.[34-40] Bile salts are an excellent alternative choice when compounds do not separate and migrate with the t_{mc} of a SDS run buffer system. Figure 8.2 shows how changing the micelle from SDS to sodium cholate can affect the resolution of corticosteroids. The effect of the hydrophobicity of different bile salt can also be seen in Figure 8.3. Resolution of the same corticosteroids can be improved when replacing sodium cholate with sodium deoxycholate. Because of their unique structure various bile salts have demonstrated the ability to separate chiral compounds.[32,35,66-68] Bile salt micelles can handle organic solvents better than the alkyl chain surfactants broadening the scope of their applications.[65] The source of the bile salts is important to ensure a clean electropherogram, since they can contain impurities that interfere with UV detection at low wavelengths, potentially decreasing the sensitivity of the assay. Therefore several sources of bile salts should be considered to ensure that the highest purity salts are selected for the run buffer. Table 8.2 shows a list of the most commonly used bile salts.

Adding an organic modifier to the run buffer[41,69-71] can enhance the separation of neutral and hydrophobic molecules. Organic solvents have been used widely to affect the selectivity of HLPC separations. They usually obtain optimal resolution, reasonable retention times, and help remove compounds that are very hydrophobic and highly retained by the stationary phase. Increasing the concentration of the organic modifier causes the compounds to elute at faster retention times. However, when an organic modifier is added to the run buffer, a reverse effect is observed with MEKC. When the organic modifier is added to the run buffer, it increases the solubility of the compounds in the aqueous portion of the run buffer, causing them to migrate at a time faster than the t_{mc}. Besides changing the resolution adding an organic modifier decreases the electroosmotic flow of the system. This creates a larger migration window, which increases the analysis time of the system. The observed migration time appears to be directly proportional to the viscosity changes that come with using different organic

Table 8.1. Long-Chain Alkyl Surfactants for MEKC

Surfactant	CMC (mM)
Cationic	
Decyltrimethylammonium bromide (DeTAB)	61
Decyltrimethylammonium chloride (DeTAc)	68
Dodecyltrimethylammonium bromide (DTAB)	15
Dodecyltrimethylammonium chloride (DTAC)	20
Tetradecylammonium bromide (TTAB)	3.5
Tetradecylammonium chloride (TTAC)	4.5
Cetyltrimethylammonium bromide (CTAB)	0.92
Cetyltrimethylammonium chloride (CTAC)	1.5
Hexadecyltrimethylammonium bromide (HTAB)	
Hexadecyltrimethylammonium chloride (HTAC)	
Anionic	
Sodium dodecyl sulfate (SDS)	8.1
sodium tetradecyl sulfate (STS)	2.1
Sodium decanesulfonate	40
Sodium dodecanesulfonate	7.2
Sodium N-lauroyl-N-methyltaurate (LMT)	8.7
Sodium polyoxythylene(3) dedecyl ether sulfate	2.8
Potassium perfluoraoheptanoate	28
Sodium tetradecene sulfonate (OS-14)	
Sodium polyoxyethylene(3) dodecyl ether acetate	
Sodium di-2-ethylhecyl sulfosuccinate	2.1
Disodium 5,12-bis(doedecyloxymethyl)-4,7,10,13 -tetraoxa-1,16-hexadecanedisulfonate	
Perfluorooctanesulfonate	
3β-glucopyranosyl-5β-cholan-12α-hydroxy-24-oic acid	
Neutral	
Polyoxyethylene(6) dodecyl ether	0.09
Polyoxyethylene(23) dodecyl ether (brij 35)	0.09
Polyoxyethylene (20) sorbitanmonolaurate (tween 20)	0.95
Polyoxyethylene (20) sorbitanmonooctadecanate (tween 60)	
Heptyl-β-D-glucopyranoside	79
Octyl-β-D-glucopyranoside	25
Nonyl-β-D-glucopyranoside	6.5
Decyl-β-D-glucopyranoside	3
Octanoyl-N-methylglucamide	58
Nonanoyl-N-methylglucamide	25
Decanoyl-N-methylglucamide	6
Octyl-β-maltopyranoside	23
n-Octanoylsucrose	24

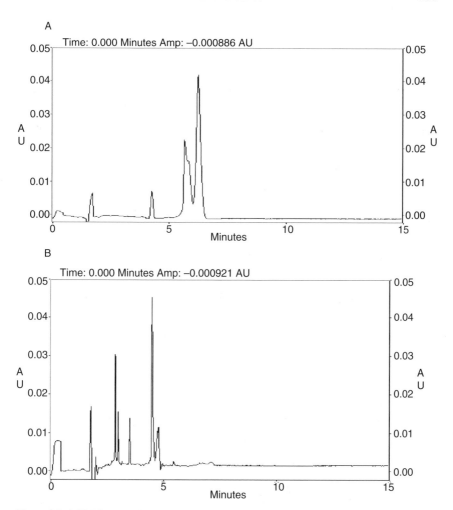

Figure 8.2. MEKC separation of some corticosteroids using (*A*) 50-mM SDS and (*B*) 50-mM sodium cholate as the surfactant in the run buffer. The solution also contained 25-mM sodium borate pH 9.0. The separation was performed with a 57-cm open tube capillary. The compounds were detected at 200 mn.

solvents.[72] Figure 8.4 shows how increasing the percentage of acetonitrile increases the resolution and migration times of the corticosteroids. The efficiency that is achieved can be more or less dependent on the organic modifier that is used in the system. Current decreases as the percent organic modifier is increased allowing for the use of higher voltages and a decrease

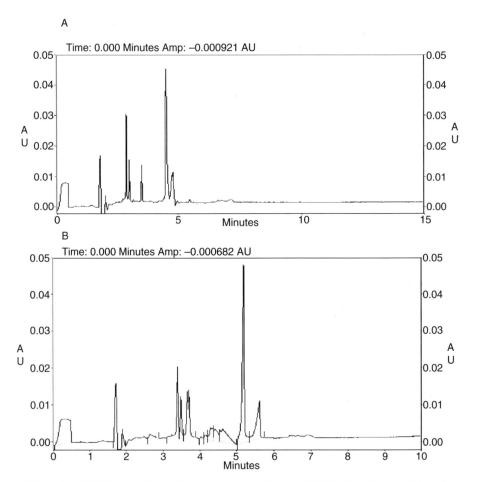

Figure 8.3. MEKC separation of some corticosteroids using (*A*) 50-mM sodium cholate and (*B*) 50-mM sodium deoxycholate as the surfactant in the run buffer. The solution also contained 25-mM sodium borate buffer pH 9.0. The separation was performed with a 57-cm open tube capillary. The compounds were detected at 200 mn.

in Joule heating. However, it should be noted that to high a concentration of organic modifier can lead to a breakdown of the micelle structure. The amount of organic modifier used is dependent on the type of organic modifier and the type of micelle that is used in the run buffer system. It should be noted that the organic modifier can precipitate the salt that is used in the system when the concentration of either gets to high. Care must be taken to ensure that the buffer used is compatible with the organic mod-

Table 8.2. Commonly Used Bile Salts in MEKC

Surfactant	CMC
Sodium cholate (SC)	15
Sodium deoxycholate (SDC)	6
Sodium taurocholate (STC)	10
Sodium taurodeoxycholate (STDC)	6
Dehydrocholate	
Ursodeoxycholate	7
Chenodeoxycholate	4
Lithocholate	0.6
Glyco-cholate	10
Glyco-ursodeoxycholate	4
Tauro-ursodeoxycholate	2
Tauro-cholate	6
Glyco-chenodeoxycholate	2
Tauro-chenodeoxycholate	3

ifier. In most cases up to 20% organic modifier in MEKC can be used without any difficulty. The typical solvents that can be used are methanol, acetonitrile, ethanol, and isopropanol. Organic modifiers are often used when one or more components have a migration time of t_{mc}. This usually happens when the analyte has a very low solubility in the run buffer. Addition of the organic modifier to the run buffer will increase the solubility of the analyte in the run buffer. This causes the analyte to partition into the bulk aqueous phase more, which leads to a decrease in migration time. Care should be taken to make sure that no compounds are eluting close to the EOF. Addition of an organic modifier to the run buffer can cause the early migrating component to elute with the EOF.

Chiral surfactants can improve the separation of chiral mixtures as well as enhance the separation of compounds that contain no chiral center. These surfactants can be neutral or charged, and a list of them is given in Table 8.3. As mentioned earlier, bile salt surfactants can be effective in separating enantiomers of several drugs.[73–76] Acylamino acids and alkoxyacylamino acid surfactants can be used to obtain chiral separations.[26,77–81] When considering an acylamino acid or an alkoxyacylamino acid, one must make sure that the molecule is not extremely hydrophobic. A highly hydrophobic analyte will have a migration time of t_{mc}, which will lead to poor or no resolution of the analyte. Some enantiomer separations can be performed using a mixed micelle system that contains two chiral surfactants or a combination of a surfactant and a chiral additive such as cyclodextrin. Cyclodex-

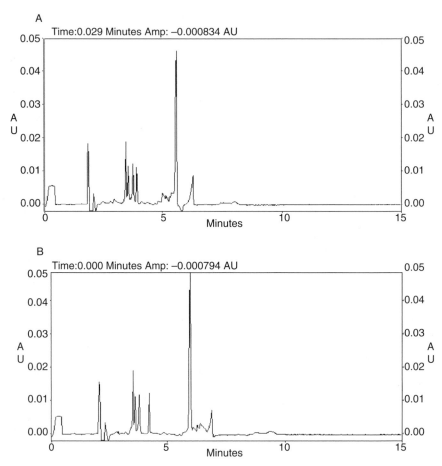

Figure 8.4. Electropherogram of some corticosteroids using a run buffer composed of (*A*) 25-mM sodium borate pH 9.0, 50-mM sodium deoxycholate, 5% acetonitrile, and (*B*) 25-mM sodium borate pH 9.0, 50-mM sodium deoxycholate, 10% acetonitrile. The capillary used was a 57-cm open tube capillary, and the compounds were detected with UV detection at 200 nm.

10 trins are oligosaccharides composed of glucose linked α 1→4 in a circular fashion. These molecules can be composed of six, seven, and eight glucose units, and the compounds are called α, β, and γ cyclodextrin, respectively. The core of the cyclodextrin molecule is hydrophobic while the outer portion is hydrophilic. The physical properties of these molecules are listed in Table 8.4. The hydrophobic core allows for the inclusion of organic molecules into the cyclodextrin cavity. The complex that is formed depends on the size of the analyte in the mixture and the size of the host cyclodextrin

Table 8.3. Chiral Surfactants Used in MEKC

Surfactants	CMC
Bile salts and saponis	
Sodium cholate (SC)	15
Sodium deoxycholate (SDC)	6
Sodium taurocholate (STC)	10
Sodium taurodeoxycholate (STDC)	6
Digitonin	
Glycyrrhizic acid	
β-Escin	
Amino acid head type	
Sodium N-dodecanoyl-L-valinate (SDVal)	6.2
Sodium N-dodecanoyl-L-alaninate	12.5
Sodium N-dodecanoyl-L-glutamate	
N-Dodecanoyl-L-serine	
Sodium N-dodecanoyl-L-threoninate	
Sodium N-undecylenyl-L-valinate	36
Poly(sodium N-undecylenyl-L-valinate)	
(R)-N-Dodecoxycarbonylvaline	
(S)-N-Dodecoxycarbonylvaline	
(S)-2-[(Dodecoxycarbonyl)amino]-3(S)-methyl-1-sulfoxypentane	
Sugar head type	
n-Heptyl-β-D-thioglucopyranoside	30
n-Octyl-β-D-glucopyranoside	24
n-Octyl-β-D-thioglucopyranoside	9
n-Dodecyl-β-D-glucopyranoside	2
n-Dodecyl-β-D-glucopyranoside monophosphate	0.5
n-Dodecyl-β-D-glucopyranoside monosulfate	1
Cyclodextrins	
α-cyclodextrin	
β-cyclodextrin	
γ-cyclodextrin	
Mono-3-O-phenylcarbamoyl-β-Cyclodextrin	
2,6-di-O-methyl-β-cyclodextrin	
Hydroxypropyl-β-cyclodextrin	
Sulfobutylether-cyclodextrin	
Trimethyl-β-cyclodextrin	

Table 8.4. Physical Properties of Cyclodextrins

Cyclodextrin	Number of Glucose Units	Modular Weight	Cavity Diameter	Water Solubility (g/100 mL)
Alpha	6	972	0.57 nm	14.5
Beta	7	1135	0.78 nm	1.85[a]
Gamma	8	1297	0.95 nm	23.2

[a] If the 2 or 3 hydroxyl groups of β-cyclodextrin are derivatized the solubility can increase.

cavity. This mechanism represents the attraction of the nonpolar molecule with the nonpolar segment of the cavity. When an aromatic group is present, its orientation in the cavity is selective due to the electron sharing of the methylene groups with those of the glycoside oxygens. This orientation can also be affected by the function groups attached to the guest molecule, which can interfere with or enhance the electronic sharing between the analyte and the cyclodextrin, and lead to chiral separations. Linear molecules occupy more random positions within the cyclodextrin cavity. Therefore the best chiral separations are obtained for molecules that contain at least one aromatic ring. In the past cyclodextrins have been added to mobile phases to improve the separation in HPLC.[82-85] Later D. W. Armstrong et al. attached the cyclodextrin molecules to silica gel and produced HPLC columns that contained native and modified cyclodextrins. These stationary phases were able to separate compounds in three different modes: reversed-phase,[86-88] normal phase,[89,90] and the polar organic mode.[91-94]

When cyclodextrins are added to the run buffer in MEKC, they can improve the separation of enantiomeric mixtures. The choice of cyclodextrin is important because the size of the cavity can affect the separation. Since the run buffer is primarily aqueous, it is assumed that the primary separation mechanism is inclusion into the cyclodextrin cavity. If the molecule is too small for the cyclodextrin cavity, then it will be included in a random order, and if the molecule is too large, then inclusion will not occur. Some of the native cyclodextrins have low solubilities in the aqueous run buffers and have to be added at low concentrations. These low concentrations may not be enough to affect the MEKC separation. Modified cyclodextrins have a higher solubility and are often used in place of the native cyclodextrins. In MEKC, cyclodextrins are considered electrically neutral and have no electrophoretic mobility. They are not assumed to be incorporated into the micelle because of the hydrophilic nature of the outside surface of the molecule. Therefore, when cyclodextrin is added to the run buffer, the separation is based on the equilibrium distribution among the analyte in the aqueous phase, the cyclodextrin cavity, and the

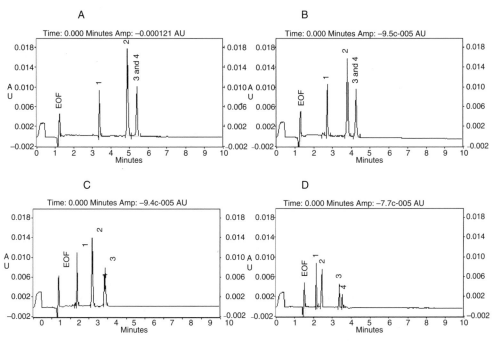

Figure 8.5. Electropherograms generated using (*A*) 0-mM hydroxypropyl-β-cyclodextrin, (*B*) 10-mM hydroxypropyl-β-cyclodextrin, (*C*) 25-mM hydroxypropyl-β-cyclodextrin, and (*D*) 50-mM hydroxypropyl-β-cyclodextrin added to the 100-mM SDS, 25-mM sodium tetraborate run buffer. All the samples were detected using UV at 254 nm.

micelle. The analyte molecule included by the cyclodextrin migrates at the same velocity as the EOF because electrophoretically cyclodextrin migrates with the bulk aqueous phase. Therefore the addition of cyclodextrin reduces the apparent distribution coefficient and enables the separation of highly hydrophobic analytes, which otherwise would be totally incorporated into the micelle in the absence of cyclodextrin.[95] This results in shorter analysis times and improved resolution when compared to run buffers that do not contain cyclodextrins. Figure 8.5 demonstrates the affect of adding hydroxypropyl-β-cyclodextrin on the separation of hydrocortisone and its derivatives. As the concentration of hydroxypropyl-β-cyclodextrin is increased from 0 to 50 mM, the resolution of the four components is also increased. In addition to the improving resolution, the analysis time can be decreased to approximately 3.5 minutes using 50-mM hydroxypropyl-β-cyclodextrin. The 50-mM concentration cannot be achieved with native cyclodextrin because of the low solubility. Modifying the hydroxyl groups

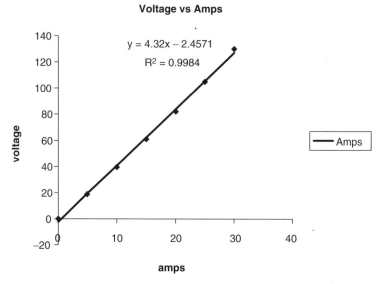

Figure 8.6. Ohm's law plot of the run buffer composed of 50-mM hydroxypropyl-β-cyclodextrin, 100-mM SDS, 25-mM sodium tetraborate pH 9.0 at 25°C.

of cyclodextrin with carboxymethyl, sulfate, heptakis, and methyl can affect the selectivity of the run buffer. Several cyclodextrins should be tried to ensure that the appropriate system is selected. It is important to note that adding cyclodextrin to the run buffer increases the viscosity. This in turn will increase the current applied across the capillary, and thus increase the Joule heating of the system. For this reason one must determine the maximum amount of voltage that can be applied to the system at a constant temperature, which might be lower than the system with cyclodextrin added to the run buffer. This can be accomplished by making an Ohm's law plot. When the applied current versus voltage is plotted, an ideal Ohm's law plot should yield a straight line. This indicates that the heat being generated by the capillary is adequately dissipated. Deviation from linearity is an indication of inadequate Joule heat dissipation, which leads to irreproducible results. Lowering the temperature can increase the linear range of the system and permit more current to be applied. Therefore, once the run buffer is selected, the Ohm's law experiment should be performed at a constant temperature to ensure that the proper voltage is being applied. The Ohm's law plot for the run buffer system used to acquire the data discussed above is shown in Figure 8.6. The graph shows that the current is linear up to 30 kV with an r^2 value of 0.9984.

Table 8.5. High Molecular–Mass Surfactants

Butyl acrylate-butyl methacrylate-methacrylic acid copolymer
Methyl methacrylate-ethyl acrylate-methacrylic acid copolymer
Poly(sodium 10-undecenyl sulfate) (polySUS)
Poly(sodium 10-undecylenate) (polySUA)
Poly(sodium N-undecylenyl-L-valinate) [poly(l-SUVal)]

High-molecular mass surfactants can be used in the run buffer for micellar electrokinetic chromatography separations. These surfactants have the unique ability to form micelles with just one molecule, and this allows them to be used at very low concentrations compared to other surfactants that must be above the CMC to form micelles. The high-molecular mass surfactants have the advantage of being more stable and rigid. It is also easier to control the size of these surfactants. Since only one molecule can be used to form the micelle, the CMC is essentially zero. The micelle concentration therefore is independent of buffer concentration, temperature and organic solvents. Table 8.5 list some of the high-molecular mass surfactants successfully used in MEKC. Terabe et al. and Yang et al. have had success with polymers composed of methylacrylate that have excellent selectivity for substituted aromatic compounds. The selectivity is better than separations achieved using SDS micelles.[96–98] When cyclodextrin is added to the run buffer the separation of dansylated amino acids improved over other cyclodextrin-SDS run buffer systems.[97] Other types of polymers that have been used as high-mass surfactants include those built from alkyidene-bridged resorcinol units[99] and sodium 10-undecylenates.[100,101] Sulfated polymers have been used and result in good separation of small molecules.[102] These polymers can form different shapes in solution and therefore change the selectivity of the system. Some form dendrimers[103] and monolayers,[99] and some can form emulsions[104] that affect the selectivity and migration times of the method.

Since these molecules have such a low CMC, they can be used for mass spectrometry analysis. Their low concentration and high molecular weight make them highly amenable to this technique. They result in much less background in the mass spectrum compared to the low molecular weight sufactants.[105,106] The only drawback is that the separations are not effective at low concentrations. However, when the concentration of the surfactant is increased the separation improves, but the trade-off is an increase in background noise. While the technique holds great promise, more experiments need to be performed to achieve optimal separation and the high sensitivity needed for mass spectrometry analysis.

Cyclodextrins can be added to the run buffers that contain high-mass surfactants. However, these surfactants can interact with the cyclodextrins, reducing the selectivity of the run buffer system and causing a loss in resolution.

8.3.1. Buffer Selection

The most important component of the run buffer system is the actual buffer itself. This controls the current and plays a big role in the selectivity of the system. Selection of the run buffer is especially important in MEKC. Using the wrong buffering system can result in poor reproducibility, and the loss of buffer capacity can shorten the amount of analysis time a certain system can use. Once the buffering capacity is lost, the EOF will cease. Therefore the pH and buffer concentration should be a value that will maximize the number of injections that can be made with the run buffer.[107] Most MEKC systems use various concentrations of sodium borate as the run buffer. This system is most effective at pH 9.0 or greater. Under these conditions the run buffer can be used to make a great number of injections. The graph in Figure 8.7 plots the migration time of hydrocortisone and its derivatives in

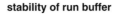

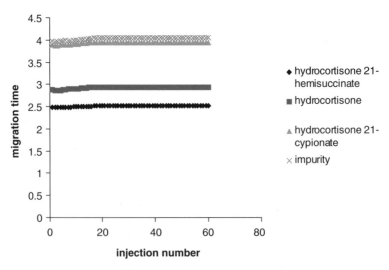

Figure 8.7. Graph of injection number versus retention time of hydrocortisone, hydrocortisone 21-hemisuccinate, hydrocortisone 21-cypionate and its impurity using a run buffer composed of 50-mM hydroxypropyl-β-cyclodextrin added to the 100-mM SDS, 25-mM sodium tetraborate run buffer. All the samples were detected using UV at 254 nm.

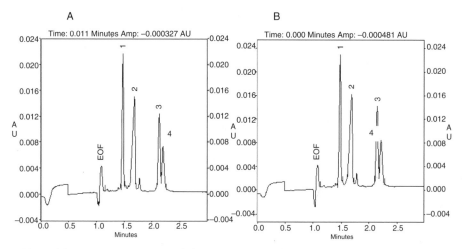

Figure 8.8. Electropherograms of (*A*) 1st injection and (*B*) 60th injection of the (1) hydrocortisone 21-hemisuccinate, (2) hydrocortisone, (3) hydrocortisone 21-cypionate, and (4) impurity using a 27 cm × 50 μm open tube capillary with 10-mM sodium tetraborate, 100-mM SDS, and 50-mM hydroxypropyl-β-cyclodextrin.

terms of the injection number. Note that after 60 injections the migration times show very little change. The electropherograms displayed in Figure 8.8 show the initial injection and the sixtieth injection. Both electropherograms show a clean baseline, which indicates that this run buffer system is good for 60 injections.[108] When the pH is lowered, the buffering capacity decreases and the number of reproducible injections decreases. It has been reported that run buffers used to perform separations at pH 7.4 were only good for three injections before the buffering capacity was lost.[109,110] The pH of the run buffer can have a dramatic affect on the EOF. Decreasing the pH slows the EOF and at pH around 2.5 the EOF is essentially suppressed.[111–114] Under these conditions the compounds must be charged to achieve a separation. Therefore they can migrate in the direction of opposite charges. Only compounds of like charge can be separated, and neutral compounds can only be separated from charged compounds and not each other. The micelles that can be used under these conditions have to be neutral, since a charged micelle will migrate in the direction of opposite charge and not provide any separation. At low pH most scientist choose cyclodextrins to perform the separations.[24–26] Since cyclodextrins are neutral, they will not migrate at the low pH, and behave the same at low pH as they do at high pH. It is wise include molecules using the same mechanism mentioned earlier, which can even help solubilize molecules that might have low solubility in the run buffer. When selecting a pH, it is important to ensure that the pH used to perform the separation does not cause any sample degradation or precipitation. The solubility of certain compounds is changed dramatically with changes in pH, and if the pI of proteins or peptides is known it is logical to avoid a similar pH.

The ionic strength of the buffer can affect the separation and the capacity of the system. At low concentrations the buffer separates more slowly, which leads to irreproducible results and decrease the amount of current generated, slowing the electroosmotic flow and increasing the analysis time. This can be used to improve the resolution. Concentrations as low as 10 mM have been reported to work well,[115-119] though, as the literature reports, the maximum number of injections one can get from these systems is three. High concentrations of buffers have the advantage of high capacities, which lead to a longer shelf life and an increased number of injections. However, increasing the buffer strength currents and greater heat production. When a very high concentration of salt is used, one needs to ensure that the proper voltage is being applied to the system by making an Ohm's law plot and examining the linearity of the line. When selecting an aqueous buffer, one must also consider the other solvents that will come in contact with the run buffer, preventing precipitation, which can clog the capillary. Samples that are dissolved in such solvents as chloroform and methylene chloride can cause the buffer to precipitate when the samples are injected. These solvents can be useful, but it is the experience from our laboratory that only a small amount of these solvents can be injected into the capillary at a given time. When these solvents were used in conjunction with a 50-cm capillary it was determined that the maximum inject time was 2 seconds. With other solvents such as aqueous mixtures of methanol/water or acetonitrile/water, the inject time could be as great as 5 seconds. Longer inject times cause a loss of current because the buffer precipitates. In addition too much organic solvent can destroy the micelles and decrease the resolution. At certain concentrations of organic solvents, the lost current is believed to be caused by precipitation of the run buffer as the sample is injected.[120]

There are important factors to consider when selecting a buffer for analysis. These factors can be applied to any CE experiment being designed.

1. Appropriate pKa
2. High solubility of the analyte in the aqueous system
3. Minimal salt effects
4. Low UV absorption
5. Low amount of Joule heating
6. Unreactive with the analytes

7. Minimal effects on dissociation due to concentration, temperature, and ionic composition
8. Nonexisting effects from mineral cations
9. Chemical stability

Some of these criteria are self-explanatory, but a discussion of others is needed. A buffer with a high ratio of water solubility to solubility in relatively nonpolar solvents can affect how much organic modifier can be used. A buffer with a low ratio can result in buffer precipitation when an organic solvent is used. Most MEKC experiments are performed at a high pH (>9), but one needs to ensure that the buffer works well in this range. Many buffers are supplied as crystalline acid or bases, and the pH of these buffers is usually not near the pKa when the buffer is placed into solution, so the pH must to be adjusted. The buffer should be selected with a pKa near the desired working pH, and if the pH needs to be adjusted, then the appropriate acid or base should be used.

There are certain cases where the mineral cations of a buffer can interfere with the compound separation. In these cases one often uses tetramethylammonim hydroxide as the buffer. This buffer has a pH around that of sodium hydroxide is effective.

Changing the temperature can affect the buffer, which can in turn affect the separation. High temperature causes a decrease in viscosity and lowers the analysis times. The current also increases with increasing temperature. Temperature can affect resolution, efficiency, and peak shape as well. It can affect the pH, especially of good buffers.

8.3.2. Voltage Effects

Increasing the voltage can increase efficiency and resolution, which is directly proportional to the change in voltage up to the point where the Joule heating is not adequately dissipated. Shorter analysis times and an increased amount of heat is produced as a product of increased voltage. As the voltage is increased, there is a greater need for temperature control. Ohm's law plots are very helpful in determining the maximum voltage that can be applied to the MEKC system.

8.3.3. Capillary Column Selection

The most widely used capillary for these types of experiments is the open, uncoated, fused silica tube. The silica groups are ionized at the pH where MEKC experiments are typically performed, allowing the EOF to occur.

There are very few references on other capillaries being used for MEKC experiments. The two parameters that are most important when selecting a column are length and diameter. Increasing the column length increases the resolution and the efficiency of the experiment. This is particularly helpful with very hard to resolve compounds or mixtures. Increasing the capillary length lowers the resistance to the current applied across the capillary and also helps dissipate the Joule heat. However, the lower current leads to longer analysis times, which can be overcome by increasing the voltage. The typical capillary lengths for MEKC experiments range from 40 to 75 cm. The length of the capillary is a factor if many components are to be separated or if the analysis time must be shorten. When a lot of components are present in the samples, a longer capillary would allow for greater migration time. Longer capillary tubes are also helpful when the sample is very dilute, because a larger volume of sample can be injected for analysis.

The capillary's diameter can also affect resolution and efficiency. Typical capillary diameters range from 5 to 100 μm. The narrower diameter capillaries have extremely high efficiency and give very good resolution because of the decrease in band broadening. The smaller diameter capillaries often give higher current values than larger capillaries when the same voltage is applied. The result is a shorter analysis time. The buffer concentration used in a small diameter column is typically less than that used for a larger column. In contrast, larger diameter capillaries have the big advantage of allowing more sample to be injected for analysis, making them a better choice for dilute samples. In addition higher concentrations of buffer and surfactant can be used with the wider capillaries. These capillaries typically require more voltage to achieve the same separation as that with a narrower diameter capillary. However, as the capillary diameter is increased, there is a decrease in resolution and column efficiency. The resolution and efficiency can be improved by using organic modifiers or by increasing the concentration of surfactant in the run buffer. Column diameter is an important aspect of the experiment, and proper column selection can give good resolution while maintaining a fast analysis time.

8.3.4. Optimization of Methods

The best place to start for method optimization is the length of the capillary. Good starting lengths are approximately 27 to 40 cm, with a diameter of 50 μm. These lengths give the best resolution and the shortest analysis times. The exact length chosen depends on how many compounds are in the sample mixture. The more compounds in the mixture, the longer the capillary may need to be to achieve a good separation. There are many different parameters besides column length that must be taken into account when one is optimizing a method. The type and concentration of surfactant are very critical, as are the pH and concentration of the buffer used. Capillary length is a primary consideration if the method being designed is going to be used for high throughput, since the length will have a big effect on the analysis time. In high-throughput separation the resolution is often compromised somewhat for speed. In optimizing a method, it is important to obtain as much information on the samples as possible. Hydrophobicity is important because this will help determine the types of surfactants that can be used. If the surfactant has a hydrophobicity that is too weak then the compounds will migrate with the t_{mc}, which will result in poor or no resolution. The micelle that is used for analysis should meet the following criteria: the surfactant must be highly soluble, and the micelle must be UV transparent, homogeneous, and compatible with the sample. If the micelle does have a UV absorbance, then it should not interfere with detection of the compounds. The CMC of the micelle should be known so that concentrations above this level are used.

The selectivity of the run buffer system can be manipulated by looking at the nature of the surfactant used. For example, using a surfactant that is positively charged like CTAB as opposed to one that is negatively charged like SDS can lead to a different type of interaction because of the differences in charges. The positively charged molecule can interact with compounds that are negatively charged in the form of an ion-pairing reagent. In addition there is a difference in the number of carbons in the surfactant, and this too will have an effect on the migration times. Mixed micelles provide a different selectivity because they have a different surface from that of the original ionic micelle.[121] Mixed micelles can provide resolution of chiral compounds without using chiral selectors.[122] When using mixed micelles, one must make sure that the CMC is reached for both micelles or the system will not work. The use of chiral additives is also of potential value. The important feature of chiral selectors is that, if neutral, they can decrease the analysis time and improve the selectivity. This is especially important when considering MEKC for a high-throughput assay.

Another parameter that needs to be optimized during method development is the capillary length. This will have a big effect on parameters such as resolution, capacity, migration time, and analysis time. The typical capillary lengths that can be used were mentioned earlier. For high-throughput assays shorter columns can be considered as long as the length provides the necessary resolution needed to obtain the purity information. Optimizing the capillary voltage is important because this affects the reproducibility of the assay. The best way to optimize the voltage is to construct an Ohm's law plot. This way one can determine how much voltage to apply and still achieve a reproducible separation. Longer capillaries can tolerate a higher voltage applied across them, whereas the shorter capillaries need sufficient cooling at the same setting.

When considering whether or not to used MEKC as a high-throughput screen, one needs to make sure that the compounds are compatible with the system. One needs to make sure that the compounds are stable at a high pH and that the compounds are somewhat hydrophobic. Very hydrophilic compounds, such as carbohydrates, will not be incorporated into the micelle and will migrate with the EOF. Compounds that are very hydrophilic might be separated using traditional CE methods, provided that the compounds contain a charge. If the compounds are compatible with MEKC, then one can prepare a run buffer that is suitable for the analysis. It is important that the run buffer selected be at the correct pH and concentration so that compound stability is maintained and the buffer system be able to generate an EOF. The concentration of the buffer will help determine what voltage can be used and how many injections can be made with one set of buffer vials. A high buffer concentration increases the viscosity of the run buffer, raising the current applied across the capillary and leading to inadequate cooling of the capillary. Constructing an Ohm's law plot determines how much voltage can be applied and achieve reproducible results. The higher the buffer concentration, the more capacity and more injections are possible. A concentration greater than 20 mM often works well. The actual separation is determined by the surfactant that is added to the run buffer. If there are any data available on the hydrophobicity of the compounds, this would help in selecting the type of surfactant that can be used. HPLC data are very helpful in this case. Compounds that are highly retained on a C18 column might require a very hydrophobic surfactant. If a highly hydrophobic surfactant is not available, the other options include addition of organic modifier to the run buffer or the use of a neutral surfactant. The important factors here are (1) the micelle must be above the CMC, (2) any organic modifier added must be compatible with the ionic portion of the run buffer, and (3) the surfactant must be UV transparent. Examining the charged portion of the surfactant for these characteristics can aid in designing the separation method. Using a micelle that has a charge similar to that of the analyte can decrease the run time of an assay by causing an ionic repulsion that reduces the amount of time the analyte spends in the micelle, therefore decreasing the analysis time. Care must be taken when using this approach to avoid a loss in resolution. Doing the reverse can enhance the resolution. The micelle creates a phenomenon known as "sample stacking," which is used to decrease peak broadening and allow greater efficiency and resolution. However, it can increase the analysis time.[123,124]

At this point one can optimize the voltage by constructing an Ohm's law plot. This experiment, performed at constant temperature, will determine the maximum voltage that can be applied across the capillary. The voltage will indicate if the assay time and if resolution is adequate for the laboratory needs. If the assay time must be decreased, one could look at the option of adding a neutral surfactant to the run buffer, which will decrease the assay time and still provide good resolution. Voltage optimization may be required, however. In addition one could increase the concentration of the buffer, which will increase the current, and thus the EOF, and reduce the analysis time. When the goal is to keep the analysis time short, an organic modifier should be considered along with a very short capillary. The use of an organic modifier will decrease the current and allow for higher voltage. In addition the organic modifier might provide the selectivity needed with the shorter capillary.

8.3.5. Applications and High-Throughput Considerations

MEKC has been used to analyze a large number of pharmaceutical compounds.[26,119,125–131] The applications have been primarily for compounds in drug formulations.[132–138] This technique has provided improved resolution over HPLC in these cases. Since the run buffering systems are transparent to UV, these methods have allowed for detection of samples that absorb light at 200 nm. Assays involving samples from biological matrices also have been developed with MEKC. These methods typically have a limit of detection around 500 ng/ml.[138–141] Recently methods have been developed for examining materials from plant extracts.[142–145] The appropriate system will depend on the compounds to be analyzed. MEKC can be used to complement HPLC analysis. For example, Figure 8.9 compares a chromatogram where the part of the sample is eluted in the void volume with an electropherogram. With MEKC the sample does not elute in the void, giving a better estimate of the purity. In some cases the electropherogram generated by MEKC can have the same order of elution as HPLC. This is very helpful when trying to compare the data with LC/MS. The chromatogram and electropherogram shown in Figure 8.10 compare the separation of hydrocortisone and its derivatives. The elution order is the same in both cases. Basic compounds often have strong interactions with unreacted silica in the HPLC stationary phases, and this causes peak broadening and peak tailing. Modifiers have to be added to the mobile phase to suppress this interaction as it occurs. Typical HPLC modifiers are trifluoroacetic acid, formic acid, and ion-pairing reagents. MEKC systems do not have this interaction and can give a better estimate of the sample purity with additives

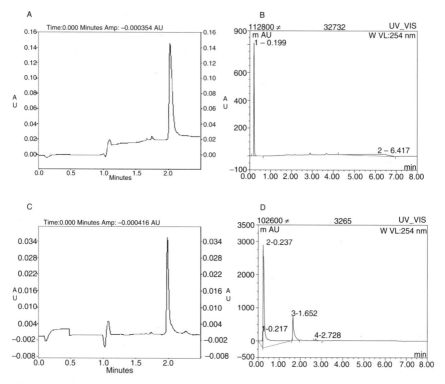

Figure 8.9. (*A*) and (*C*) electropherograms compared with (*B*) and (*D*) chromatograms of combinatorial chemistry samples generated from the synthesis of four 400 compound libraries. The samples elute close to the void in HPLC and were retained by MEKC. The MEKC samples were analyzed using a 27 cm × 50 μm open tube capillary with a 100-mM SDS, 50-mM hydroxypropyl-β-cyclodextrin, 10-mM sodium tetraborate run buffer. The chromatograms were generated using a 50 × 4.6 mm C18 column along with an acetonitrile/water/trifluoroacetic acid gradient at a flow rate of 3.5 mL/min. The gradient conditions were as follows: 10% acetonitrile to 40% acetonitrile in 0.5 minute, from 0.5 to 1.5 minute hold 40% acetonitrile. Then ramp to 90% acetonitrile in 3 minutes and hold 90% for 1 minute. Ramp back to initial conditions in 0.1 minute and hold for 2.5 minutes. The compounds were detected at 254 nm with both methods.

such as ion-pairing reagents. Figure 8.11 shows several electropherograms that were generated on combinatorial chemistry samples, using a 27 cm × 50 μm open tube capillary along with a run buffer composed of 50-mM hydroxypropyl-β-cyclodextrin, 100-mM SDS, and 10-mM sodium tetraborate. This demonstrates that the method has good selectivity for these different samples and can give a good estimate of the purity.

A   B

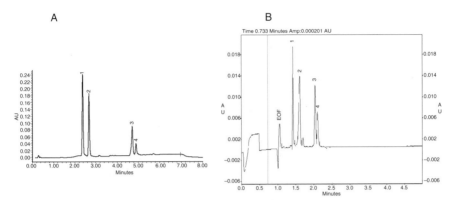

Figure 8.10. (*A*) Reversed-phase HPLC chromatogram and (*B*) electropherogram of (1) hydrocortisone 21-hemisuccinate, (2) hydrocortisone, (3) hydrocortisone 21-cypionate, and (4) impurity in 3. The chromatogram was generated using a 50 × 4.6 mm C18 column along with an acetonitrile/water/trifluoroacetic acid gradient at a flow rate of 3.5 mL/min. The gradient conditions were as follows: 10% acetonitrile to 40% acetonitrile in 0.5 minute, from 0.5 to 1.5 minute hold 40% acetonitrile. Then ramp to 90% acetonitrile in 3 minutes and hold 90% for 1 minute. Ramp back to initial conditions in 0.1 minute and hold for 2.5 minutes. The MEKC samples were analyzed using a 27 cm × 50 μm open tube capillary with a 100-mM SDS, 50-mM hydroxypropyl-β-cyclodextrin, 10-mM sodium tetraborate run buffer. The compounds were detected at 254 nm for both HPLC.

The most important item for a high-throughput assay is the stability and reproducibility of the system. There should be very little variability in the assay when it is performed in different laboratories or on different instruments. The number of samples that can be analyzed using the buffering system needs to be considered. Most CE instruments require approximately 4 mL of run buffer to perform several sample analyses. One needs to ensure that the buffer can support a large number of samples. After repeated injections the run buffer will lose its capacity to generate a current. As this interferes with the generation of the EOF, separation will not occur. Most run buffers use a sodium borate buffer around pH 9.0, which is quite stable. Many injections are possible under these conditions. In some cases as many as 60 injections were made from the same run buffer vials before the buffer lost its capacity. The method time under these conditions was approximately 3.5 minutes, and approximately 400 samples could be analyzed using approximately 50 mL of run buffer. MEKC has clear advantages for high-throughput analysis. It requires few reagents and little sample. The use of certain systems can allow the analysis to be completed as quickly as 3 minutes. Often the migration profile is similar to the elution profile generated by reverse-phase HPLC. This technique nicely complements the many separation techniques used to analyze combinatorial chemistry samples.

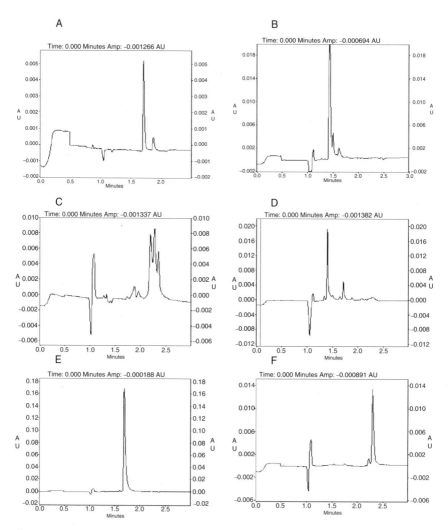

Figure 8.11. Electropherograms of compounds generated from four 400 compound libraries using a 27 cm × 50 μm open tube capillary with a 100-mM SDS, 50-mM hydroxypropyl-β-cyclodextrin, 10-mM sodium tetraborate run buffer. UV detection at 254 nm. The electropherograms show the different levels of purity of each sample and resolution that can be obtained in a short period of time.

REFERENCES

1. R. J. Linhart, A. Pervin, *J. Chromatogr.* **720**, 323–335 (1996).

2. J. Zhongjiang, T. Ramstad, M. Zhong, *J. Pharm. Biomed. Anal.* **30** (3), 405–413 (2002).

3. C. F. Silverio, A. Plazas, J. Moran, F. A. Gomez, *J. Liq. Chromatogr. Related Technol.* **25** (10–11), 1677–1691 (2002).

4. P. R. Brown, C. S. Robb, S. E. Geldart, *J. Chromatogr.* **965** (1–2), 163–175 (2002).

5. N. Suzuki, Y. Ishihama, T. Kajima, N. Asakawa, *J. Chromatogr.* **829**, 411–415 (1998).

6. D. Heiger, R. Majors, A. Lumbardi, *LC-GC* **15** (1), 14–23 (1997).

7. S. Honda, *J. Chromatogr.* **720**, 337–351 (1996).

8. K. Raith, R. Wolf, J. Wagner, R. H. H. Neubert, *J. Chromatogr.* **802**, 185–188 (1998).

9. S. Terabe, K. Otsuka, T. Ando, *Anal. Chem.* **57** (4), 834–831 (1985).

10. P. V. Van Zomeren, M. J. Hilhorst, P. M. Coenegracht, J. G. DeJong, *J. Chromatogr.* **867** (1–2), 247–259 (2000).

11. S. Mekaeli, G. Thorsen, B. Karlberg, *J. Chromatogr.* **907** (1–2), 267–277 (2001).

12. S. Bellini, M. Uhrova, Z. Deyl, *J. Chromatogr.* **772**, 91–101 (1991).

13. S. Surapaneni, K. Ruterbories, T. Lindstrom, *J. Chromatogr.* **761**, 249–257 (1997).

14. R. D. Tippetts, *US At. Energy Comm.* 115 (1965).

15. S. Hjerten, *Methods Biochem. Anal.* **18**, 55–79 (1970).

16. G. Browwer, G. A. Postema, *J. Electrochem. Soc.* **117** (3), 874–878 (1970).

17. K. Tsuji, *J. Chromatogr.* **662** (2), 291–299 (1992).

18. T. LeBricon, E. Launay, P. Houze, D. Bengoufa, B. Bousquet, B. Gourmel, *J. Chromatogr. B.* **775** (1), 63–70 (2002).

19. F. J. Fernandez, G. Rivas, J. Salgado, I. Montero, A. Gonzalez, C. Mugueta, I. Monreal, *Am. Clinical Lab.* **21** (4), 26–29.

20. D. Rochu, F. Renault, P. Masen, *Electrophoresis* **23** (6), 930–937 (2002).

21. H. Cottet, M. P. Struijk, J. L. Van Dongen, H. A. Claessens, C. A. Cramers, *J. Chromatogr.* **915** (1–2), 241–251 (2001).

22. J. Tjornelund, S. H. Hansen, *J. Biochem. Biophys. Methods* **38** (2), 139–153 (1999).

23. J. Torneland, S. H. Hnasen, *Chromatographia* **44** (1–2), 5–9 (1997).

24. L. Doowon, S. A. Shamsi, *Electrophoresis* **23** (9), 1314–1319 (2002).

25. K. W. Phinney, J. W. Jackson, L. C. Sander, *Electrophoresis* **23** (9), 1308–1313 (2002).

26. K. L. Rundlett, D. W. Armstrong, *Electrophoresis* **22** (7), 1419–1427 (2001).

27. M. P. Malloy, N. D. Phadke, H. Chen, R. Tyldesley, D. E. Garfin, J. R. Maddock, P. C. Andrews, *Proteomics* **2** (7), 899–910 (2002).

28. H. W. Towhin, O. Ozbey, O. Zingel, *Electrophoresis* **22** (10), 1887–1893 (2001).

29. X. Zuo, L. Echan, P. Hembach, H. Y. Tong, K. D. Speicher, D. Santoli, D. W. Speicher, *Electrophoresis* **22** (9), 1603–1615 (2001).

30. A. S. Rathore, C. Horvath, *J. Chromatogr.* **781** (1–2), 185–195 (1997).

31. C.-X. Zhang, Z-P. Sun, D.-K. Ling, *J. Chromatogr.* **655** (2), 309–316 (1993).

32. Y. He, K. Lee, *J. Chromatogr.* **793**, 331–340 (1998).

33. A. Peterson, J. P. Foley, *J. Chromatogr.* **695**, 131–143 (1997).

34. C. Zhang, C. Zhu, X. Lin, F. Gao, Y. Wei., *Anal. Sciences* **18** (5), 595–597 (2002).

35. H. Yarabe, R. Hyacinthe, J. K. Rugutt, M. E. McCarroll, I. M. Warmer, *Electrophoresis* **21** (10), 2025–2032 (2000).

36. E. S. Okerberg, S. Elshihabi, P. T. Carmichael, K. A. Woody, T. A. Barckhotz, J. A. Burke, M. M. Buskey, *J. Microcolumn Separ.* **12** (3), 391–397 (2000).

37. A. L. Crego, M. J. Gonzalez, M. L. Marina, *Electrophoresis* **19** (12), 2113–2118 (1998).

38. J. M. Herrero-Martinez, M. Fernandez-Marti, E. Simo-Alfonso, G. Ramis-Ramos, *Electrophoresis* **22** (3), 526–534 (2001).

39. S. E. Lucangioli, C. N. Caducci, V. P. Tripodi, E. Kenndler, *J. Chromatogr.* **765** (2), 113–120 (2001).

40. M. Aguilar, A. Farran, C. Serra, M. J. Sepaniak, K. W. Whitaker, *J. Chromatogr.* **778** (1–2), 201–205 (1997).

41. M. G. Khaledi, J. G. Bumgarner, M. Hodjmohammadi, *J. Chromatogr.* **802** (1), 35–47 (1998).

42. J. G. Clothier, Jr. L. M. Daley, S. A. Tomellini, *J. Chromatogr.* **683** (1), 37–45 (1996).

43. J.-M. Lin, M. Nakagawa, K. Uchiyama, T. Hobo, *Chromatographia* **50** (11–12), 739–744 (1999).

44. S. K. Poole, F. Colin *Anal. Commun.* **34** (2), 57–62 (1997).

45. E. Dabek-Zlotoszynaka, E. Lai, *J. Capillary Electrophoresis* **3** (1), 31–35 (1996).

46. A. E. Bretnall, G. S. Clarke, *J. Chromatogr.* **716** (1–2), 49–55 (1995).

47. P. L. Desbene, C. Rony, B. Desmazieres, J. C. Jacquier, *J. Chromatogr.* **608** (1–2), 375–383 (1992).

48. C.-X. Zhang, Z.-P. Sun, D.-K. Ling, *J. Chromatogr.* **655** (2), 309–316 (1993).

49. K. Ghousi, J. P. Foley, R. J. Gale, *Anal. Chem.* **622** (4), 2714–2721 (1990).

50. P. G. Mujselaar, H. A. Claessens, C. A. Cromers, *J. Chromatogr.* **764** (1), 127–133 (1997).

51. S. K. Poole, C. F. Poole, *Analyst* **122** (3), 267–274 (1997).

52. Y. S. Wu, H. K. Lee, F. Y. Sam, *Electrophoresis* **19** (10), 1719–1727 (1998).

53. P. A. Bianco, E. Pramauro, M. Gallarate, M. E. Carlotti, G. Orio, *Analytica Chimca Acta* **412** (1–2), 141–148 (2000).

54. J. Lin, J. F. Banks Jr., M. Novotny, *J. Microcolumn Sep.* **1** (3), 136–141 (1989).

55. C. P. Ong, L. C. Ng, H. K. Lee, Y. F. Sam, *Electrophoresis* **15** (10), 1273–1275 (1994).

56. J. G. Bumgarner, M. G. Khaledi, *Electrophoresis* **15** (10), 1260–1266 (1994).

57. T. Kaneta, S. Tanaka, M. Taga, H. Yoshida, *J. Chromatogr.* **653** (2), 313–319 (1993).

58. T. Kaneta, S. Tanaka, H. Yoshida, *J. Chromatogr.* **538** (2), 385–391 (1991).

59. D. Lukkari, H. Tiren, M. Pantsar, M. L. Riekkola, *J. Chromatogr.* **632** (1–2), 143–148 (1993).

60. J. T. Smith, Z. El Rassi, **685** (1), 131–143 (1994).

61. C.-E. Lin, Y.-T. Chen, T. Z. Wang, *J. Chromatogr.* **837** (1–2), 241–252 (1999).

62. M. J. Hilhorst, G. W. Somsen, G. J. De Jong, *J. Pharm. Biomed. Anal.* **16** (7), 1251–1260 (1998).

63. A. Dworschak, U. Pyell, *J. Chromatogr.* **848** (1–2), 387–400 (1999).

64. J. T. Smith, Z. El Rassi, *J. Microcolumn Sep.* **6** (2), 127–138 (1994).

65. J. T. Smith, W. Nashach, Z. El Rassi, *Anal. Chem.* **66** (7), 1119–1133 (1984).

66. R. O. Cole, M. J. Sepaniak, W. L. Hinze, J. Gorse, K. Oldiges, *J. Chromatogr.* **557** (1–2), 113–123 (1991).

67. H. Nishi, T. Fukuyama, M. Matsuo, *J. Microcolumn Sep.* **1** (5), 234–241 (1989).

68. X. Z. Qin, S. T. Diem, P. Dominic, *J. Chromatogr.* **16** (17), 3713–3734 (1993).

69. A. L. Crego, M. A. Garcia, M. L. Marina, *J. Microcolumn Sep.* **12** (1), 33–40 (2000).

70. T. S. K. So, L. Jin, C. W. Huie, *Electrophoresis* **22** (11), 2159–2166 (2001).

71. S. H. Hamen, G. Gabel-Jansen, S. Pedersen-Bjergaurd, *J. Sep. Science* **24** (8), 643–650 (2001).

72. J. M. Herrero-Martinez, E. F. Simo-Alfonso, C. Mongay-Fernandez, G. Ranis-Ramos, *J. Chromatogr.* **895** (1–2), 227–235 (2000).

73. Z. Liu, H. Zou, M. Ye, J. Ni, Y. Zhang, *Sepu.* **17** (2), 147–152 (1999).

74. E. Szoko, J. Gyimesi, Z. Szakacs, M. Tasnui, *Electrophoresis* **20** (13), 2754–2760 (1999).

75. L. A. Crego, J. M. Gonzalez, L. M. Marina, *Electrophoresis* **19** (12), 2113–2118 (1998).

76. H. Nishi, T. Fukuyama, M. Matsuo, *J. Microcolumn Sep.* **1** (5), 243–241 (1989).

77. S. Terabe, M. Shibata, Y. Miyashita, *J. Chromatogr.* **480**, 403–411 (1989).

78. A. Dobashi, M. Hamada, J. Yamaguchi, *Electrophoresis* **22** (1), 88–96 (2001).

79. K. Otsuka, J. Kawahara, K. Tatekawa, S. Terabe, *J. Chromatogr.* **559** (1–2), 209–214 (1991).

80. J. R. Mazzeo, E. R. Grover, M. E. Swartz, J. J. Petersen, *J. Chromatogr.* **680** (1), 125–135 (1994).

81. J. R. Mazzeo, M. E. Swartz, E. R. Grover, *Anal. Chem.* **67** (17), 2966–2973 (1995).

82. K. Otsuka, S. Terabe, *J. Chromatogr.* **875** (1–2), 163–178 (2000).

83. N. Thaud, B. Schillc, E. Renard, *J. Biochem Biophys Methods* **54** (1–3).

84. Y. C. Guillaumc, J. F. Robert, C. Guinchard, *Ann. Pharm. Pr.* **59** (6), 392–401 (2001).

85. P. Jandera, S. Buncekove, J. Planeta, *J. Chromatogr.* **871**, 139–152 (2000).

86. P. K. Zarzycki, R. Smith, *J. Chromatogr.* **912**, 45–52 (2001).

87. T. Vaisar, T. Vanek, *J. Chromatogr.* **547**, 440–446 (1991).

88. M. L. Hilton, S.-C. Chang, M. P. Gasper, M. Palwoswka, D. W. Armstrong, A. M. Stalcup, *J. Chromatogr.* **16** (1), 127–147 (1993).

89. J. W. Ryu, H. S. Chang, Y. K. Ko, J. C. Woo, D. W. Kim, *Microchem. J.* **63**, 168–171 (1999).

90. M. WoBner, K. Ballschmiter, *Fcesencius J. Anal. Chem.* **366**, 346–350 (2000).

91. S. L. Abidi, T. L. Mounts, *J. Chromatogr.* **670**, 67–75 (1994).

92. C. Chacau, A. Thienpont, M.-H. Soulard, G. Felix, *Chromatographia* **44** (7–8), 411–416 (1997).

93. G. Farkas, L. H. Ingerns, G. Quintero, M. D. Besson, A. Al-Saeed, G. Vigh, *J. Chromatogr.* **645**, 67–74 (1993).

94. S. C. Chang, G. L. Reid III, S. Chen, C. D. Chang, D. W. Armstrong, *Trends Anal. Chem.* **12** (4), 144–153 (1993).

95. S. Chen, *J. Chinese Chem. Soc.* **46**, 239–244 (1999).

96. S. Terabe, Y. Miyashita, O. Shibata, E. R. Barnhart, L. R. Alexander, D. J. Patterson, B. L. Karger, K. Horaoya, N. Tanaka, *J. Chromatogr.* **516**, 23–31 (1990).

97. H. Ozaki, S. Terabe, *J. Chromatogr.* **794**, 317–325 (1998).

98. L. Yang, C.-S. Lee, *J. Chromatogr.* **780** (1–2), 207–218 (1997).

99. H. Ozaki, S. Terabe, A. Ichihara, *J. Chromatogr.* **680** (1), 117–123 (1994).

100. H. Ozaki, A. Ichihara, S. Terabe, *J. Chromatogr.* **709** (1), 3–10 (1995).

101. S. Yang, J. G. Bumgarner, G. Jefferson, M. G. Khaledi, *J. High Resolut. Chromatogr.* **18** (7), 443–445 (1995).

102. K. Baechmann, A. Bazzenalla, I. Haag, K.-Y. Han, R. Arnethe, V. Bochmer, W. Vogt, *Anal. Chem.* **67** (10), 1722–1726 (1995).

103. C. P. Palmer, M. V. Khalad, H. M. McNair, *J. High Resolut. Chromatogr.* **15** (11), 756–762 (1992).

104. C. P. Palmer, H. M. McNair, *J. Microcolumn Sep.* **4** (6), 509–514 (1993).

105. C. P. Palmer, S. Terabe, *J. Microcolumn Sep.* **8** (2), 115–121 (1996).

106. C. P. Palmer, *J. Chromatogr.* **780** (1–2), 75–92 (1997).

107. C. P. Palmer, S. Terabe, *Anal. Chem.* **69** (10), 1852–1860 (1997).

108. L. Paugam, R. Menard, J.-P. Larue, D. Thauvenot, *J. Chromatogr.* **864** (1), 155–162 (1999).

109. P. J. Simms, C. T. Jeffries, Y. Huang, L. Zhang T. Arrhenius, A. Nadzan, *J. Comb. Chem.* **3** (5), 427–433 (2001).

110. B. J. Herbert, J. G. Dorsey, *Anal. Chem.* **67** (4), 685–776.

111. J. T. Smith, D. V. Vinjamoori, *J. Chromatogr.* **669**, 59–66 (1995).

112. E. F. Hilder, C. W. Klampfl, W. Buchberger, P. R. Haddad, *J. Chromatogr.* **922** (1–2), 293–302 (2001).

113. M. J. Hilhorst, A. F. Derksen, M. Steringa, G. W. Somsen, G. J. DeJong, *Electrophoresis* **22** (7), 1337–1344 (2001).

114. G. M. Janin, G. M. Muschek, H. J. Issaq, *J. Chromatogr.* **683** (1), 29–35 (1996).

115. M. A. Garcia, M. L. Marina, J. C. Diez-Masa, *J. Chromatogr.* **732** (2), 345–359 (1996).

116. D. Michalke, S. Kobb, T. Welsh, *J. Chromatogr.* **916** (1–2), 113–122 (2001).

117. C.-E. Lin, W.-C. Lin, W.-C. Chiou, *J. Chromatogr.* **722** (1–2), 333–343 (1996).

118. U. Pyell, U. Buetehorn, *Chromatographia* **40** (3–4), 175–184 (1995).

119. Ph. Morin, J. C. Archanbault, P. Andre, M. Dreux, E. Gaydou, *J. Chromatogr.* **791** (1–2), 289–297 (1997).

120. H. Nishi, S. Terabe, *J. Biomed. Anal.* **11** (11–12), 1277–1287 (1993).

121. N. Onyewuenyi, P. Hawkins, *J. Chromatogr.* **749** (1–2), 271–277 (1996).

122. M. Roses, C. Rafols, E. Bosch, A. M. Martinez, M. H. Abraham, *J. Chromatogr.* **845**, 217–226 (1999).

123. J. G. Clothier Jr., S. A. Tomellini, *J. Chromatogr.* **723** (1), 179–187 (1996).

124. H. Nishi, S. Terabe, *J. Chromatogr.* **735**, 3–27, (1996).

125. V. Revilla, L. Alma, M. Vagas, G. Maria, *Revista Mexicana de Ciencias Farmaceuticas* **33** (1), 18–25 (2002).

126. K. Otsuka, S. Terabe, *J. Chromatogr.* **875** (1–2), 163–178 (2000).

127. M. R. Taylor, S. A. Westwood, D. Parrett, *J. Chromatogr.* **768** (1), 67–71 (1997).

128. M. R. Taylor, S. A. Westwood, D. Parrett, *J. Chromatogr.* **745** (1–2), 155–163 (1996).

129. F. Tagliaro, F. P. Smith, S. Turrina, V. Esquisetto, M. Marigo, *J. Chromatogr.* **735** (1–2), 227–235 (1996).

130. Y. M. Li, A. Van Schepdael, E. Roets, J. Hootzmartens, *J. Pharm. Biomed. Anal.* **15** (8), 1063–1070 (1997).

131. M. Steppe, M. S. Prado, M. F. Tavares, E. R. Kedor-Hackmann, M. I. R. M. Santoro, *J. Capillary Electrophor. Microchip Technol.* **7** (3–4), 81–86 (2002).

132. M. K. Srinivasu, N. A. Raju, G. O. Reddy, *J. Pharm. Biomed. Anal.* **29** (4), 715–721 (2002).

133. C. Lucas, M. J. Gliddon, M. M. Safarpour, S. Candaciotto, E. S. Ahuja, J. P. Foley, *J. Capillary Electrophor. Microchip Technol.* **6** (3–4), 75–83 (1999).

134. R. H. H. Neubert, Y. Mrestani, M. Schwarz, B. Colin, *J. Pharma. Biomed. Anal.* **16** (5), 893–897 (1998).

135. K. N. Altria, *J. Chromatogr.* **646** (2), 245–247 (1993).

136. M. Korman, J. Vindevogel, P. Sandra, *Electrophoresis* **15** (10), 1304–1309 (1994).

137. C. Fang, J.-T. Liu, C.-H. Liu, *J. Chromatogr.* **775** (1), 37–47 (2002).

138. B. Fan, J. T. Stewart, *J. Liq. Chromatogr. Related Technol.* **25** (6), 937–947 (2002).

139. R. Wang, H. Fan, W. Ma, *J. Liq. Chromatogr. Related Technol.* **25** (6), 857–864 (2002).

140. B. X. Mayer, U. Hollenstein, M. Brunner, H.-G. Eidler, M. Muller, *Electrophoresis* **21** (8), 1558–1564 (2000).

141. C. A. Ogawa, C. A. Diagone, F. M. Lancas, *J. Liq. Chromatogr. Related Technol.* **25** (10–11), 1651–1659 (2002).

142. C. Simo, C. Barbas, A. Cifuentes, *J. Agri. Food Chem.* **50** (19), 5288–5293 (2002).

143. X. Shang, Y. Zhuobin, *Analy. Lett.* **35** (6), 985–993 (2002).

144. C.-E. Lin, Y.-C. Chen, C.-C. Chang, D.-Z. Wang, *J. Chromatogr.* **775** (1–2), 349–357 (1997).

145. C.-E. Lin, W.-C. Lin, W.-C. Chiou, *J. Chromatogr.* **722** (1–2), 333–343 (1996).

CHAPTER

9

CHARACTERIZATION OF SPLIT-POOL ENCODED COMBINATORIAL LIBRARIES

JING JIM ZHANG and WILLIAM L. FITCH

9.1. INTRODUCTION

The split-pool approach for solid-phase synthesis has been proved an effective way to generate large, diverse chemical libraries. Libraries prepared on beaded resins by the split-pool approach are characterized by the "one bead–one compound" rule. Such libraries, in principle, combine the advantages of split-pool synthesis with those associated with the screening of discrete compounds, since individual beads may be assayed for biological activity. After the synthesis of a split-pool combinatorial library, it is desirable to verify the success of the synthesis before screening the library against biological targets. However, since the beads are randomized in each pooling operation, tracking the reagents and building blocks to which each bead is exposed becomes troublesome, and the identity of the chemical ligand on an active bead may be very hard to determine.

Encoding provides a general solution to this problem, whereby a surrogate analyte, or "tag," is attached to the beads to allow determination of the reaction history of a certain bead and thus the identity of the final product attached to it. A variety of encoding strategies have now been proposed; they share the critical property that the code is much easier to analyze than the associated compounds.

Encoded combinatorial libraries are a key component of our strategy for accelerating drug discovery,[1–3] and the technology required to realize such libraries has evolved continuously in our organization over the last decade. The development of DNA-based encoding[4] was a milestone in combinatorial science and led to the invention of polyamide "hard tags,"[5–9] which have been shown to be compatible with a wide range of synthetic chemistries and of demonstrable utility in drug discovery.[10–16]

Analysis and Purification Methods in Combinatorial Chemistry, Edited by Bing Yan.
ISBN 0-471-26929-8   Copyright © 2004 by John Wiley & Sons, Inc.

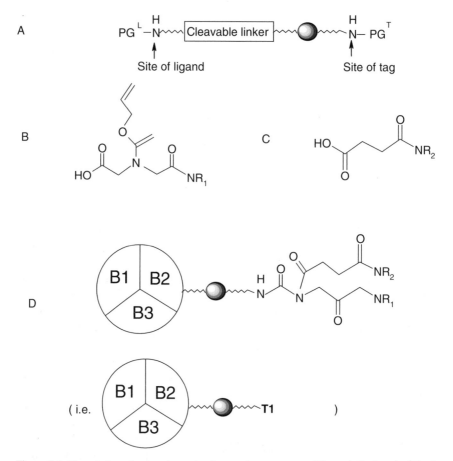

Figure 9.1. Encoded synthesis using secondary amine tags on a differentially functionalized polymer support: (*A*) Functionalized resin; (*B*) structure of Alloc tagging monomer unit; (*C*) structure of succinoyl monomer tagging unit; (*D*) schematic representation of the product of a two-step encoded synthesis.

Figure 9.1 shows the general structure of a compound prepared on a polymer bead using polyamide tags. It shows the construct for a three-step synthesis where the first two steps are encoded; the differentiated bead (*A*), the tagging monomer units (*B* and *C*), and the generic final encoded bead (*D*). The cleavable linker contains a photo- or acid-cleavable group optimized for screening; the tag will stay with the bead until mineral acid hydrolysis frees it for structure determination. The protecting groups PG^L and PG^T are normally Fmoc and Boc, respectively. The Alloc protecting group of (B) masks the branch point and is readily removed for subsequent encoding steps. Building blocks are encoded by mixtures of secondary amines HNR_1R_2. Decoding entails release of the tags, determination of the tags, and thereby (with reference to a coding table) determining the structure of the ligand that was carried by the bead.

A key assumption for the utility of encoded libraries is that the code is predictive of the structure of the attached compound. Decoding must be unambiguous, and the correct compound must have been produced. Obtaining this assurance presents significant analytical challenges,[17,18] but it is critical to obtain, as poor quality libraries can lead to significant wasted time in the screening and re-synthesis of spurious hits.

The analytical challenges posed by encoded libraries are largely the result of the small amount of analyte that is available. The beads used in these libraries typically have a few hundred picomoles of synthesis sites. We differentiate the beads such that approximately 90% is ligand and 10% is tag (in more recent libraries we have switched this to 80/20). While larger beads would make analysis easier, it would be more difficult to prepare libraries of large numbers of compounds, as it is necessary to ensure that many beads bearing each compound will be produced. In addition larger beads often suffer from mechanical instability during synthesis.

We will describe the processes by which we determine the quality of encoded libraries and decode biological hits, including the evolution of our code-reading process, and quantitative analysis of libraries.

9.2. METHOD FOR ENCODING AND DECODING LIBRARIES

9.2.1. Encoding

Our encoding approach relies on the use of secondary amines as tags. The number of these amines required is dependent on the number of compounds in the library. The simplest use of these tags is binary encoding, whereby the presence or absence of an amine is recorded as a 1 or 0. Thus the code {101} contains two of the three amines, and may, for instance, correspond to the incorporation of a particular building block. In the binary encoding scheme the number of possible codes is given by $2n - 1$ where n is the number of amine tags that are available. A maximum of three amines is used to encode each building block synthesis. The individual amines will be present at levels from a few picomoles to a few tens of picomoles per analysis.

We originally chose 18 tags for encoding three position libraries of up to 250,000 members.[5] For many libraries we do not need this many tags; for instance, if the final position is not chemically encoded but, instead, is positionally encoded by simply keeping the pools separate after incorporation of the last building block, then only two sets of codes and 12 amines are needed. In practice, it has been found that it is often not necessary to encode more than two positions.

Any unbranched dialkylamine can be used in our encoding scheme. Anilines and benzylamines are excluded. Alpha-branched amines give anomalous encoding ratios due to slower reaction rates. Our initial tags (Table 9.1) were all commercially available, and were chosen for the ability of their dansylated derivatives to be separated on a C18 column. Separation of all 18 tags required 80 minutes because (1) some amines eluted close together, requiring a rather slow gradient to separate them, and (2) several amines were very hydrophobic and needed prolonged elution with neat acetonitrile for elution from the column.

To achieve a high-throughput method, we had to significantly reduce the decode cycle time. We employed two main strategies. First, systematic analysis of the retention patterns of available amines by experiment and modeling led to predictive models of amine behavior, allowing an "ideal" set of tags to be identified with evenly spaced elution characteristics. This allowed uniform compression of the gradient with minimal loss of resolu-

Table 9.1. Original Tag Set

Number	Amine Name
1	ethylbutylamine
2	methylhexylamine
3	dibutylamine
4	methylheptylamine
5	butylpentylamine
6	dipentylamine
7	butylheptylamine
8	dihexylamine
9	pentyloctylamine
10	propyldecylamine
11	methyldodecylamine
12	bis(2-ethylhexyl)amine
13	dioctylamine
14	butyldodecylamine
15	pentyldodecylamine
16	hexyldodecylamine
17	heptyldodecylamine
18	didecylamine

Table 9.2. Optimized Tag Set

Number	Amine Name	Molecular Weight	Abbreviation
1	2,2,2-trideuteroethylpentylamine	118.2	depn
2	2,2,2-trideuteroethylhexylamine	132.3	dehx
3	2,2,2-trideuteroethylheptylamine	146.3	dehp
4	2,2,2-trideuteroethyloctylamine	160.3	deoc
5	2,2,2-trideuteroethylnonylamine	174.3	denn
6	2,2,2-trideuteroethyldecylamine	188.4	dedc
7	pentyloctylamine	199.4	pnoc
8	diheptylamine	213.4	hphp
9	heptyloctylamine	227.4	hpoc
10	dioctylamine	241.4	ococ
11	pentyldodecylamine	255.5	pndo
12	hexyldodecylamine	269.5	hxdo
13	heptyldodecylamine	283.5	hpdo
14	didecylamine	297.6	dcdc

tion. Second, changing to a less retentive cyano column allowed all amines to be eluted by the gradient and permitted chromatographic compression to be achieved. In addition the column proportions were changed to allow much faster flow. These factors, and the need for tags of unique molecular weight, led to our second generation tag set as illustrated in Table 9.2. The enhanced throughput is illustrated in Figure 9.2, where all 14 tags are separated in a 4-minute run.

9.2.2. Decoding

The tags can be released for analysis with strong acid. Single beads were placed in individual tapered glass microvials (National Scientific, decontaminated before use by heating at 400°C for 2 hours) by an automated bead-picker.[19,20] The ligand was removed by prior cleavage, 25 µL of HCl (6N constant boiling grade) was added, and the sealed vials were heated at 135°C for 12 to 16 hours. The HCl was removed from the cooled vial in vacuum.

Mass spectroscopy offers the obvious advantage of providing mass information on the molecular species present. If used in conjunction with chromatography (e.g., LC/MS) an additional dimension of data is provided that can assist in decoding. For example, amines are often used as building blocks in library assembly. If for some reason they remain associated with beads during the tag analysis process, they may elute near one of the tags,

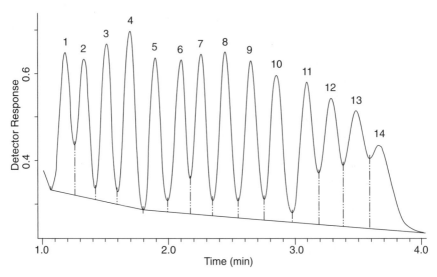

Figure 9.2. Gradient reverse phase HPLC/fluorescence analysis of dansylated tags from a single-bead encoded with an equimolar mixture of 14 tags. Tags are identified in Table 9.2.

interfering with tag assignment. Mass information would eliminate such interferences and thus increase the reliability of decoding.

We initially expected that the dialkylamine tags could be analyzed as the underivatized protonated amines by high-throughput, flow injection, electrospray MS. Despite considerable effort, adequate detection limits were never achieved due to ion suppression effects of the residual HCl and other impurities in the samples. Lane and Pipe describe the CEC/MS/MS of dansylated tags.[21] We have now developed a robust LC/MS method for the analysis of underivatized tags.

Strong cation exchange chromatography in a highly organic solvent mixture is ideally suited to the task of determining basic amines. We identified isocratic conditions whereby all the amine tags were eluted in a close packet separated from the solvent front but with overall a remarkably short run time of 3 minutes. By having unique masses for each tag, we could set up selective ion monitoring channels such that although the tags are not well-separated in time, they are individually analyzable by mass. This led to the chromatogram shown in Figure 9.3, with 14 separate selected ion monitoring channels overlaid. In this example an undifferentiated bead has been hydrolyzed and analyzed. This bead contains a nominal 400/14 = 28 picomoles per tag. In normal encoding practice a 10/1 differentiated bead (90% ligand and 10% tag) sample will never contain more than 3 tags per

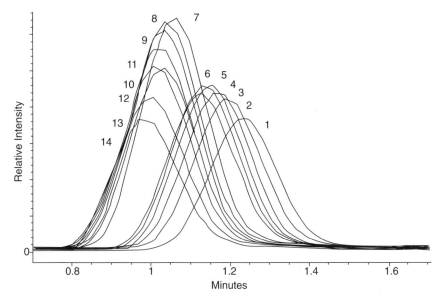

Figure 9.3. Isocratic cation exchange LC/MS analysis of underivatized tags from a single-bead encoded with an equimolar mixture of 14 tags. Tags are identified in Table 9.2. Each tagging molecule generates a single ion, which is detected by a selected ion monitoring chromatogram. The 14 chromatograms are overlaid in the figure.

synthesis step, for a nominal 400/10/3 = 13 picomoles for the smallest analyte.

LC/MS was conducted with an Agilent 1100 Series LC/MSD system. LC autosampling from the 96-well glass microvial plates was done with a CTC PAL autosampler. The sample was dissolved in 200 μL of acetonitrile. The column used is a Supelco LC-SCX column, 1 × 50 mm in a column heater at 40°C. Injection volume is 20 μL. Flow rate is 0.2 mL/min isocratic with 85% acetonitrile/15% 0.4% aqueous TFA with a total run time of 3.5 minutes. Electrospray drying gas is set at 7 L/min and 350°C. Capillary voltage is held at 2000 V. Fragmentor is ramped from 70 to 120 V, based on the masses sampled. Detection is by selected ion monitoring of the appropriate M + H values. A precolumn in-line 0.45 micron filter protects the column from plugging. The ion exchange capacity of the column slowly changes under isocratic conditions, causing a slow reduction in retention times. The drifting retention times are readily recalibrated by running standards every 24 samples. After every few hundred samples the column is regenerated by switching to 50% acetonitrile/50% 0.4% TFA for 15 minutes, followed by a 15-minute re-equilibration.

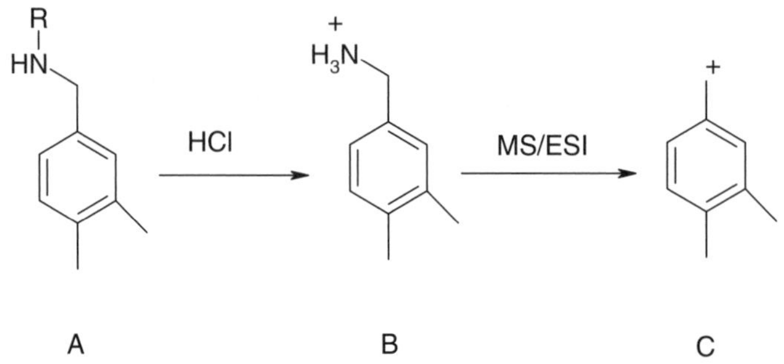

Figure 9.4. Interference of benzylamine libraries to the deuteroethylpentylamine tag. (*A*) Dimethylbenzylamine compounds in the libraries; (*B*) dimethylbenzylamine ion; (*C*) tropylium ions, m/z 119, which interferes with the target M + H ion for the deuteroethylpentylamine tag.

The high-throughput, specificity, and simplified sample preparation (relative to derivatization) make this an excellent alternative to HPLC methods. And yet no analytical method is trouble free. One rare error we have detected is in the encoding of benzylamine libraries, such as (*A*) in the Figure 9.4. A variety of solid phase chemistries, including simple benzylamides, will liberate the free benzylamine on strong acid hydrolysis. These will be retained on the cation exchange column, eluting close to the tag molecules. Certain of these molecules can fragment in the mass spectrometer to give stabilized tropylium ions, which interfere with tags. This is illustrated for a dimethylbenzylamine (*B*) which will yield an ion of $m/z = 119$ (*C*). This is also the target M + H ion for deuteroethylpentylamine. So in this case we get a false decode. This has been so rare as to be mainly an interesting novelty.

9.2.3. Library Analysis and Quality Control

The ligand cleaved from a single bead is analyzed by an LC/MS method. Obtaining a representative mass spectrum of compound from a single bead of a split/pool library presents a combination of challenges to the analytical chemist. The actual amount of compound can vary from less than 50 picomoles to more than 1 nanomole depending on the size and type of bead used as the solid support. Bead size is selected to optimize many factors; ease of analysis being only one. For this library we used 130-μm Tentagel beads with a theoretical loading near 500 picomoles per bead. The actual amount released from a given bead is often less due to variations in bead size, ease of compound synthesis, and efficiency of cleavage. The bead will contain a pure single component or, more often, a mixture of product, starting material and by-products. Once separated from the bead all of the common problems (contamination, adsorption, and transfer of small volumes) associated with small dilute samples are observed.

Photo-linked libraries are convenient for split/pool synthesis and screening. Cleavage is effected by adding methanol and subjecting the bead to intense UV light. 25 μL of methanol was added to each vial and the vial sealed. The vial was placed on an orbital shaker and irradiated for 120 minutes with 365 nm UV light from an arc lamp ($10 mW/cm^2$). The resultant solution is ready for mass spectrometry.

A second common linker is the acid cleavable linker. Treatment with 95% TFA cleaves the ligand. Evaporation and re-dissolution in acetonitrile prepares the sample for MS. For example, 50 μL of TFA was added to each vial, and the vials were left at room temperature for 60 minutes. The TFA was evaporated to dryness, and the sample re-dissolved in 25 uL of acetonitrile. Methanol is to be avoided at this point because methanolic TFA, even if dilute, is an effective methylating milieu. The mass spectrum is most readily obtained by positive ion electrospray ionization. But we also employ negative ion electrospray when advantageous. Flow injection analysis can be used for sample introduction but conventional[22] or capillary[17,18] LC provide better sensitivity and the opportunity to compare LC/UV and LC/MS profiles for more complete characterization.

An Agilent 1100 Series LC/MSD system was used to analyze ligand at the single-bead level. The cleaved ligand was dissolved in 25 μL of either acetonitrile or methanol in a glass microvial of 96-well plate. A CTC Leap model 200 autosampler was used to inject 25 μL of water (0.1% formic acid) into each sample vial. The premix of organic solvents with water is to reduce the hydrophobicity of sample solutions; therefore the sample solutions are more compatible with optimal HPLC separation requirements. A total of 50 μL of the sample mixture was then injected into HPLC column. The column used was Polaris 2000 C18 columns packed with 3 μm particles, 30 × 3 mm (i.d.) (MetaChem Technologies, Torrance, CA), in a column heater at 60°C. The mobile phase consisted of water (0.1% formic acid) and acetonitrile (0.1% formic acid) delivered in a linear gradient from 0% to 100% acetonitrile in 4 minutes with 1 mL/min flow rate. The total analysis time (time between injections) for each sample was 5.5 minutes. UV detection was performed at 220 nm. The mass spectrometer scanned a window of 150 to 1000 m/z in 1 second. After the library product was analyzed by LC/MS, the tags were cleaved from the bead, and analyzed by LC/MS.

The main application of these single-bead analytical techniques is in library quality control. For every 1 biology hit bead decoded to reveal an active structure, we may decode 10 or more beads to ensure the quality of the library prior to screening. There is some debate[18] about the number of single beads that must be analyzed to generate meaningful statistics regarding library quality. Our present view is to pick five beads for each final pool. For a $36 \times 36 \times 36$ library this equals $5 \times 36 = 180$ beads. Each of the final set of building blocks will then be analyzed exactly five times, while building blocks from the first two sets (having been randomized by pooling) will be observed on average on five beads. While more information may be desired, we believe this sampling protocol is a good compromise between analysis time and gathering the information necessary for library evaluation.

The process for evaluating the success of a library begins by picking the appropriate sampling of beads into clean glass vials. The compound is released by the appropriate cleavage protocol, and its mass spectrum measured. The residual bead is then subjected to acid hydrolysis to liberate the secondary amine tags that may be analyzed by the appropriate technique. Decoding the amine data allows the determination of which monomers were used in the synthesis, and thence, by knowledge of the chemistry involved, the structure and mass of the compound that should be present on the bead. Comparison with the MS data for the ligand allows pass/fail determination for that bead.

Figures 9.5 and 9.6 show typical results from the analysis of a single thiazolidinone bead. An encoded thiazolidinone library was produced as previously described[23] using the photocleavable linker.[24] The peak at 1 minute in the UV is due to a mismatch in formic acid concentration between the sample and the mobile phase. The lower trace is the total ion current for the entire scanned mass range (150–1000), while the upper trace is the total extracted ion current for the mass range of interest (400–700). This library product contains a Boc (tertiary-butyloxycarbonyl) protecting group that is partially lost from the compound during electrospray. We observe ions corresponding to MH^+, MNa^+, and MH^+-100 for the code-predicted compound. Thus the bead depicted in Figure 9.5 contained only the product compound. The results for another bead, shown in Figure 9.6, indicate that in addition to the desired product being present, an impurity was present on the bead. A total of 20 beads were analyzed in this way. Decoding failed on one of these beads, a typical failure rate (as often as not due to bead-picking error). Of the remaining 19 beads, 13 gave correct M + H values in the LC/MS data. The other 6 were either weak or wrong. This is a fairly typical success rate in library quality control.

We can determine how successful the synthesis was by comparing the observed products to the expected products. Usually the expected product is observed as the major component; otherwise, if another compound in the correct mass range is observed, an identifiable (and generally correctable)

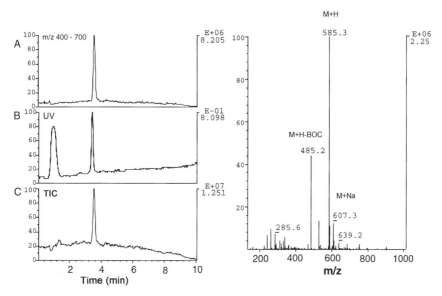

Figure 9.5. LC/MS analysis of the product from a single bead. Shown are (*A*) the *m/z* 400–700 chromatogram, (*B*) the UV chromatogram, (*C*) the total ion current chromatogram, and (*D*) the mass spectrum of the peak from (*A*).

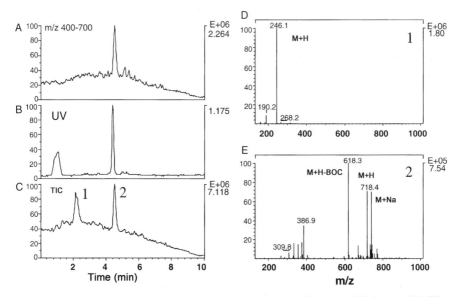

Figure 9.6. LC/MS analysis of the product from a single bead. Shown are (*A*) the *m/z* 400–700 chromatogram, (*B*) the UV chromatogram, (*C*) the total ion current chromatogram, and the mass spectra from peaks (*D*) 1 and (*E*) 2 from (*C*).

human error is often involved. These include erroneous entry of structures in the database and mis-assignment of decode data. Sometimes, though, the synthesis results in a product other than that predicted. We can use this information to better understand the library chemistry and improve the quality of subsequent libraries. The LC/MS analysis also yields a measure of purity (although because we cannot reliably predict the UV or MS response, we do not obtain a quantitative measure of how much material comes off each bead). In short, LC/MS of single beads gives us information both on product identity and purity. Combined with decoding, this provides the information necessary to give a detailed insight into the quality of the library produced.

9.2.4. Hardware for Library QC and Decoding Automation

With the LC/MS decoding method in hand, the decoding bottleneck became the slow, manual, sample preparation steps. This was especially true in the library quality control process, where specific beads are explored both for ligand and for tags. The previous method required manual (using a glass capillary and a microscope) transfer of the bead after ligand cleavage, to a screw cap vial, addition of HCl, heating, lyophilization, derivatization, and filtration. We have now incorporated an automated beadpicker[19,20] and a new hydrolysis device into the process. The beadpicker reliably picks and transfers hundreds of beads per hour eliminating the most tedious manual step of library analysis.

The new apparatus for automating this whole process in a 96 well format is shown disassembled in Figure 9.7A. The three metal pieces are made from anodized aluminum. The bottom plate sets the correct micro-titer plate footprint. The middle piece holds the glass vials in place. The top piece has holes appropriate for a 96-well autosampler. Tapered glass vials are used for their cleanliness, acid resistance, and volume efficiency. A Teflon insert between the middle metal piece and the lips of the vials prevents vial break-age when the device is sealed with a Teflon/silicone membrane and screw clamps. Assembled, the apparatus has a standard 96-well plate footprint, ready to be used as illustrated in Figure 9.7B, in commercial autosamplers.

This apparatus has allowed us to adopt many features of automation such as multichannel pipetting and direct autosampling. Appropriate negative controls for ensuring contamination control, positive controls for ensuring hydrolysis, and reference standards for optimizing peak identifi-cations and ensuring instrument performance are readily included in the sample queue of the hydrolysis plate. Once the beads are placed in the vials, the steps of ligand cleavage, ligand LC/MS, code hydrolysis, acid evapora-tion, and code LC/MS can be automated with no further bead manipula-

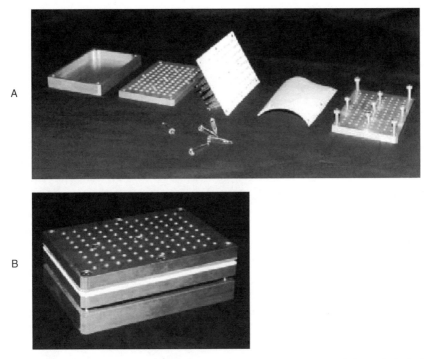

Figure 9.7. (*A*) Tag hydrolysis automation device shown disassembled; (*B*) tag hydrolysis device after assembly, in a microtiter type autosampler.

tion during the library quality control process. For biology hits the beads are manually picked into these same vials and the apparatus used for hydrolysis and LC/MS. The fast MS run times allow completion of several 96-well plates in a day.

9.3. SOFTWARE FOR ANALYSIS OF ENCODED LIBRARIES

The improvements of analytical methods, hardware, and automation allow for great throughput for decoding. It is now practical to perform quality control sampling and analysis on a good proportion of the members of an encoded library. This analysis has the benefit of increasing confidence in the encoded library synthesis-and-screening process. The analytical work can have the negative effect of generating a large pile of paper data, whose study can distract the synthetic chemist from the more fundamentally important duties of library design and evaluation of biological data. An

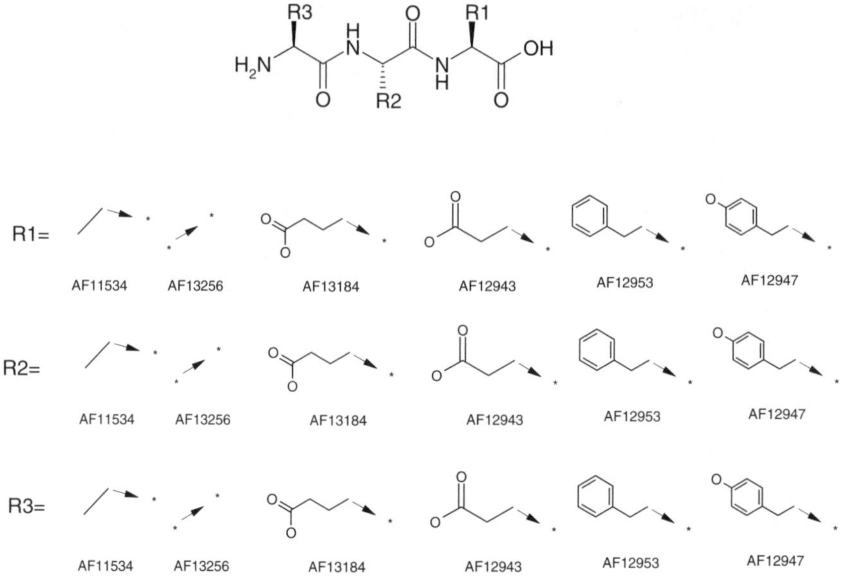

Figure 9.8. Generic structure and building blocks for the virtual library.

information management system is needed to streamline this process. Many of these issues have been previously discussed by us and by others.[25–34] We describe herein our customized software for automating the decoding of encoded combinatorial libraries.[35]

To illustrate the issues, a $6 \times 6 \times 6$ tripeptide library synthesis will be simulated. We will use the same six amino acid building blocks at each position. Figure 9.8 shows the generic structure of the cleaved product as well as the structures and names of the building blocks. We will use 1, 2, or 3 tags to encode the first building block and 1, 2, or 3 tags to encode the second. The third position will not be encoded; the final pools, each containing 36 tripeptides will be kept separate. Table 9.3 is the building block to tag encoding scheme.

Three tags could have encoded for up to 7 building blocks at each position. We never use more than 3 tags in a single code to maintain good tag detectability. Indeed, because we have a total of 14 tags in our tag set, the preferable codes for this simple library would be to use 6 different single tags at each position. We have not seen the need for introducing parity checking in encoding.[36] We illustrate all of the different 1, 2, and 3 tag combinations in this scheme to explain all of the software issues. Of course, the

Table 9.3. Building Blocks and Tags of Tripeptide Library

Building Block		First-Position Codes			Second-Position Codes		
Number	Structure	depn	dehx	dehp	deoc	denn	dedc
1	ala	×			×		
2	gly		×			×	
3	glu			×			×
4	asp	×	×		×	×	
5	phe	×		×	×		×
6	tyr	×	×	×	×	×	

numbers of building blocks do not have to be the same in a library synthesis.

The process steps of encoded library synthesis are as follows (detailed descriptions can be found in the indicated references):

1. Differentiate the resin with BOC protected tag attachment point and cleavable linker.[37]
2. Split the resin into 6 reaction vessels.
3. Remove the tag protecting group and couple the first code as the Alloc-protected versions.[37]
4. Couple building block 1 to the resin.
5. Pool and resplit to 6 reaction vessels.
6. Remove the Alloc group and couple the second code mixture as succinoyl derivatives.[38]
7. Couple building block 2 to the resin.
8. Pool and resplit to 6 reaction vessels.
9. Couple building block 3.

Figure 9.9 shows the structure of the first member of this library. The structure implies a 1:1 ratio of ligand and codes, but remember that we started with differentiated resin so the actual ratio is 10:1. PL is the photocleavable linker.[24] Cleavage will yield trialanine. Mineral acid cleavage of the tags will yield a mixture of depn and deoc.

The CodeGen and Decode programs are applications written for Microsoft Access 97, using the programming language Visual Basic for Applications. Proper installation is required because the application uses three different MS Access databases and several Oracle database tables used by MDL (Molecular Design Ltd.) ISIS/Host and ISIS Base. To access

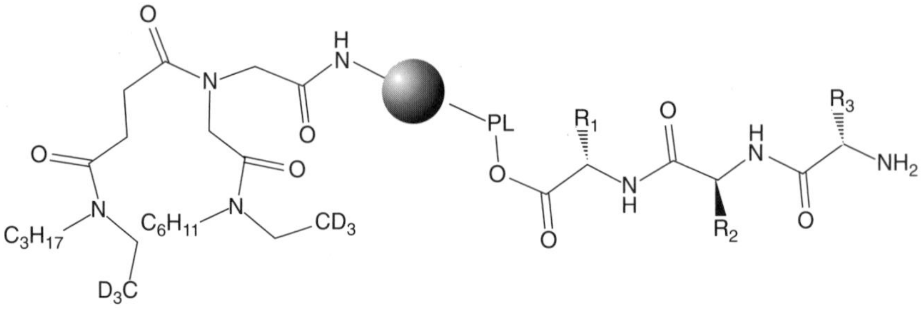

Figure 9.9. Structure of the first member of the tripeptide library.

data from the Oracle databases, Open DataBase Connectivity (ODBC) drivers are used. CAPTURE (Comprehensive Analysis Program That Ultimately Reveals Everything, a GlaxoSmithKline proprietary application), is written in Microsoft Visual C++. CAPTURE retrieves plate maps created by CodeGen and the structures from Code-to-Structure using MDL ISIS/Object Library. Mass spectral data is read from files created by the ChemStation macros. Code to structure software is written in C/C++.

MS acquisition computers should not be encumbered with complex data processing software. We do preliminary processing on the LC/MS computer but subsequent processing on desktop PCs and network servers. Postacquisition, all raw data and text reports are transferred to the decoding server.[32]

9.3.1. Code Generation

Prior to developing the Microsoft Access applications, decoding raw data were converted to structures manually. By reference to the encoding scheme, the detected tags were assigned to either R_1 or R_2 and then the best matches were made. For example, if a bead from the fifth peptide pool yielded codes depn, dehp, and deoc, the first two tags would be assigned to R_1 and decoded as structure 5, phe, and the last tag assigned to R_2 and decoded as structure 1, ala. Combining the decoded information with the pool information gives the full structure as consisting of building blocks 5–1–5. For peptides these are easily put together and adequately described as "phe–ala–phe." For more complex libraries it is necessary to use a chemical drawing program to assemble the chemical structure. Finally the molecular weight of the assembled compound can be calculated for comparison to the ligand LC/MS data.

In automating decoding, the first step is to get the tagging information

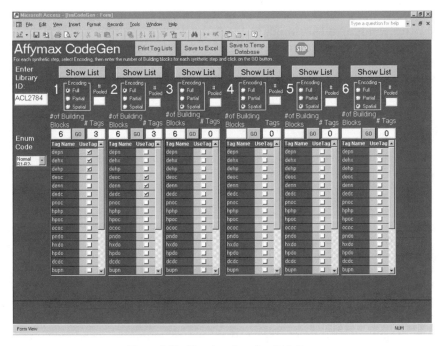

Figure 9.10. User interface for CodeGen.

(e.g., Table 9.3) into a database. CodeGen was written to generate lists of tag codes to identify each building block in a multi-step synthesis. Figure 9.10 shows the user interface for CodeGen. For each library the user enters the type of encoding for each step in the synthesis: full, partial, or spatial.

After selecting the type of encoding for each step, the user enters the number of building blocks for each synthetic step and clicks on the GO button, starting with step 1. The program computes the minimum number of tags required to encode the building blocks without using more than three tags per building block,[38] or the user can select any higher number of tags by checking the boxes next to the tag names. The program will use all of the one-tag assignments first, then the two-tag, and finally the three-tag combinations.

The main CodeGen algorithm is called for each synthesis step. All of the tag codes start with the number of tags used, 1, 2, or 3. Then the tag codes are listed in alphabetical order separated by commas. The tag assignments are reviewed with the show list button. Once the user is satisfied with the tag lists an Excel spreadsheet formatted as Table 9.2 can be obtained and the tag information is saved to a temporary file.

At this point the chemist can go to the lab and start generating the combinatorial library. The building block structures are then entered into ISIS/Host in the standard way. These codes can then be loaded into the appropriate Oracle table in the TAG field. When the codes are loaded, the encoding type for each step along with the number of partially pooled building blocks are placed in a second Oracle table.

9.3.2.   Decode

Decode is used to identify the building blocks for each set of tag codes. The program has three parts: BeadSetup, Chromread, and Decode.

BeadSetup

The process of picking single beads from encoded library pools and assessing both the codes and the ligand on each bead were described in literature.[27] This process has been automated through the use of a robotic beadpicker.[19,20,39] Experiments were conducted on an Agilent LC/MSD instrument controlled by ChemStation Rev. A08.04.

The first step in the decode process is to place beads on to a 96-well plate. The BeadSetup program allows the user to enter the bead and pool numbers in two ways for quality control (Regular QC) and hit decoding (Irregular) as shown in Figure 9.11. For quality control, the number of pools to be placed on the plate(s) is entered along with the number of beads per pool (typically 4 or 5) and the starting pool number.

By default, the first 11 wells in each row are designated as sample wells. The wells in the twelfth column are reserved to hold controls, blanks, and standards. The Pick Beads button sets up the wells to each contain one bead for the correct pool. This information is stored with the plate and ultimately can be viewed in the CAPTURE program, so any additional information about a specific well can be entered at this time. The Print Bead Report button can print out the plate information. The Save button saves this plate information in file.

If Irregular Hits is selected, an entry box and button are displayed. The user enters the number of pools and then each pool number and the associated number of beads for that pool number. The plate is then set up and can be saved.

ChromRead

The decoding analytical method involves acid hydrolysis to release the covalently bonded amines, followed by their LC/MS detection, and by ref-

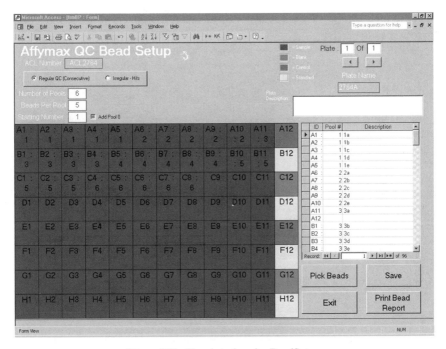

Figure 9.11. User interface for BeadSetup.

erence to a calibration table, their quantitative determination. The LC/MS instrument software reduces the raw selected ion monitoring data into a report.txt file. The standard ChemStation (Agilent) report output of this analysis lists the code abbreviations, the MS signal, the semiquantitative information on the number of picomoles detected, and the expected and measured retention times as in the example of Table 9.4. ChromRead assigns these data to the appropriate well of the 96-well test format as shown in Figure 9.12. This program has a list box to select the plate information. The Browse button is used to select the files holding the chromatographic data for the first well. Standard Microsoft NT tools are used to navigate to the server where the LC/MS instrument has sent the datafiles. Next the number of files to load is entered and the first well number. The Get Data button populates the Access chromatography data table. Well colors change to reflect the status of each well. The chromatographic data can be viewed either by clicking on the well or as a list by using the View Table of Injections or View Table of Peaks buttons. If the chromatographic data are correct, the Save button will save the data and their connection with the selected plate.

Table 9.4. LC/MS Decoding Report

Tag	MS Signal	Area	Picomoles	Expected RT	Measured RT
ococ	MSD1 242,	0.000	0.000	1.277	0.000
hpoc	MSD1 228,	0.000	0.000	1.287	0.000
dcdc	MSD1 298,	0.000	0.000	1.255	0.000
hphp	MSD1 214,	7.510e3	1.879	1.296	1.267
hpdo	MSD1 284,	0.000	0.000	1.277	0.000
pndo	MSD1 256,	0.000	0.000	1.305	0.000
pnoc	MSD1 200,	0.000	0.000	1.320	0.000
hxdo	MSD1 270,	0.000	0.000	1.289	0.000
denn	MSD1 175,	0.000	0.000	1.442	0.000
deoc	MSD1 161,	3.816e4	14.586	1.460	1.409
dedo	MSD1 217,	0.000	0.000	1.418	0.000
dedc	MSD1 189,	0.000	0.000	1.431	0.000
dehp	MSD1 147,	0.000	0.000	1.488	0.000
depn	MSD1 119,	5.028e4	25.957	1.595	1.527
dehx	MSD1 133,	1.258e4	6.245	1.529	1.476
debu	MSD1 105,	0.000	0.000	1.695	0.000

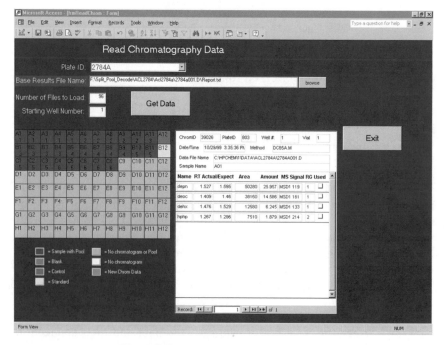

Figure 9.12. User interface for ChromRead.

Decode

The primary screen of the main Decode application is shown as Figure 9.13. When a plate number is selected, the software organizes the plate pool numbers from the Beadsetup step, the building block information from Oracle and the chromatographic results for each well. The left half of this screen shows the numbers of building blocks that were used for each library and the tags that were used in each step. The right side of the screen will change as each well is checked. The top half shows the 96-well graphical user interface. The bottom half shows the chromatographic data for each well. A new column in the chromatographic data now also shows the R group (RG) with which the code is associated: in this case 1 for R_1, 2 for R_2, and X for the codes that were present in the chromatographic data but not used in the library synthesis. The X can only be present due to lab contamination or the presence of an interfering ligand-derived molecule in the bead hydrolyzate. Note that within each group (1, 2, or X) the amines are sorted by the number in the amount column.

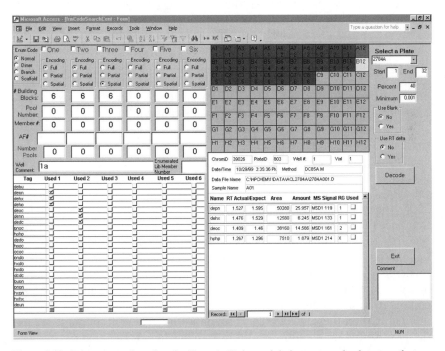

Figure 9.13. Primary user interface for Decode. Debu and dedo are two obsolete tags that are included for backward compatibility.

By clicking on the 96-well diagram, the chromatographic data can be viewed. The analyst can now review the analytical quality of the data. For example, viewing the blank wells C12 or G12 will reveal contamination during analysis. Viewing the reference standard wells B12, D12, F12, and H12 will assure the analyst that adequate detection of all tags was maintained across the plate. Viewing the positive control wells A12 and E12 will assure the analyst that the plate was subjected to adequate acid hydrolysis to release the amines.

Two contamination correction tools are provided for the analyst, the blank and the minimum amount buttons. The chromatographic data from a blank can be subtracted from all of the wells by using the Yes radio button in the Use Blank selection. A "well to use" entry box will appear, with the C12 well used as default. Any well can be entered as the blank well. The peak amounts for each tag in the blank well will be multiplied by 2 and subtracted from all of the same peaks in all other wells as the decode is performed. The Minimum button provides another way to assure that very small peaks will not be used for decoding.

The LC/MS decoding chromatography is isocratic and low resolution. The specificity of the method is primarily due to the mass selection of the selected ions. The retention time window in the primary Agilent peak identification protocol is left relatively wide to allow for normal isocratic chromatographic drift across a set of analyses. In some cases interferences are seen in the data. A tighter retention time discrimination can be applied during the Access analysis of Decode by hitting the Use RT Delta radio button. An absolute retention time window is then entered and the Decode software will exclude identifications outside that window.

The user now can select a Percent value for the chromatographic peaks to be used in the decode process. The default is 40%. Empirically this has been found to be a good compromise value to adjust for variations in encoding chemistry and decoding analysis. But more or less stringent percentages can be useful in special cases.

The decode algorithm is run with the Decode button. The key logic is that once the tags used for a specific synthesis step are known, one is simply required to determine if one, two, or three of those tags are in the chromatography data. When the chromatography data for a well have been read, the tag peaks, which are used in synthesis step 1, are sorted by amount and corrected for blanks or retention time. The first peak amount is then compared to the minimum amount. If the peak amount is greater than the minimum, then there is one tag. If there are no tags with amounts greater then the minimum, then the synthesis step has failed. The next peak is next checked to see if it is at least the Percent of the first peak. If it is, then there is a second code, and the program looks for a third tag. If there is a fourth

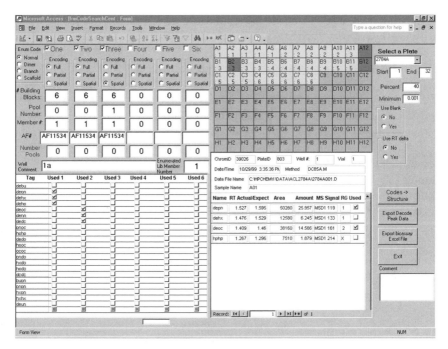

Figure 9.14. Decode user interface demonstrating the Code-to-Structure step.

peak, which is higher than the Percentage, this is considered a failure for the synthesis step.

If one, two, or three tags are found, then the tags are placed in alphabetical order and then the tag code is searched in the Oracle R-Group table. If there is a match, then the AF number (AF is the Affymax compound numbering system) is retrieved for the R-group along with the member number. These tags are marked as used. The pool number is used to retrieve a building for spatially or partially encoded steps. The screen is updated to show the Decode information, as in Figure 9.14.

The well colors in the plate graphical user interface are adjusted to reflect success or failure of the Decode. Obviously the control wells will always come out red, as they do not represent coded molecules. At this point in the analysis, the user can review the data for completeness and study the failures. Failures are usually due to multiple beads or no beads in a well.

The Export Decode Peak Data feature exports to Microsoft Excel a worksheet that can be used to track quantitative results of the decode process. The worksheet includes building block identifiers and the amounts of tag detected in each well. These results are used in quality control of bead differentiation and encoding.

9.3.3.   Code to Structure

Prior to structure generation, it is assumed that a generic structure describing the combinatorial library exists in the Affymax ACL database. The ACL database is an ISIS/Host v3. × RCG database that contains generic structures described with root structures and corresponding named building blocks (see Figure 9.9). Here each building block has a previously assigned name (e.g., AF11534). After the decode is complete, the Code to Structure button appears as shown in Figure 9.14. This button causes the enumeration string for positive decodes (in this case ACL2784-AF11534-AF11534-AF11534) to be generated for each positive well and appends them to the AFFYBANK_ACL_DECODE table. An Oracle procedure is then launched, and it generates the structure from each enumeration string. Structures for the positive decodes are generated and placed in an ISIS database.

9.3.4.   Capture

We use the Windows NT application, CAPTURE,[32] to correlate the decoded structures with the mass spectrum from each well in library quality control. For the example peptide library we would pick five beads from each of the six pools. Each bead yields a ligand LC/UV/MS datafile and a decoding LC/UV/MS datafile. The user interface in CAPTURE represents a 96 well microtiter plate, as shown in Figure 9.15. For the green wells the predicted structure from decoding has matched a spectrum from the LC/MS data. This application uses simple isotopic composition and electrospray ionization rule sets to predict mass spectra and judge the concordance of a structure-mass spectrum data set. The white well is the case where no structure is available because the decoding has failed. The red well is the case with a good decoded structure but the LC/MS data do not match.

The data "behind" the green light of well A01 is shown as Figure 9.16. The ligand LC/UV/MS is conducted "full scan," treating the ligand as an unknown. In the figure the top trace is a base peak chromatogram (BPC), the middle trace is the UV chromatogram, and the bottom trace is the spectrum of the major or expected compound. The weakness of these signals is typical for single-bead mass spectra. Although the nominal amount of ligand cleavable from a single bead is in the hundreds of picomoles, variations in synthetic yield, cleavage yield, and ionization efficiency can cause several orders of magnitude differences in actual LC/UV/MS peak sizes.

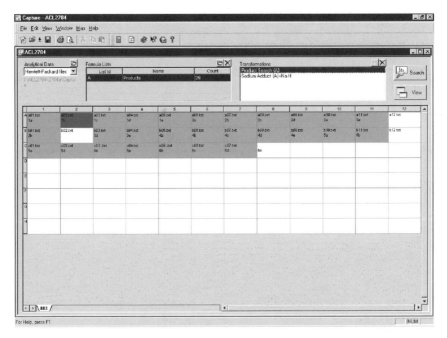

Figure 9.15. Main CAPTURE user interface.

This simple peptide has insufficient UV absorbance to be seen at this concentration.

As seen in Figure 9.16, three of the peaks in the BPC are labeled. These are the three peaks detected by the integration parameters set in the ChemStation software. These parameters are optimized for each combinatorial library to minimize the total number of peaks but detect the desired peak in most cases. A slider bar in CAPTURE sets the criteria for what constitutes a green light as predicted M + H must be >50% of the base peak in any of the detected spectra. In the Figure 9.16 example, the predicted M + H is 100% of the base peak.

9.3.5. EIC for Capture

Encoded library quality control using ligand single-bead mass spectrometry is often complicated by a very weak signal observed in CAPTURE. Figure 9.17 gives an example. Both the UV trace (bottom) and the BPC (middle) show nothing above background; no spectra are available for green/red analysis. This situation arises because of weak UV chromophores,

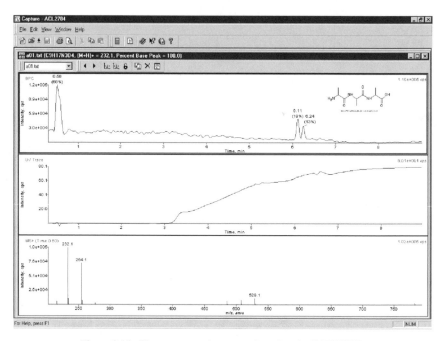

Figure 9.16. Chromatogram/spectrum interface in CAPTURE.

poor electrospray ionization, inadequate cleavage from the bead, a failure of synthesis, badly contaminated solvents used in sample preparation, or inadequate sensitivity of the mass spectrometer. An additional data analysis tool in this case is to look for the presence of the predicted mass spectrum in the full scan data. The technique for doing this is called *extracting an ion chromatogram*. The result is called an extracted ion chromatogram (EIC). For example, if the predicted tripeptide has a monoisotopic molecular weight of 231.1, the predicted positive ion electrospray mass spectrum of this compound would have a dominant ion at M + H = 232.1. The three-dimensional LC/UV/MS data set is then queried for the presence of ion current at this mass and this current plotted against time. The top curve of Figure 9.17 shows the EIC for this structure in this data set. You can see that the focused data give evidence for the presence of a molecule of the predicted properties. Just as important would be an EIC that shows no peak in the predicted EIC. This would provide strong evidence that the predicted molecule is not present.

Manually obtaining the EIC for each well of a 96-well analysis can be painful. First the decoded structure is evaluated for molecular weight. Then

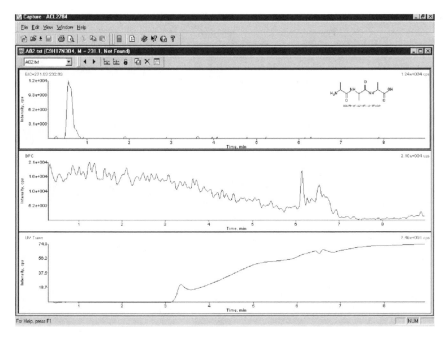

Figure 9.17. Chromatogram/structure interface when using EIC for CAPTURE.

the M + H is calculated. Then the raw LC/MS datafile is opened in the instrument control software, ChemStation. Finally an EIC is created for the predicted molecule. Doing this for 96 data files is prohibitive. A new piece of Access software has been created to automate this task. The user interface is shown as Figure 9.18. For each plate the user associates the raw datafile of the ligand LC/MS with each well. Then the appropriate type of EIC is chosen depending on the chemistry of the ligands. For positive ion mass spectrometry of small molecules, the default choice is to create M + H and M + Na (sodium). For negative ion mass spectrometry, the default is M − H. An option to look at doubly charged positive ions is offered as is an option to create a user-defined EIC.

The process is then initiated, each structure is queried, and a set of target ions is created. ChemStation is opened and a macro is run that opens each raw LC/MS datafile and extracts the predicted ion. The macro creates new data transfer text file for each well and places them in a separate EIC-CAPTURE folder next to the existing CAPTURE text files. This EIC process is of critical importance in encoded library quality assurance,[34] and it has saved our chemists considerable effort previously expended on doing these steps manually.

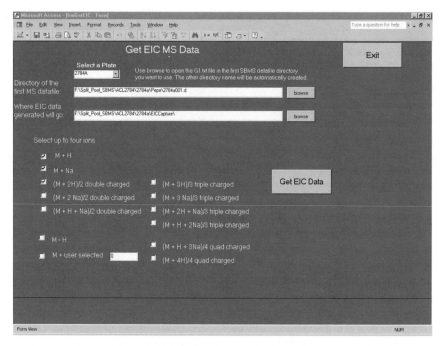

Figure 9.18. User interface for EIC CAPTURE.

9.3.6. Summary of Software

The new process for analyzing decoding and library quality control information is summarized in Figure 9.19. The Access software resides only on the analytical chemist's computer. The structure database resides on the corporate UNIX server. The LC/MS data are collected on dedicated data acquisition computers and transferred to NT servers for processing. CAPTURE and Isis Base reside on the end-user's desktops. Starting in the upper left corner, first the synthetic chemist plans and produces a library. The MDL software is used to enter the building blocks in an Oracle database. The Access application CodeGen is used here to set up the building block/code correspondence and load this information into the Oracle tables of the ISIS/Host ACL database.

When the chemist is done with the synthesis, samples are submitted to analytical for quality control. The analyst runs the Decode *Beadsetup* software to arrange single beads from each pool of the library into 96-well plates. Samples are analyzed for ligands using LC/UV/MS, and these datafiles are stored on the Decode NT server along with the CAPTURE

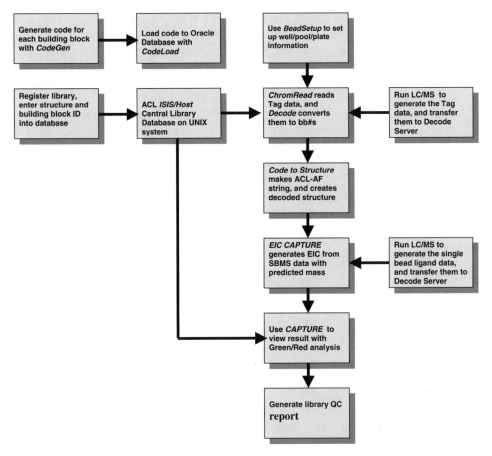

Figure 9.19. Summary chart of software applications.

.txt files created by the ChemStation macro. The same samples are then decoded by LC/MS to generate datafiles and report.txt files that are organized onto the same NT server.

The analyst next starts the Decode *ChromRead* application to associate a decode.report.txt file with each well of each plate. The Decode main application is run to convert the data for each well into an enumeration string ACLxxxx–Afaaaa–Afbbbb–Afcccc. The Code-to-Structure application creates an ISIS database of these enumerated structures. As a final step the analyst uses the EICCAPTURE application in Access to generate the EIC-CAPTURE text files for use in CAPTURE.

The chemist is then informed that the analysis is complete and the data can be viewed. Each chemist has the CAPTURE application on his or her desktop. They first load the appropriate structures from the ISIS database of decoded structures onto their 96-well template for their quality control. Next they find the LC/UV/MS text files from the server. They have a choice of looking at the data without EICs (in the CAPTURE folder) or with EICs (in the EICCAPTURE folder).

For the decoding of hit structures from biological screening, the process is simpler. The beads are delivered to the analytical laboratory with a 96-well hardcopy diagram that shows the library number and the pool number. The analyst uses the Beadsetup software to arrange the plate for analysis. After decoding LC/MS, the ChromRead, Decode, and Code-to-Structure applications are used to create the structures. These are then communicated back to the originating scientist using the feature of CAPTURE which creates an Excel spreadsheet of the structures.

Encoded combinatorial chemistry is a young technology. The formats for applying this technology to drug discovery change on a regular basis; chemists can invent new formats much faster than software algorithms can be adapted and tested. We have found it valuable to use flexible Access software for application development in this area.

This software and its application are complex. It requires diligence to avoid errors in mismatching codes and structures or datafile names and beads. However, the process has been shown to be a tremendous time saver over manual methods of decoding data handling. Only after this system was put in place, has Affymax been able to realize the full potential of encoded library chemical diversity.

9.4. QUANTITATIVE ANALYSIS OF COMBINATORIAL LIBRARIES

Encoding and decoding of combinatorial libraries provides a general solution to verify whether the desired compound is present in the solid-phase synthesis. The next step is to determine how pure the sample is and how much was made. However, it is always a challenge for analytical chemists to develop generic analytical techniques to quantitatively analyze a diversity of single compounds released from single beads. The lack of quantitation methods has hindered the method development for single-bead high-throughput screening as well as chemical synthesis of split-pool libraries and compound cleavage from beads.

In large-scale synthesis, quantitation is done by purifying and weighing the product. In solid-phase synthesis, there is often a submilligram yield that cannot be quantified by weighing. LC/UV/MS is often used to quantify libraries. However, when working with small amount of combichem compounds, reference standards are often not available. UV cannot detect many salts, aliphatic reagents, and solvents, and the unpredictability of UV response makes it a questionable choice for relative quantitative analysis. Relative peak heights in electrospray mass spectra have been proposed for purity assessment in peptide synthesis.[41] Generally, the ion peak heights in mass spectra are not proportional to the concentration of small molecule compounds.

Evaporative light scattering detection (ELSD) often in conjunction with HPLC is chosen by many companies for quantitation of combinatorial synthesis products.[42,43] This method is based on the assumption that all molecules give equal response to this detector. While promising for crude quantitation, this approach will not be appropriate for high-accuracy property measurement.

IR methods can be used to support a proposed intermediate in a reaction[44] or to follow the incorporation of a distinctive functional group (e.g., CHO) onto a resin.[45,46] IR is especially useful for surfaces such as pins or crowns where NMR techniques are not useful. High-quality IR spectra can be obtained from single beads.[47] The UV spectrophotometric quantitative measurement of Fmoc release from derivatized amino groups is still a very common method for measuring loadings.[37] Qualitative color tests are frequently used to follow reactions to assure completion. The Kaiser ninhydrin test is the best known of these.[48] An improved method for detection of secondary amines has been reported.[49]

Here we will describe several approaches of quantitative analysis of combinational libraries in Affymax, including QC of libraries by chemiluminescent nitrogen detector (CLND), quantitative analysis by standard bead LC/MS method, and estimation of concentration from UV response.

9.4.1. Quality Control of Libraries by CLND

The CLND is a unique method for directly measuring the yield and purity of solid and solution phase reactions. This detector oxidizes nitrogen compounds to NO, converts the NO to excited NO_2 by reaction with ozone, and detects the emitted light in a photomultiplier. A nitrogen detector is a remarkably universal detector for pharmaceuticals. The limitation is that the compound must contain nitrogen. Of the compounds in the commercial database of developmental and marketed drugs, MDDR (MDL Information Systems Inc. San Leandro, CA), 91% contain nitrogen. Many of the scaffolds that Affymax has described for solid-phase synthesis of combinatorial libraries contain nitrogen.

The usefulness of CLND is based on the hypothesis that all nitrogen compounds give equal combustion response. We, and others, have confirmed this with most molecules tested. The exception is molecular nitrogen, which requires higher combustion temperatures to oxidize. Molecules that contain N–N bonds may have thermal decomposition pathways that yield N_2 and thus will not give equimolar CLND response.[50] This limitation of the CLND detector is important for certain classes of molecules (tetrazoles) but is not a huge problem.

The CLND has been developed for use with HPLC flows up to $300 \mu L/min$. The LC/CLND measures the total nitrogen content of a chromatographic peak. It can be calibrated with any nitrogen containing compound, and thus allows quantitative analysis when authentic reference standards are not available. For direct purity estimation of a reaction mixture, CLND is complementary to UV. Many reaction by-products or starting materials do not contain nitrogen; others lack a UV chromophore. The CLND is especially useful in conjunction with splitting to a mass spectrometer. The electrospray mass spectrometer is likely to detect many of those peaks that do not contain nitrogen, offer structural proof for all components, and confirm the purity of each peak.

In the split/pool method of combinatorial synthesis, mixtures of compounds are made that are difficult to characterize. The LC/CLND of a nominally equimolar pool (based on nitrogen) should yield equal-sized chromatographic peaks of compounds. In the early stages of a lead development project, weighable quantities of authentic pure samples of a compound are not available, and yet quantitative measurements such as IC50, solubility, or plasma stability need to be made. LC/CLND can be used to calibrate solutions made from submilligram synthetic samples. LC/CLND is an important new technique to add to the arsenal of the organic analytical laboratory.

The limitation of the LC/CLND detector and the reason for its slow acceptance in the industry is that it is not as trivial to use as is a UV or ELSD detector; it is still a specialist's instrument. We have introduced an alternative, direct inject CLND (DI-CLND), that is easy to use. This detector measures the combustible nitrogen in a sample with two limitations relative to the LC/CLND; the samples are not chromatographically separated and on-column concentration is not possible, so sensitivity is compromised. We find these limitations are more than balanced by the ease of use for certain applications. Following the synthesis of a small molecule library, compounds analyzed using this technique were characterized by mass spectrometry, and an accurate concentration of the compound was assessed by CLND. Characterization of one compound is completed in 60 s, allowing for up to 1000 compounds to be analyzed in a single day. The data are summarized using pass/fail criteria in internally developed software. Therefore very fast 100% quantitation of parallel synthesis libraries is achievable.[33,51]

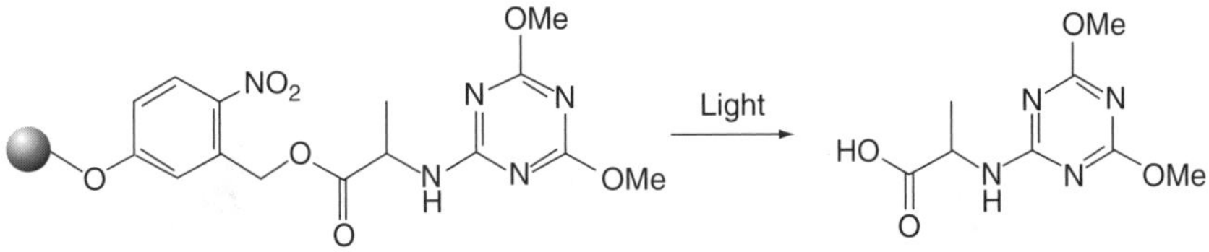

Figure 9.20. Compound A prepared on TentaGel resin with a photolinkage.

To illustrate these principles, compound A in Figure 9.20 was prepared on TentaGel resin with a photolinkage.[24,51] The resin was characterized by quantitative solid-phase NMR, traditional elemental analysis, and solids CLND. By NMR the loading was 0.38 mmole/g while the CHN analysis showed 0.34 mmole/g, and the solids CLND showed 0.40 mmole/g. While not strictly relevant to the cleavage question, the solids CLND is a convenient, high-throughput way to characterize solid-phase synthesis samples.

The photolytic cleavage of this compound was measured using direct inject CLND. A weighed sample of the resin (3–5 mg) was placed in a small glass vial and slurried with 200 μL of solvent. The sealed vial was then exposed to 10 mW/cm^2 light energy, measured at 360 nm.[24] After an appropriate time, the vial was removed from the photolysis and diluted for analysis. We have previously shown that the DI-CLND is sensitive to solvent and must be calibrated with the standards dissolved in the same solvent as the samples are dissolved in. The time course for cleavage of the compound in four different solvents is shown in Figure 9.21.

9.4.2. Quantitative Analysis by Standard Bead LC/MS Method

It is very useful to quantitatively determine the amounts of compound released from single beads. The CLND, ELSD, and NMR methods are not applicable for the subnanomole amounts of sample released from single beads, so these measurements will need the more sensitive UV or MS detectors. For single compounds an LC/MS run in selected ion mode is an extremely sensitive and specific analytical method.

To conduct our quantitative LC/MS single-bead analysis, we developed a so-called standard bead LC/MS method.[52] A standard set of compounds with different cLogP's was selected. They are Mono-Fmoc-Lysine, Bis-Cbz-Lysine, Fmoc-Cbz-Lysine, and Bis-Fmoc-Lysine. All compounds were synthesized on both TentaGel beads with the Affymax photo-cleavage linker and Wang acid-cleavable linker. The cLogP varied from 0.9 to 7.2, representing the cLogP diversity of common split/pool libraries. The standard compounds were simple and commercially available (Advanced

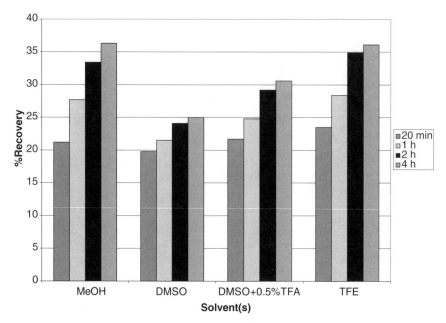

Figure 9.21. Time course for cleavage of compound A from resin in four solvents.

ChemTech, Louisville, KY). Figure 9.22 shows the molecular structures of the standard compounds.

The standard bead synthetic procedure is summarized in Figure 9.23. ABI 433 Amino Acid Synthesizer was used to synthesize all compounds on both TentaGel beads with the Affymax photo-cleavage linker and Wang acid-cleavable linker. First, the TentaGel photolinker or Wang resin was washed with dichloromethane (DCM) and N-methylpyrrolidone (NMP). Then 4 equiv (relative to the resin) of R_1-R_2-Lys-OH, 4 equiv of N,N-dicyclohexylcarbodiimide, and 4 equiv of 1-hydroxybenzotriazole anhydrous in NMP were added to the resin, where R_1 is Fmoc or Cbz, and R_2 is Boc, Cbz or Fmoc. The reaction mixture was gently shaken for 2 hours at room temperature, and then washed with NMP and DCM 6 times. The resin was lyophilized for 2 days. For mono-Fmoc-Lysine, Boc was removed with TFA after the coupling of Fmoc-Lys-OH was done.

After synthesis and drying, the standard beads were ready for use. Single beads were picked into individual tapered glass microvials. For the beads with the Affymax photo-cleavage linker, 25 μL mixture of 75% IPA, 25% DMSO, and 0.3% TFA was added to a single bead, and the microvial was sealed. The vial was placed on an orbital shaker and irradiated with 360 nm

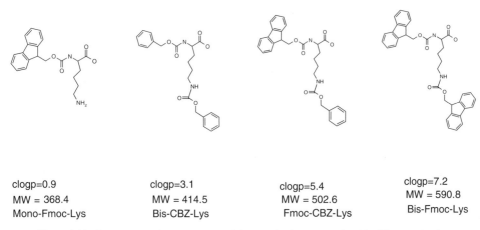

clogp=0.9
MW = 368.4
Mono-Fmoc-Lys

clogp=3.1
MW = 414.5
Bis-CBZ-Lys

clogp=5.4
MW = 502.6
Fmoc-CBZ-Lys

clogp=7.2
MW = 590.8
Bis-Fmoc-Lys

Figure 9.22. Structures and mass spectra of the standard compounds with different cLogPs.

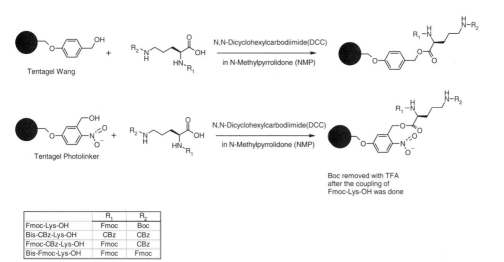

Figure 9.23. Synthetic scheme of standard compounds.

UV, $10\,mW/cm^2$, for 2 hours. The sample was then ready for LC/MS analysis. For the beads with the Wang acid-cleavable linker, $50\,\mu L$ of 95% TFA was added to a single bead, and the microvial was left at room temperature for 1 hour to cleave the compound. After evaporating the TFA, $25\,\mu L$ of ACN was added into the bead vial. The whole sample cleaved from a single bead was injected into LC/MS for analysis after a premix with an equal volume of water (0.05% formic acid).

LC/MS experiment was conducted with an Agilent 1100 LC/MSD system. A CTC PAL autosampler from LEAP Technologies was used to introduce samples to the LC/MSD. Polaris 2000 C18 columns packed with 3 μm particles, 30 × 3 mm (i.d.) at 60°C were selected for the study. Flow rate was 1 mL/min, linear gradient starting form 100% water (0.05% formic acid) to 100% acetonitrile (0.05% formic acid) in 4 minutes. ESI source with positive selected ion monitoring mode (SIM) was employed to detect the corresponding ions of standard compounds. The positive ions (m/z) of 369, 371, 459, and 591 were selected to monitor Mono-Fmoc-Lysine, Bis-Cbz-Lysine, Fmoc-Cbz-Lysine, and Bis-Fmoc-Lysine separately. The mass spectra of standard compounds are shown in Figure 9.24. Four levels of calibration—25 picomolar, 50 picomolar, 100 picomolar, and 200 picomolar of standard compounds—were used in the quantitation method.

The total amount of the compounds on a single bead can be measured by Fmoc number method if the compounds contain Fmoc. The Fmoc group can be removed from the compounds on the beads by adding base solution (30% piperdine in DMF) to the beads. The absorbance of the sample solution at 300 nm of UV was used to determine the concentration of Fmoc in samples. The total amount of Fmoc removed from the compounds on the resin represents the total amount of compounds on the resin (assume that all of Fmoc can be cleaved from the compounds on the resin).

The CLND method described in Section 9.4.1 provides another option to measure both the total amount of the compounds (by solid CLND) and the cleavable compounds on a single bead. Since both Fmoc and CLND methods are not sensitive enough to measure the amount of compounds on a single bead, multi-milligram samples of beads were used to measure the amount of compounds per milligram of standard beads. By measuring the numbers of bead/mg for an aliquot, we can convert this number to the picomoles/bead that is the amount of compounds on a single bead.

The standard beads were first used to study the cleavage efficiency. The results of standard compounds with different cLogPs on both TentaGel beads with the Affymax photo-cleavage linker and Wang acid-cleavable linker are summarized in Table 9.5 as well as in Figure 9.25. The average results of Fmoc and CLND were used as the total amount of the compounds on a single bead. For the compound without Fmoc, Bis-Cbz-Lys-OH, we used CLND results only. The % theoretical is the percentage of the cleavable compound measured by LC/MS over the total amount of compound on a single bead. For both Affymax photo-cleavage linker and Wang acid-cleavable linker resins we found that about half of total amount of compounds on resin can be cleaved off by using the cleavage conditions in the

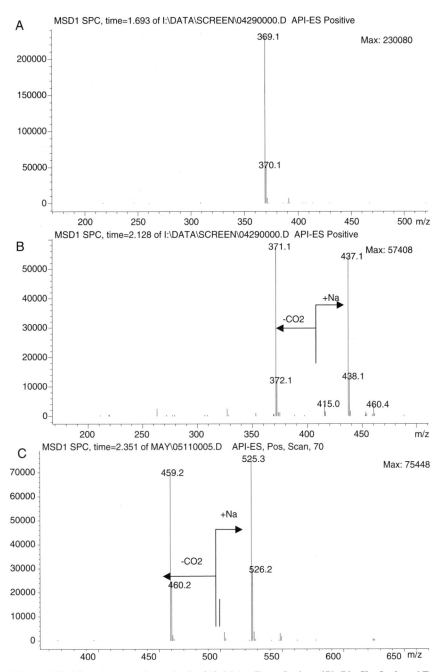

Figure 9.24. Mass spectra of standards: (*A*) Mono-Fmoc-Lysine; (*B*) Bis-Cbz-Lysine; (*C*) Fmoc-Cbz-Lysine.

Table 9.5. Cleavage Results of Standard Compounds on Affymax Photolinker and Wang Acidlinker Resin

Linker	Compound	Total Fmoc/CLND (pmol/bead)	Released CLND (pmol/bead)	Released LCMS (pmol/bead)	% Theoretical
Photolinker	Fmoc-Lys	240	200	126	53
	Bis-CBz-Lys	200	150	109	55
	Fmoc-CBz-Lys	—	200	116	—
	Bis-Fmoc-Lys			< 2	
Acidlinker	Fmoc-Lys	270	180	166	61
	Bis-CBz-Lys	310	120	103	33
	Fmoc-CBz-Lys	110	100	60	55
	Bis-Fmoc-Lys			< 2	

experiment. Because of its poor synthesis yield and poor solubility in tris buffer, Bis-Fmoc-Lysine was dropped from the standard list.

Magic-angle spinning (MAS) NMR is a very useful tool in solid-phase chemistry development. MAS is used to remove the inhomogeneities in the magnetic field immediately surrounding the bead and, therefore, to reduce the broadening of the resonances. We found that MAS is useful for obtaining high-resolution spectra for any liquid-phase samples, present as either a solution or a suspension, that might otherwise have acceptably broad line shapes due to susceptibility discontinuities. We use the hig-resolution proton MAS NMR to check the purities of standard beads in the study. Figure 9.26 is the high-resolution proton MAS NMR spectrum of Fmoc-Lysine on the TentaGel beads with Affymax photolinker. Presaturation of the PEG resonance at 3.6 ppm is critical to spectral quality. The complex peaks from 3.4 to 3.9 ppm are due to residual PEG, ^{13}C satellites and the terminal CH_2 of the PEG chain. We also took the MAS NMR spectrum of the same resin after photo cleavage. By comparing the spectra of the resin beads before and after cleavage, we were not surprised to find that significant amounts of Fmoc-Lysine still attach on the resin even after the 4-hour photo cleavage.

The standard bead LC/MS method proved useful in optimizing the cleavage and recovery conditions for HTP biology screening.[52] For example, Fmoc-Lysine and Cbz-Lysine standard beads were selected to examine how compounds with different cLogPs behave at different cleavage conditions. Two different 864-well setups,[53] 864-bead cup plate and 864 master plate, were tested in the experiment. Methanol or tris buffer solution was used to recover the cleaved compounds. The results are summarized in Figure 9.26.

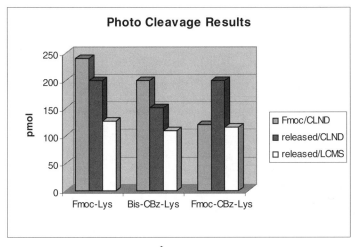

A

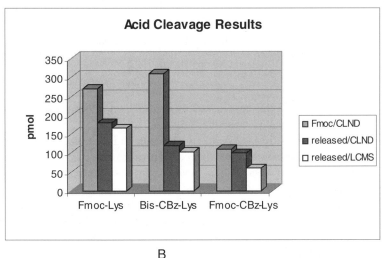

B

Figure 9.25. (A) Cleavage results of standard compounds on TentaGel beads with (*A*) Affymax photo-cleavage linker and, (*B*) Wang acid-cleavable linker.

It was found that much more Fmoc-Lysine can be cleaved and recovered from beads in the 864-well master plate than the 864-bead cup plate, while for Cbz-Lysine, the recovery amounts are about the same for both 864 master plate and 864-bead cup. We also can conclude from the experiment

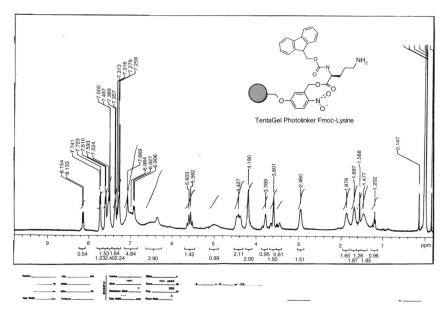

Figure 9.26. High-resolution proton MAS NMR spectrum of Fmoc-Lysine on the TentaGel beads with Affymax photolinker.

that for both Fmoc-Lysine and Cbz-Lysine more compound can be recovered by using methanol as a recover solvent.

In conclusion, the standard bead LC/MS method was successfully developed to quantitatively measure the concentration of the standard compounds cleaved from the single beads. The results from CLND, UV response, and NMR measurements of standard compounds cleaved from single beads were investigated and compared to the LC/MS results. The standard bead method proved to be very useful in development of new HTP screening methods, in improving cleavage reliability in biology, and in monitoring the cleavage efficiency of library compounds before biological screening.

9.4.3. Estimation of Concentration from UV Response

In the modern drug discovery laboratory, analysts are asked to quantify the amount of target compound in hundreds of novel samples each day. These molecules are made in submilligram amounts and have never been synthesized before. The only information available is the structure of the molecule and properties that can reliably be calculated from the structure. The preceding section showed the classic approach to this problem: synthesize a few standard compounds that might represent all compounds. We believe that the prediction of UV spectral intensity would be a useful alternative for high-throughput organic synthesis. We recently published the first attempt to do this.[54] In the process we created a large database of spectra, developed a sophisticated QSPR using quantum chemical descriptors, and demonstrated a good correlation for a target set of compounds.

Additional work will be needed to bring this method to the mainstream. This could include the general improvement of UV databases and parameterization of organic molecules, but for the case of split pool library synthesis, it may be simpler to use a set of the molecules synthesized during library rehearsal as a learning set to create simple QSPRs that are adequate for *relative* quantitation of specific libraries. Inclusion of a known compound/bead combination in a library quality control experiment could then become a simple and elegant tool for increasing the quantitative quality assurance of a combinatorial drug discovery process. Such experiments are in process at Affymax.

ACKNOWLEDGMENTS

Many current and former employees of Affymax, and its collaborators, contributed to this work. The authors would like to especially acknowledge Derek Maclean and Ken Lewis for much of the early work on decoding and single-bead MS, Mario Geysen for the beadpicker, Rob Wilgus for CAPTURE; Frank Holden for the invention of the maize and blue 96-well hydrolysis plate, Nikhil Shah for much of the MS infrastructure, Liang Tang for the elegant sensitivity of the single-bead LC/MS approach, and Glenn Ouchi for the Decode software.

REFERENCES

1. M. A. Gallop, R. W. Barrett, W. J. Dower, S. P. A. Fodor, E. M. Gordon, *J. Med. Chem.* **37**, 1233 (1994).
2. E. M. Gordon, R. W. Barrett, W. J. Dower, S. P. A. Fodor, M. A. Gallop, *J. Med. Chem.* **37**, 1385 (1994).
3. M. Needels, J. Sugarman, in J. F. Kerwin, E. M. Gordon, eds., *Combinatorial Chemistry and Molecular Diversity in Drug Discovery*, Wiley, New York 339 (1998).
4. M. C. Needels, D. G. Jones, E. H. Tate, G. L. Heinkel, L. M. Kochersperger, W. J. Dower, R. W. Barrett, M. A. Gallop, *Proc. Nat. Acad. Sci. USA* **90**, 10700 (1993).

5. Z. Ni, D. Maclean, C. P. Holmes, M. M. Murphy, B. Ruhland, J. W. Jacobs, E. M. Gordon, M. A. Gallop, *J. Med. Chem.* **39**, 1601 (1996).

6. Z. J. Ni, D. Maclean, C. P. Holmes, M. A. Gallop, *Methods Enzymol.* 267(Comb. Chem), 261 (1996).

7. J. W. Jacobs, Z. J. Ni, in E. M. Gordon, J. F. Kerwin, eds., *Combinatorial chemistry and molecular diversity in drug discovery*, Wiley, New York, 271 (1998).

8. M. A. Gallop, E. M. Gordon, Z. Ni, D. Maclean, C. P. Holmes, W. L. Fitch, N. Shah, *Methods for hard-tagging an encoded library*, US Patent **5**, 846, 839 (Dec. 8, 1998).

9. W. L. Fitch, T. A. Baer, W. Chen, F. Holden, C. P. Holmes, D. Maclean, N. Shah, E. Sullivan, M. Tang, P. Waybourn, S. M. Fischer, C. A. Miller, L. R. Snyder, *J. Comb. Chem.* **1**, 188 (1999).

10. B. Ruhland, A. Bhandari, E. M. Gordon, M. A. Gallop, *J. Am. Chem. Soc.* **118** (1), 253 (1996).

11. D. Maclean, J. R. Schullek, M. M. Murphy, Z. J. Ni, E. M. Gordon, M. A. Gallop, *Proc. Nat. Acad. Sci. USA* **94** (7), 2805 (1997).

12. A. Ateugbu, D. Maclean, C. Nguyen, E. M. Gordon, J. W. Jacobs, *Bioorg. Med. Chem.* **4** (7), 1097–1106 (1996).

13. J. R. Schullek, J. H. Butler, Z.-J. Ni, D. Chen, Z. Yuan, *Anal. Biochem.* **246**, 20 (1997).

14. M. K. Schwarz, D. Tumelty, M. A. Gallop, *J. Org. Chem.* **64**, 2219 (1999).

15. J. L. Silen, A. T. Lu, D. W. Solas, M. A. Gore, D. Maclean, N. H. Shah, J. M. Coffin, N. S. Bhinderwala, Y. Wang, T. Tsutsui, G. C. Look, D. A. Campbell, R, L. Hale, M. Navre, C. R. Deluca-Flaherty, *Antimicrob. Agents Chemother.* **42**, 1447 (1998).

16. O. Lorthioir, R. A. E. Carr, M. S. Congreve, M. H. Geyesn, C. Kay, P. Marshall, S. C. McKeown, J. N. Parr, J. J. Scicinski, S. P. Watson, *Anal. Chem.* **73**, 963 (2001).

17. K. C. Lewis, W. L. Fitch, D. Maclean, *LC/GC*, **16**, 644 (1998).

18. R. E. Dolle, J. Guo, L. O'Brien, Y. Jin, M. Piznik, K. J. Bowman, W. Li, W. J. Egan, C. L. Cavallaro, A. L. Roughton, Q. Zhao, J. C. Reader, M. Orlowski, B. Jacob-Samuel, C. Caroll, *J. Comb. Chem.* **2**, 716 (2000).

19. H. Kedar, R. M. Gavin, J. H. Sugarman, D. T. Roth, *Bead dispensing device and methods*, US Patent **5**, 722, 470 (March 1998).

20. G. Karet, *Drug Discovery and Development*, 55 (1999).

21. S. J. Lane, A. Pipe, *Rapid Commun. Mass Spectrom.* **12**, 667 (1998).

22. T. Tang, W. L. Fitch, P. Smith, D. Tumelty, K. Cao, S. Ferla, *Comb. Chem. High Through. Screen.* **4**, 287–293 (2001).

23. C. P. Holmes, J. P. Chinn, G. C. Look, E. M. Gordon, M. A. Gallop, *J. Org. Chem.* **60**, 7328 (1995).

24. C. P. Holmes, D. G. Jones, *J. Org. Chem.* **60**, 2318 (1995).

25. D. J. Daley, R. D. Scammell, D. James, I. Monks, R. Raso, A. E. Ashcroft, A. J. Hudson, *Am. Biotechnol. Lab.* **15**, 24 (1997).

26. K. C. Lewis, W. L. Fitch, D. Maclean, *LC/GC*. **16**, 644 (1997).

27. G. I. Ouchi, *LC/GC* **16**, 362 (1998).

28. R. Richmond, E. Gorlach, J. Seifert, *J. Chromatogr. A* **835**, 29 (1999).

29. H. Tong, D. Bell, K. Tabei, M. M. Siegel, *J. Am. Soc. Mass Spectrom.* **10**, 1174 (1999).

30. K. Koch, *LC/GC* **18**, 500 (2000).

31. N. Shah, S. Teeter, W. L. Fitch, R. L. Wilgus, C. Koch, *Scientific Comput. Instrument.*, August (2000).

32. N. Shah, K. Tsutsui, A. Lu, J. Davis, R. Scheuerman, W. L. Fitch, *J. Comb. Chem.* **2**, 453 (2000).

33. R. E. Dolle, J. Guo, L. O'Brien, Y. Jin, M. Piznik, K. J. Bowman, W. Li, W. J. Egan, C. L. Cavallaro, A. L. Roughton, Q. Zhao, J. C. Reader, M. Orlowski, B. Jacob-Samuel, C. Caroll, *J. Comb. Chem.* **2**, 716 (2000).

34. K. C. Nicolaou, J. A. Pfefferkorn, H. J. Mitchell, A. J. Roecker, S. Barluenga, G.-Q. Cao, R. L. Affleck, J. E. Lillig, *J. Amer. Chem. Soc.* **122**, 9954 (2000).

35. W. L. Fitch, J. J. Zhang, N. Shah, G. I. Ouchi, R. L. Wilgus, S. Muskal, *Comb. Chem. High Through. Screen.* **5**, 531 (2002).

36. L. W. Dillard, J. A. Connelly, J. J. Baldwin, E. G. Horlbeck, G. L. Kirk, G. Lauri, *PCT Int. Appl.* (2000).

37. Z. Ni, D. Maclean, C. P. Holmes, M. M. Murphy, B. Ruhland, J. W. Jacobs, E. M. Gordon, M. A. Gallop, *J. Med. Chem.* **39**, 1601 (1996).

38. W. L. Fitch, T. A. Baer, W. Chen, F. Holden, C. P. Holmes, D. Maclean, N. Shah, E. Sullivan, M. Tang, P. Waybourn, S. M. Fischer, C. A. Miller, L. R. Snyder, *J. of Comb. Chem.* **1**, 188 (1999).

39. H. Kedar, R. M. Gavin, J. H. Sugarman, D. T. Roth, US, *Glaxo Group Limited*, 12 (1998).

40. X.-Y. Xiao, *Front. Biotechnol. Pharm.* **1**, 114 (2001).

41. S. S. Smart, T. J. Mason, P. S. Bennell, N. J. Meaji, H. M. Geysen, *Int. J. Pept. Protein Res.* **47**, 47–55 (1996).

42. B. H. Hsu, E. Orton, S. Tang, R. A. Carlton, *J Chromatogr. B* **725**, 103 (1999).

43. L. Fang, M. Wan, M. Pennacchio, J. Pan, *J. Comb. Chem.* **2**, 254 (2000).

44. M. F. Gordeev, D. V. Patel, E. M. Gordon, *J. Org. Chem.* **61**, 924 (1996).

45. B. Cherera, J. A. Finklestein, D. E. Veber, *J. Am. Chem. Soc.* **117**, 11999 (1995).

46. J. R. Hauske, P. Dorff, *Tetrahedron Lett.* **36**, 1589 (1995).

47. B. Yan, G. Kumaravel, H. Anjaria, A. Wu, R. C. Petter, C. F. Jewell Jr., J. R. Wareing, *J. Org. Chem.* **60**, 5736 (1995).

48. E. Kaiser, R. L. Colescott, C. D. Bossinger, P. I. Cook, *Anal. Biochem.* **34**, 595 (1970).

49. T. Vojkovsky, *Peptide Res.* **8**, 236 (1995).

50. K. C. Lewis, D. Phelps, A. Sefler, *Am. Pharm. Rev.* **3**, 63–68 (2000).

51. N. Shah, G. Detre, S. Raillard, W. L. Fitch, in B. Yan, A. Czarnik, eds., *Optimization of Solid-Phase Combinatorial Synthesis*, Marcel Dekker, New York, 251 (2002).

52. J. J. Zhang, W. L. Fitch, K. Wright, Z. Taiby, in *Proceedings 50th ASMS Conference on Mass Spectrometry and Allied Topics* No. A022025 (2002).

53. J. M. Dias, N. Go, C. P. Hart, L. C. Mattheakis, *Anal. Biochem.* **258**, 96 (1998).

54. W. L. Fitch, M. McGregor, A. R. Katritzky, A. Lomaka, R. Petrukhin, M. Karelson, *J. Chem. Info. Comput. Sci.* **42**, 830 (2000).

PART
III

HIGH-THROUGHPUT PURIFICATION TO IMPROVE LIBRARY QUALITY

STRATEGIES AND METHODS FOR PURIFYING ORGANIC COMPOUNDS AND COMBINATORIAL LIBRARIES

JIANG ZHAO, LU ZHANG, and BING YAN

10.1. INTRODUCTION

The *absolute purity* requirement of combinatorial library compounds delivered for biological screening has been raised. Improving compound purity is the most effective way to remove any ambiguity in the screening data. Even with the rapid advances in solid-phase and solution-phase synthesis and the intensive reaction optimization, excess reagents, starting materials, synthetic intermediates, and by-products are often found along with the desired product. Furthermore the strong solvents used to swell the resin bead for solid-phase synthesis and the scavenging treatment in solution-phase reactions often introduce additional impurities leached from resins and plastic plates. Therefore high-throughput purification has become an indispensable technology in all combinatorial chemistry and medicinal chemistry laboratories.

Throughput is a main consideration in purifying combinatorial libraries. Parallel synthesis often produces large numbers of samples, ranging from hundreds to thousands per library. Parallel processes are therefore preferred as productivity is multiplied by the number of channels. A 10-channel flash column chromatography system is presented by Isco, and 96-channel systems of solid-phase extraction (SPE) and liquid-liquid extraction (LLE) are also reported. The off-line process is often used as a time-saving measure in preparative HPLC where parallel processing is difficult. Column re-equilibrating and samples loading can be done off-line to reduce the cycle time.

Cost is a deciding factor in conducting high-throughput purification. Lengthy purification, scale-up in library production, low-purification recovery yield, plus all the reagents and accessories used for purification boost

Analysis and Purification Methods in Combinatorial Chemistry, Edited by Bing Yan.
ISBN 0-471-26929-8   Copyright © 2004 by John Wiley & Sons, Inc.

the cost of the purified products. With other factors optimized, purification recovery is the primary concern in every high-throughput purification protocol.

Automation is another key factor in considering purification strategy and efficiency. Purifying a combinatorial library is a highly repetitious process, especially when the library size is large. Robotics provide the best precision for repetitive processes, and thus reduce the chance for human error. Unattended processes can work around the clock to improve the daily throughput. However, mechanical failure can also be a major drawback in unattended processes.

Resolution is another factor for a purification process. Low-resolution method such as LLE can only remove impurities with a major difference from the product in terms of hydrophobicity. High-resolution methods such as HPLC and SFC can often separate compounds of close structural similarities. However, high-resolution methods are often more costly and time-consuming. Resolution is also related to the scale of sample loading, and it may decrease significantly as loading increases. The resolution decreases when the throughput increases, so it is often sacrificed for speed.

A "general"purification method should be sufficient to purify at least a major portion of a library. Reverse-phase HPLC is generally method of choice. Affinity methods apply only to compounds with specific structural features. Nevertheless, a successful purification strategy always involves identifying the properties of the target compounds as well as those of the impurities.

Finally solvent removal from the aqueous solution is not trivial. As an integral part of the whole purification process, solvent removal strategy needs to be considered in choosing and designing the process. Unlike organic solvents the removal of aqueous solvent involves a lengthy lypholyzation process or centrifugal evaporation. An additional SPE step can be added to exchange the aqueous medium with organic solvent.

In this chapter we review various purification strategies, factors that impact on the purification efficiency, and recent progresses in high-throughput purification of combinatorial libraries.

10.2. REVERSED-PHASE SEMIPREPARATIVE HPLC

In the last 15 years' 60% to 90% of the analytical separations was done in reverse-phase HPLC. The preference for HPLC can be attributed to its relative simplicity and its economic solvent systems in the reverse-phase HPLC. The another advantage of reverse-phase HPLC is its capability of separating different classes of compounds, ranging from aromatic hydrocarbons and fatty acid esters to ionizable or ionic compounds such as carboxylic acids, nitrogen bases, amino acids, proteins, and sulphonic acids. The recent advances in automation, detection, and method development have made it possible to use semipreparative reverse-phase HPLC to purify 200 to 250 compounds a day per instrument.[1,2] It has been reported that an parallel automatic HPLC system is capable of purifying dozens to hundreds of samples in unattended mode. For example, 200 mg of sample can be purified in 5 minutes by the fast gradient and very short column reverse-phase HPLC method.[3,4]

10.2.1. Effects of Stationary Phase

When choosing a stationary phase, we have to consider the chemical properties (bonded-functional groups) and physical properties, such as pore size, column dimensions, and particle size for the solid stationary phase. The silica packing with surface covalently bonded hydrophobic octadecylsilyloxy group (C-18) is the most popular stationary phase in both analytical and preparative separations. For preparative HPLC methods described in the literature, the columns packed with spherical C-18 media with various dimensions were mostly used for small organic molecules.[2–5] An experimental study of the relationship between the purification recovery and sample loading using various columns was reported (Table 10.1).[1]

While an examination of the chromatogram, shows that the 10-mm diameter column was overloaded at the 50-mg sample; the data in Table 10.1 indicate excellent recovery independent of sample or column size. In the preparative chromatography nonlinear effects caused by column overload are often observed,[6] and this affects the separation resolution as sample

Table 10.1. Recovery of Preparative HPLC Samples

Sample	Percent Recovery From		
	10×100 mm	20×100 mm	30×100 mm
10 mg, component 1	95 ± 3	99 ± 1	94 ± 7
10 mg, component 2	92 ± 3	90 ± 13	94 ± 7
50 mg, component 1	92 ± 6	94 ± 2	91 ± 6
50 mg, component 2	89 ± 3	92 ± 3	86 ± 9
100 mg, component 1	—	91 ± 3	85 ± 1
100 mg, component 2	—	83 ± 3	82 ± 1
200 mg, component 1	—	92 ± 1	93 ± 2
200 mg, component 2	—	94 ± 1	91 ± 2

Note: Component 1: p-nitrobenzoic acid; component 2: 1-(4-chlorophenyl)-1-cyclobutanecarboxylic acid. Results are from triplicate experiments.

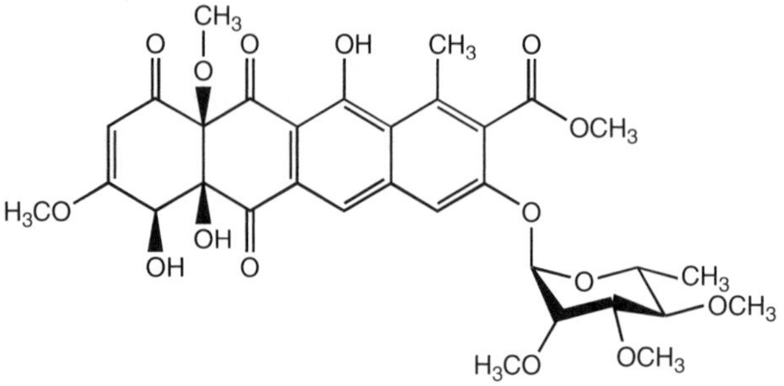

Figure 10.1. Structure of elloramycin.

Table 10.2. Purity and Recovery of Elloramycin by Column Particle Size

Particle Size (μm)	Original Purity (%)	Final Purity (%)	Recovery (%)
10	96	100	93
	20	100	69
15–25	96	97	92
	20	80	83

loading is increased. A study of the percentage of recovery for pharmaceutical compounds in overloaded column circumstances has been carried out and reported.[7] When there is enough separation resolution (e.g., $\alpha > 1$), the recovery of a desired product nevertheless turns out to be close to 100%.

For the purification of hydrophobic anthraquinone antibiotics, such as elloramycin (structure in Figure 10.1), the influence of particle size of the C-18 stationary phase on the purification efficiency has been studied.[8] The separation resolution, product purity, and recovery were compared with use of 10 μm and 15–25 μm Nucleosil C-18 column. The results shown in Table 10.2 demonstrate that with small and homogeneous particles used as the stationary phase, the separation resolution and product purity increases dramatically, though the recovery is not significantly affected.

The C-8 column has been studied for automatic purification of reaction mixtures of the amines and aldehydes after the parallel solution-phase reaction.[9] The typical column size is 20 × 50 or 20 × 75 with 5-μm particle size for 50-μmol materials. The yield of the desired products varied from 20% to 90% with purity >95%.

10.2.2. Effects of the Mobile Phase

Combinatorial compounds are highly diverse, although the choice of solid phase is usually limited. The separation of different kinds of the compounds can nevertheless be accomplished by choosing the right mobile phase. The solvent type, flow rate, gradient slope, and chemical modifiers can influence the separation efficiency, product recovery, product purity, purification speed, and the purification cost.

Generally, the best solvents for preparative LC mobile phase have the following characteristics:

- Low boiling point for easy and economical sample recovery.
- Low viscosity for minimum column back pressure and maximum efficiency.
- Low levels of nonvolatile impurities.
- Chemically inertness so as not to cause modification of sample and stationary phase.
- Good solubility properties for sample.
- Low flammability and toxicity for safety in storage and handling.

The theoretical studies for condition optimization of the preparative chromatograph has been published.[10,11] The theoretical models will not be discussed here, but the results from the studies will simplify the process of method development. They can be used as guidelines, as summarized below:

- The column should be operated at the highest flow rate to maximize the purification speed.
- The loading factor, which is the ratio of the total amount of sample to the column saturation capacity, is higher in gradient elution than in isocratic elution condition.
- The average concentration of the collected fractions and the purification speed are higher in gradient elution than in isocratic.
- The recovery yield achieved under optimum conditions is the same in gradient and in isocratic elution.
- The optimum gradient steepness depends mostly on the elution order. It is higher for the purification of the less retained component than for that of the more retained one.
- The volume of the solvents required to wash and to regenerate the column after a batch separation will always be larger in gradient than in isocratic elution.

- The gradient retention factor is a more significant parameter than the gradient steepness because the former incorporates the retention factor at the initial mobile phase composition.
- The gradient elution may use less efficient columns than isocratic elution.
- The performance in gradient mode is very sensitive to the retention factor of the two components to be separated. Optimizing their retention factors would improve the recovery yield and the purity of the final products.

In the methods for the high-throughput purification reported in the literature,[1–4,12–16] the steep and fast (4–6 minutes) gradient modes were employed for reverse-phase preparative HPLC. For purification of small organic molecules, water/acetonitrile or ware/methanol are the most commonly used solvent systems as the mobile phase. Offer 0.05% to 0.1% TFA is added to the mobile phases as a modifier. However, TFA is not a desirable chemical in the final compound. It may decompose some compounds and is detrimental to the biological screening. Other additives such as formic acid, acetic acid, or propanol may be used instead. The addition of triethylamine or ammonium acetate is to reduce the tailing of basic components in the samples. Using the acidic aqueous mobile phase can make all of the ionized groups protonated and avoid the formation of multiple forms of ions in the column. For separation of the acid labile compounds, the neutral or slightly basic conditions can be used.

10.2.3.   Effects of Other Factors

The scale of a combinatorial library is often on the order of tens of milligrams. In order to work on this scale, a larger diameter column (typically 20-mm internal diameter) is needed. The mobile phase linear velocity (u) is expressed as

$$u = \frac{4F}{\pi \varepsilon_0 d^2},$$ (10.1)

where

F = flow rate
ε_0 = column porosity
d = column diameter

To maintain the same linear velocity (and thus the retention time), the solvent flow rate should be scaled proportional to the square of the diameter ratio as

$$\frac{F_{prep}}{F_{analytical}} = \left(\frac{d_{prep}}{d_{analytical}}\right)^2. \tag{10.2}$$

Since sample retention time is proportional to the column length, the overall scaling equation is

$$\frac{rt_{prep}}{rt_{analytical}} = \frac{L_{prep}}{L_{analytical}} * \frac{F_{analytical}}{F_{prep}} * \left(\frac{d_{prep}}{d_{analytical}}\right)^2 * \frac{\varepsilon_{prep}}{\varepsilon_{analytical}}. \tag{10.3}$$

To increase purification throughput, all four factors in (10.3) need to be considered. First, columns length can be reduced for faster elution.[17] A benefit to reducing-column length is that the backpressure is also reduced, and this makes it possible to increase the mobile phase flow rate, as this will further shorten the run. The third parameter has largely to do with sample loading. However, choosing a smaller diameter column and running sample under lightly overloaded condition is the preferred way to maintain high throughput. The fourth factor is often adjusted to improve separation. Since running high flow rates on reduced column lengths degrades separation, narrower bore columns are often chosen to compensate for separation efficiency.

Since it is necessary to remove solvent from the product, the mobile phase buffer must be considered. Some popular reverse-phase HPLC buffers, such as phosphates or zwitterion organic buffers, are nonvolatile. They must be replaced by a volatile buffer such as formic acid or ammonium acetate. Otherwise, a desalting step must be added. Trifuouroacetic acid is another common buffer. Although it is fairly volatile, it forms a salt with the basic product and therefore cannot be completely removed from the final product.

HPLC can conveniently interface with various on-line detection techniques that are used to direct fraction collecting. The most common detection interfaces are the ultraviolet (UV) detector,[1,9] the evaporative light-scattering detector (ELSD), and the mass spectrometer (MS). Both UV and ELSD generate an intense analog signal over time. An intensity threshold is set, and the fraction collector is triggered to start collecting once the signal intensity exceeds the threshold. Neither method can distinguish products from impurities, and therefore all substances with certain concentration are collected. A follow-up analysis, most likely flow injection

MS, must be performed to determine the product location. In contrast, mass spectrometers are better triggering tools for compound specific fraction collection. In the select ion monitor (SIM) mode the mass spectrometer can selectively trigger fraction collection when the specific mass-to-charge ratio that corresponds to the molecular ion of the desired product, leaving impurities of different molecular weight uncollected.

10.2.4. High-Throughput Purification

Semipreparative HPLC is the most popular method for purifying combinatorial libraries. This is largely due to the relatively high resolution of HPLC, the ease with which HPLC instruments can be interfaced with automatic sampling and fraction collecting devices for unattended operation, and the possibility to develop a "generic" method for a whole library or even many libraries.

Zeng and co-workers assembled an automated "prepLCMS" system[18] using MS-triggering technique to collect fractions. Among the 12 samples tested, the average purity improved from about 30% to over 90%. Two switching valves allowed the system to select either analytical or preparative applications. Based on a similar principle, several commercial MS-triggered systems are now available.

Although the MS-triggered purification has advantages, mass spectrometry is a destructive detection method, and it can only be used in conjunction with a flow-splitting scheme. Flow splitting has negative effect on chromatography: the signals are delayed, and peaks can be distorted. The nondestructive UV detector, on the other hand, can be used in-line between HPLC column and fraction collector to record real peak shapes in real time. Ideally the fraction triggering must take advantage of both MS selectivity and UV real peak shape reporting.

Efforts that focus on parallel processing to accelerate the process have been made by various groups. The high-throughput preparative HPLC system with four parallel channels, commercially known as Parallex,[12] is based on UV-triggered fraction collection. A postpurification process is used to identify the product location. The sheath dual sprayer interface doubles the capacity of the MS-triggered system. However, the samples for two channel must be of different molecular weights for the system to be able to distinguish between the two sprayers.[19] Recently a four-channel MUX technology[20] was used and provided rapid switching to sample four HPLC channels for parallel purification.

Our group has established a high-throughput purification system based on the UV-triggered fraction collection technique. High-throughput parallel LC/MS technology is the foundation of our system due to its capacity to

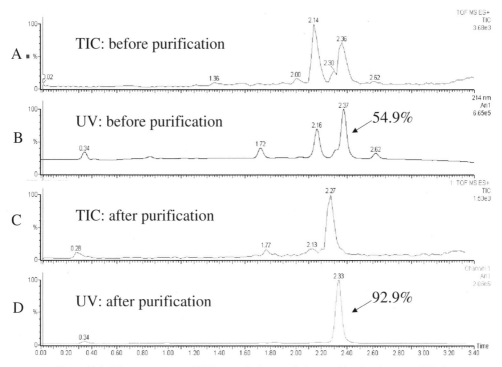

Figure 10.2. Mass spectro- and UV traces before and after purification by prep-HPLC.

provide LC/MS results for all samples before and after purification. The UV-triggered purification process is based on a prediction of the preparative retention time from the analytical retention time. An example of one sample in this pyrrolopyrrole library is shown in Figure 10.2.

As Figure 10.2 shows, before purification, there are four major chromatographic components in the UV trace (Figure 10.2*B*), and the purity of the desired compound is 54.9%. After purification, impurities at 1.7 and 2.1 are significantly reduced, while the one at 2.6 is eliminated as well as the front shoulder of the target (Figure 10.2*A*).

Figure 10.3 shows the purification results of this three diversity pyrrolopyrrole library. Figure 10.3*A* shows the purity distribution of samples before purification. Each sample was dissolved in 800 μL of DMSO, before it was loaded on a 50 × 21.2 mm C-18 HPLC column. A binary gradient of water and acetonitrile with 0.05% TFA as modifier was used to elute the samples at a flowrate of 24.9 mL/min. Fractions were collected based on peak height of UV signal. In case of multiple fractions, computer software was used to pick the fraction using a predictive calculation based

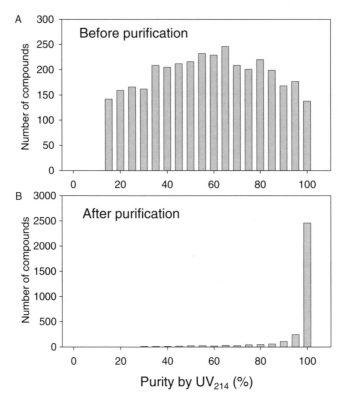

Figure 10.3. Purity distributions of a pyrrolopyrrole library (*A*) before and (*B*) after purification by prep-HPLC. This figure is a summary of high-throughput parallel LC/MS results. LC/MS was carried out using a MUX-LCT LC/MS system with eight parallel channels.

on analytical data or manual picking based on the peak eluting order and relative height in accordance with the analytical run.

Figure 10.3*B* gives the purity distribution of the same samples after purification. About 77% of samples, 2707 out of 3490, had purity higher than 90% and a reasonable weight recovery. The rest of samples failed due to three good reasons:

1. Early eluting. Under the chromatographic condition, samples with strong basic side chain eluted at solvent front along with DMSO, and were not collected.

2. Co-eluting impurity. Our prep HPLC method was of lower resolution than that of analytical. Impurities eluting closely to the target compound may get collected in the same fraction.

3. Picking fraction. Incorrect fractions were picked by the software. This was mainly due to the limited analytical capacity for postpurification LC/MS analysis. For each sample purified, only about 1 to 2 fractions were selected for LC/MS analysis.

10.3. NORMAL-PHASE PREPARATIVE HPLC

The big advantage of normal-phase LC is that the solvents are easily evaporated. Therefore the process of sample recovery is less time-consuming and the degradation of the purified products is minimal. Recently there has been reported an automation of the normal-phase preparative HPLC.[21]

The most popular stationary phases applied in normal phase preparative HPLC are alumina, silica, nitril-bonded silica, aminopropyl-bonded silica, or diol-bonded silica. Although the stationary phase doesn't vary very much in the separation application, the variation of the mobile phase is vast, ranging from the most polar solvent (e.g., water) to the most nonpolar solvent (e.g., pentane). Therefore the advantages of normal-phase over reverse-phase separation include (1) ability to provide wide range of solvent strength to increase the selectivity of the separation, (2) high solubility of lipophilic compounds in the organic solvents that allows higher loading capacity for purification, and (3) high loads of hydrophobic impurities that can be easily washed out by nonpolar solvents. For a separation in a certain stationary phase, the selection of the mobile phase strength can be theoretically predicted by Snyder theory for binary and ternary solvent systems.[22,23]

The purification of pneumocandin analogues, a class of natural lipopeptides with antifungal activity, has been carried out using normal phase preparative HPLC.[24] The silica column with a ternary solvent system of EtOAc–MeOH–H_2O gave a better resolution between the desire product B and the impurities A and C than reverse-phase HPLC. The product solubility and the resolution are affected by the percentage of MeOH and H_2O in the mobile phase. The optimum solvent composition is 84:9:7 of EtOAc:MeOH:H_2O. The recovery yield of B is a function of the flow rate and the sample loading. There is a trade-off between the recovery and the purification speed. The purification speed is increased fourfold when achieving an 82% recovery compared with achieving a 100% recovery.

The effect of the mobile phase temperature on the recovery and purification speed of the desired product was also studied.[25] Raising the temperature of the mobile phase from 25°C to 55°C increased the loading capacity of the sample. However, the resolution decreased with the overloaded column, and the recovery of the product decreased from 80–85% to 75% with the elevated temperature.

As in reverse-phase preparative HPLC, the fast gradient mode can also be applied in the normal-phase preparative HPLC. With use of a short column such $125 \times 25 \, mm$ or $125 \times 50 \, mm$ and a high flow rate, the separation of eight organic molecules can be achieved in 12 minutes.[21]

10.4. LLE AND SLE

Liquid–liquid extraction (LLE) is based on a simple principle that a compound will be partitioned between two immiscible solvents with concentration at a distribution ratio proportional to its solubility in each of the solvents. LLE is a common method of working up organic reaction mixtures. A conventional LLE application is to separate compounds between water and an organic solvent such as diethyl ether, ethyl acetate, or methylene chloride. Acidic or basic buffers are often used to control the distribution ratio of a certain substance.

However, conventional LLE requires precise removal of the aqueous layer, which is not amenable to large number of samples. To solve this problem, solid supported liquid-liquid extraction (SLE) was developed.[26] Instead of using separation funnels, the reaction mixture is loaded on a cartridge packed with diatomaceous earth, which is pretreated with an aqueous buffer and contains an aqueous layer. A water-immiscible solvent, usually methylene chloride or ethyl acetate, is then applied to elute the products off the cartridge, leaving more water-soluble impurities on the column.

Like conventional LLE, SLE is also based on partitioning of compounds between two liquid phases. Hydrophilic amines ($c \log P < 3.1$) were removed with an acid buffer of 1N HCl, but most hydrophobic amines ($C \log P > 3.1$) were retained (Table 10.1). All acids were removed with a basic buffer of 1N NaOH. Since all the acids used in this study were hydrophilic ($c \log P < 3.1$), it remains unclear how hydrophobic organic acid would respond to basic SLE. One would expect a $c \log P$ threshold for acidic compounds as well.

One distinctive advantage for SLE over the more traditional LLE is its ease of automation and parallel processing. Diatomaceous earth can be packed into a 96-deep-well filter plate. The plate is frozen to prevent leaking during the transfer. The eluent was directly collected to a 96-well microtiter plate. With the help of robotic liquid handlers, the SLE process can be automated with a throughput of four plates per hour.[27] Since the aqueous phase is immobilized on the cartridge, any water-immiscible organic solvent can be used regardless of its density. The elution is usually driven by gravity, while a slight negative pressure can facilitate this process.

The introduction of a third phase—fluorous phase—is an effective way to purify compounds with a fluorous tag.[28] This method can remove both hydrophilic and organic impurities from the reaction mixture. Studies have shown that only the desired product is taken by the fluorous solvent while impurities remain in organic and water phases. The fluorous tag can be removed later by a simple reaction such as desilylation. The purity of all final products is higher than 95%.

LLE and SLE also suffer from some limitations. There are often hydrophobic by-products, such as that from an incomplete removal of a protecting group, in combinatorial samples. These impurities will not be removed by SLE. This will affect the product purity. On the other hand, hydrophilic samples with low log Ps may get lost during the process.

10.5. SOLID-PHASE EXTRACTION

Solid-phase extraction (SPE) is the method of sample preparation that concentrates and purifies analytes from solution by sorption onto a disposable solid-phase cartridge, followed by elution of the analyte with an appropriate solvent. The SPE technique was developed in the mid-1970s as an alternative means of liquid-liquid extraction[29] but become particularly attractive for its automation, parallel purification, and pre-concentration. Since 1995, SPE has been applied in various fields, environmental, food sciences, biomedical analyses, pharmaceutical analyses, and organic synthesis.[30–34] There are a numbers of publications and reviews on the subjects of development of new solid-phase supporting materials,[26,35] instrumentation and device,[37] techniques,[38–40] and theoretical aspect.[41]

In general, the procedures of SPE consists of the following four steps:

- Conditioning the sorbent by passing through the column with a small volume of the appropriate solvent to solvate the functional groups on the surface of the sorbents. The sorbent can also be cleaned at this point to remove any impurities present in the sorbent.
- The liquid sample is applied to the column with the aid of a gentle vacuum. The interested analyte and some interfering impurities will retain in the sorbent. In this retention step the analyte is concentrated on the sorbent.
- Rinse the column with some mixed solvents to remove the interfering impurities, and let the interested analyte retain on the sorbent.

• Elute the analyte completely from the sorbent by an appropriate solvent.

Depending on the mechanism of the interaction between the analyte and the sorbent, the SPE can be classified into three modes: reversed-phase SPE, normal-phase SPE, and ion-exchanged SPE. Like liquid chromatography, the sorbents used in the reverse-phase SPE are more hydrophobic, more hydrophilic in the normal-phase SPE, and ionic in ion-exchange SPE. Unlike HPLC, where the analyte is eluted continuously with mobile phase, and collected when the detected signal appears, the analytes collected in SPE process have no monitoring signal. Therefore, no matter what kind of mechanism, the retention of the interested analytes on the sorbents has to be very specific and selective. A limitation of SPE in high-throughput purification of combinatorial libraries is the carryover of impurities with similar chemical properties. The selection of solid sorbents and the elution solvents will largely determine the recovery and purity of the desired products. The effects of sorbent properties and the elution solvents on the extraction efficiency for different classes of molecules with different modes of SPE will be discussed in the sections below.

10.5.1. Reverse-Phase SPE

The reverse-phase SPE involves the partitioning of organic solutes from a polar mobile phase, such as water, into a nonpolar solid phase, such as the C-18 or C-8 sorbent. The interaction between solute and sorbent involves van de Waals and dispersion forces. The specificity of the extraction depends on the difference in chemical potential or the solubility of the solutes between the two phases.

The most common reverse-phase sorbent used in SPE is C-18 silica with various particle sizes (40–70 μm) and pore sizes (55–125 Å). Other reverse-phase sorbents include C-8, C-4, C-2, cyclohexyl, and phenyl-bonded silica, as well as polymeric sorbents such as polystyrene-divinylbenzene (PS-DVB) and graphitized carbon.

For organic samples, there is a good correlation between the retention on C-18 silica sorbent and the octanol-water partition coefficient (log P) of the analytes. The more hydrophobic the analytes are, the higher the retention factor and the extraction recovery. For nonpolar or moderately polar analytes with log P values higher than 3, the extraction recovery can reach >95%.[40]

The effects of various solid supports on the extraction recovery for a set of polar carbamates have been studied, and the results are shown in Table 10.3. It is apparent that the extraction efficiency is lower by using C-18/OH

**Table 10.3. Comparison of Recoveries for Polar
Carbamates with Different Extraction Sorbents**

Compound	Recovery (%)			
	(1)	(2)	(3)	(4)
Aldicarb sulfone	9	18	16	29
Oxamyl	12	25	22	49
Methomyl	9	19	16	39
Aldicarb	50	102	97	89
Carbofuran	102	98	106	94

Note: Sample volume 25 mL, extraction column size: 10 × 2 mm. (1) C-18/OH from Varian, (2) standard C-18 from J. T. Baker, (3) standard C-18 from Varian, (4) PLRP-S PS-DVB from Polymer Labs.

sorbent. The OH- group in the C-18/OH sorbent can introduce the secondary H-bonding interaction with the polar carbamate, which lowers the primary hydrophobic interaction between the majority C-18 and carbamate. Therefore the total recovery is lowered by the net effect of weaker binding interaction between the sorbent and the compound. The PS-DVB polymeric sorbent gives highest recovery values due to the larger specific areas and its high carbon content (~90%).

Although C-18 silica having high carbon content provides the stronger retention for hydrophobic analytes, it also traps more interference. Other silica-based sorbents, like C-2, can extract the highly hydrophobic analytes more specifically.

The most widely used carbon-based SPE sorbent graphitized carbon black (GCB) with specific surface areas up to $210 \, m^2/g$. Applications of the GCB sorbent in SPE have been extensively studied for polar pesticides in water.[36] Table 10.4 shows the results of recovery values for extraction of 2 L of water samples using 1 g of GCB comparing with recoveries using C-18 sorbent and liquid–liquid extraction (LLE) with methylene chloride. The recoveries for GCB reach 90% to 100% for most of the compounds.

Chemically modified polymeric sorbents have also been introduced in the recently years.[35] The introduction of polar groups such as alcohol, acetyl, and the sulfonate group into PS-DVB greatly increases the retention of polar organic compounds. The comparative recovery study was performed for extraction of polar compounds such as phenols, aromatic, and pyridinic compounds with three types of PS-DVB-based sorbents and C-18 silica; the results are shown in Table 10.5.[42] The recoveries by using PS–DVB–CH_3OH

Table 10.4. LLE and SPE Recovery Data for Extraction of Polar Pesticides in 2 L Water

Compound	Recovery (%) ± RSD (%)		
	LLE	C-18	GCB
Omethoate	58 ± 8	3 ± 45	83 ± 6
Butocarboxim sulfoxide	13 ± 14	3 ± 42	102 ± 5
Aldicarb sulfoxide	16 ± 17	4 ± 29	93 ± 5
Butoxycarboxim	74 ± 7	4 ± 32	98 ± 3
Aldicarb sulfone	58 ± 11	6 ± 21	75 ± 8
Oxamyl	51 ± 10	24 ± 12	101 ± 2
Methomyl	64 ± 11	10 ± 20	100 ± 2
Monocrotophos	68 ± 5	42 ± 16	98 ± 3
Deisopropylatrazine	87 ± 4	15 ± 14	102 ± 4
Fenuron	60 ± 8	19 ± 12	99 ± 3
Metamitron	79 ± 4	28 ± 12	95 ± 5
Isocarbamid	74 ± 10	78 ± 6	97 ± 3
Deethylatrazine	85 ± 4	30 ± 13	97 ± 4
Chloridazon	75 ± 4	31 ± 11	100 ± 3
Dimethoate	78 ± 6	22 ± 14	98 ± 4
Cymoxanil	89 ± 9	28 ± 11	94 ± 4
Butocarboxim	82 ± 5	63 ± 9	95 ± 4
Aldicarb	68 ± 12	55 ± 9	99 ± 4
Metoxuron	83 ± 5	101 ± 3	97 ± 3
Hexazinone	75 ± 11	88 ± 4	98 ± 3

Table 10.5. SPE Recovery of Phenols and Aromatic Compounds

Compound	Recovery (%)			
	C-18	PS-DVB	PS-DVB-CH$_3$OH	PS-DVB-COCH$_3$
Phenol	6	91	94	100
p-Cresol	16	91	98	101
Anisole	78	91	94	98
Nitrobenzene	54	92	96	100

and PS–DVB–COCH$_3$ are slightly higher than that using PS–DVB, and significantly higher than those for C-18 silica sorbent.

The high recovery of PS-DVB for these aromatic compounds could be due to additional strong π-π interaction between the analytes and phenyl group in the polymeric sorbents besides the hydrophobic interaction. The lightly sulfonated PS-DVB sorbent (5–8 μm and 400 m^2/g) displays excellent

Table 10.6. Comparison of Extraction Recovery of Sulfonated PS-DVB and Unsulfonated PS-DVB Sorbent

Compound	Recovered Sulfonated PS-DVB (%)		Recovered PS-DVB (%)	
	Not Wetted	Wetted	Not Wetted	Wetted
Anisole	94	96	83	89
Benzaldehyde	90	89	87	96
Nitrobenzene	96	95	88	96
Hexylacetate	94	94	84	82
Benzylalcohol	90	98	78	81
Phenol	98	95	77	89
Catechol	59	34	ND	ND
m-Nitrophenol	98	99	89	95
Mesityl oxide	98	97	93	99
tert-2-Hexenyl acetate	93	90	79	89

Note: Wetting solvent is methanol; data are the average of three runs; ND = not detectable.

hydrophilicity and improved extraction efficiency for polar organic compounds over underivatized PS-DVB.[35] Table 10.6 shows the comparison of recovery of several analytes on sulfonated and unsulfonated PS-DVB cartridge. The data indicate that it is not necessary to pre-treat the sorbent before applying the sample with use of sulfonated PS-DVB sorbent.

10.5.2. Normal-Phase SPE

Normal-phase solid-phase extraction refers to the mechanism by which the analyte is adsorbed onto the polar surface of sorbent from a nonpolar solvent. The mechanism of interaction between the analyte and sorbent is a polar interaction, such as hydrogen bonding, dipole-dipole interaction, π-π interaction, and induced dipole-dipole interaction. The sorbents widely used in normal-phase SPE are silica, alumina, and magnesium silicate (Florisil), and the silica chemically modified with polar groups like amino, cyano, or diol. The samples for normal-phase SPE are typically dissolved in hexane or isooctane. Step elution with solvents of increasing polarity allows the separation into fractions on the basis of difference in polarity.

Normal-phase SPE has been in the purification of a library of N-alkylated L-amino acids.[30,43] In the synthesis of this library, the final N-alkylated L-amino acid products were usually contaminated with small amount of alcohols. These alcohols are less polar than N-alkylated L-amino acids and could be removed by SPE using silica cartridge, washing with 9:1

Table 10.7. Solvent Eluotropic Strength ($E^\varnothing$) and polarity P'

Solvent	$E^\varnothing$	P'
Acetic acid, glacial	>0.73	6.2
Water	>0.73	10.2
Methanol	0.73	6.6
2-Propanol	0.63	4.3
Pyridine	0.55	5.3
Acetonitrile	0.50	6.2
Ethyl acetate	0.45	4.3
Acetone	0.43	5.4
Methylene chloride	0.32	3.4
Chloroform	0.31	4.4
Toluene	0.22	2.4
Cyclohexane	0.03	0.00
Hexane	0.00	0.06

CH_2Cl_2/MeOH. The final products can be eluted by 65:35:5 of CH_2Cl_2/MeOH/H_2O.

The elution of the analytes from a normal-phase sorbent is a function of the eluotropic strength ($E^\varnothing$) of the solvent.[31] Table 10.7 shows the values of eluotropic strength and polarity of the organic solvents used in normal-phase SPE. The compounds are usually dissolved in the solvents with $E^\varnothing$ values less than 0.38 for silica sorbent, and eluted with solvents of $E^\varnothing$ values greater than 0.6.

10.5.3. Ion Exchange SPE

Ion exchange mechanism is a fundamental form of chromatography that was invented in 1935 by Adams and Holmes.[44] The synthetic polymers or resins that contain ionizable groups are capable of exchanging ions in the solutions. The ionizable groups are either acidic or basic. The applications of ion exchange SPE include conversion of salt solutions from one form to another, desalting a solution, trace enrichment, and removal of ionic impurities or interferences.

The sorbents used in ion exchange SPE are PS-DVB based or silica based with bonding of different ionizable functional groups such as sulfonic acid, carboxylic acid for cation exchange, or aminopropyl group for anion exchange.

Ion exchange SPE has been frequently used as a purification method for solution-phase combinatorial chemistry.[30] In the report for synthesis of

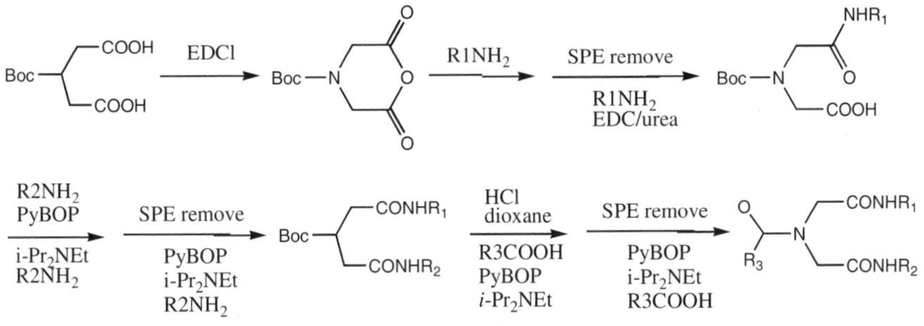

Scheme 10.1

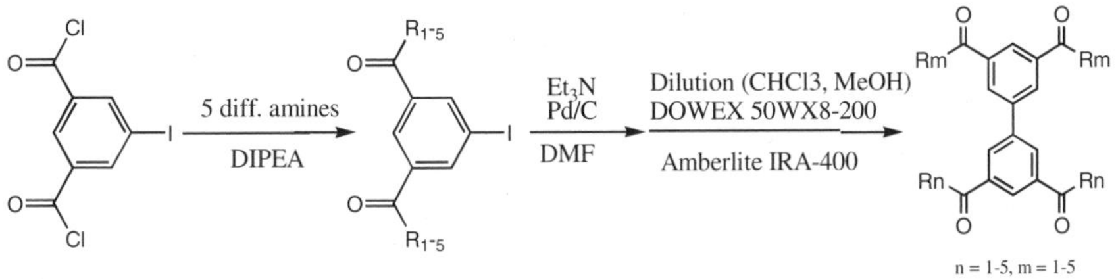

n = 1-5, m = 1-5

Scheme 10.2

dipeptidomimetics libraries, shown as Scheme 10.1,[45] ion exchange SPE was employed to remove excess reagents in each step of the reactions. Each final library product was purified in a 30 to 150 mg scale with a purity >90%.

In the library of biarys, as shown in Scheme 10.2,[30] the strong acidic Dowex 50W-X8-200 and strong basic Amberlite IRA-400 sorbents are added simultaneously to the crude reaction mixtures to remove the triethylamine and hydrogen iodide, and give the pure biaryls in 75% to 95% yields.

Examples of automatic purification of the library product using ion exchange SPE have been reported by Lawrence et al.[46] As shown in reaction Scheme 10.3, the strong cation exchange cartridge (SCX from Varian) was used to extract the final products and gave pure amides (88–98% HPLC purity) in 70% to 95% yields.

Gayo and Suto reported the condition optimization for purification of amide/ester library in 96-well format,[47] as shown in Scheme 10.4, the weakly basic Amberlite IRA-68 sorbent and EtOAc as elution solvent for extraction provided the highest yield (84–100%) and purity (98–99%) of the products.

A 96-well format SPE process for purifying carboxylic acids was developed by Bookser and Zhu.[48] The anion exchange resin Dowex 1×8–400

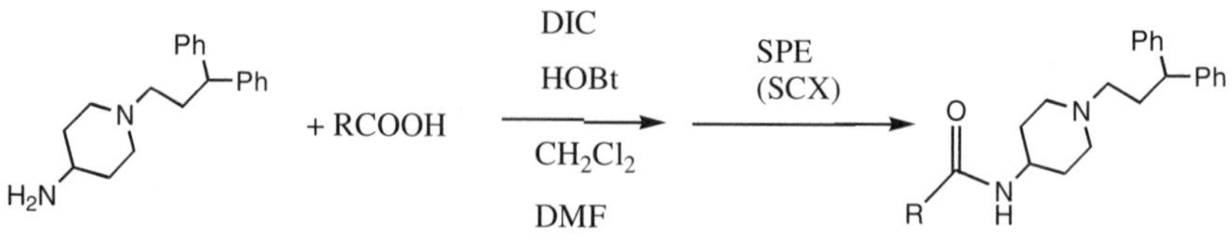

Scheme 10.3

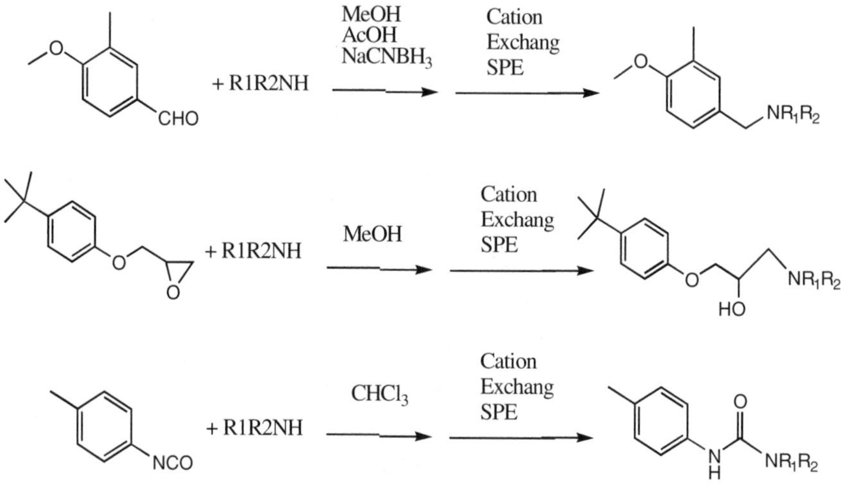

Scheme 10.4

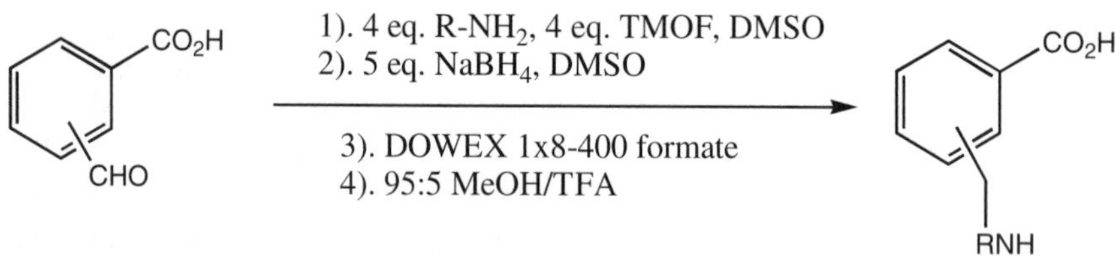

1). 4 eq. R-NH$_2$, 4 eq. TMOF, DMSO
2). 5 eq. NaBH$_4$, DMSO

3). DOWEX 1x8-400 formate
4). 95:5 MeOH/TFA

Scheme 10.5

formate was used to capture carboxylic acids from the reaction mixture. As shown in Scheme 10.5, resin, pretreated with formic acid, was allowed to exchange with organic acids. Volatile formic acid was released in the process, while carboxylic acids formed ammonium salt with the resin. Methanol was then used to remove impurities, and the 95:5 methanol/formic acid mixture was then used to recover carboxylic acids with an average purity of 89% (Scheme 10.6). The purity and yields the final

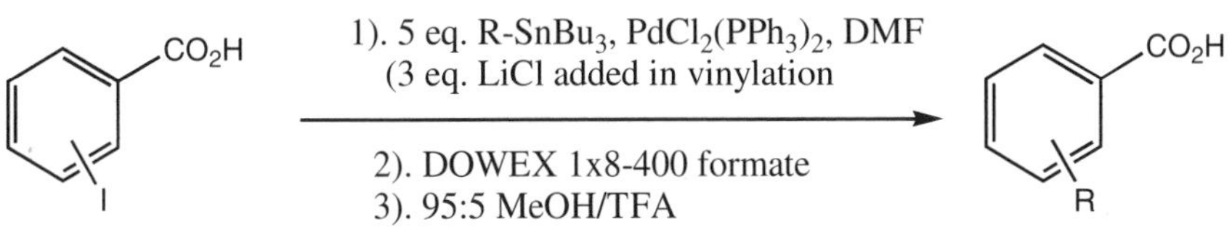

1). 5 eq. R-SnBu$_3$, PdCl$_2$(PPh$_3$)$_2$, DMF
(3 eq. LiCl added in vinylation

2). DOWEX 1x8-400 formate
3). 95:5 MeOH/TFA

Scheme 10.6

products from reductive amination and after SPE purification are 38–98% and 23–88%, respectively. The purity and yields for Stille coupling are 91–98% and 22–61%, respectively. The extraction efficiency is pK$_a$ dependent. An acid with pK$_a$ lower than the conjugate acid of the anion on the resin can be effectively exchanged in acid or salt form.

10.6.   COUNTERCURRENT CHROMATOGRAPHY

Countercurrent chromatography (CCC), also known as centrifugal partition chromatography (CPC) is the support-free liquid–liquid chromatography. The principles of separation is based on the selective partition of the samples between two immiscible liquid phase, which eliminates the irreversible adsorption of the sample onto the solid support like in other preparative LC process and gives a higher recovery yield. High-speed countercurrent chromatography has been applied in the preparative separation of both natural and synthetic products.[49–55]

The selection of the solvent systems is important to achieve the goals of separation in CCC process. The criteria for choosing a solvent system are the polarity of the samples and its solubility, hydrophobicity, charge state, and ability to form complexes. The strategies for solvent optimization have been comprehensively reviewed by Foucault et al.[52,56] In general, the sample is dissolved in a "best solvent," and this "best solvent" partition into two other solvents to build a biphasic system. Table 10.8 gives some samples of the "best solvents" and two other more or less polar solvents.[52]

The countercurrent chromatography is used in the separation and purification of the natural products of notopterol and isoimperatorin from notopterygium forbessi Boiss, a Chinese medicinal herb used as an antifebrile and anodyne.[53] The stepwise elution of two solvent systems, 5:5:4.8:5 and 5:5:5:4 of light petroleum–EtOAc–MeOH–water, respectively, was employed. The separation process took several hours and gave pure fractions of notopterol and isoimperatorin with the purity of ≥98%.

Countercurrent chromatography has also been applied in purification of antibiotic analogues.[54–55] The comparison of product purity and yield

Table 10.8. Common Solvents for CCC Separation

Less Polar Solvent	Best Solvent	More Polar Solvent
Heptane, CHCl₃	THF	Water
Heptane, CHCl₃, toluene, MiBK, EtOAc	acetone	Water
Heptane	Methyl ethyl ketone	Water
THF	DMSO	Water
Toluene, MtBE, MiBK, EtOAc	MeCN	Water
Heptane, toluene, EtOAc, CHCl₃	BuOH	Water
Heptane, toluene, EtOAc, CHCl₃	PrOH	Water
Heptane, EtOAc, CHCl₃	EtOH	Water
Heptane, toluene, EtOAc, BuOH, CHCl₃	MeOH	Water
Heptane, toluene, EtOAc, BuOH, MiBK, CHCl₃	HOAc	Water
CHCl₃	HCOOH	Water
Nonaqueous system		
Heptane	THF, EtOAc, PrOH, EtOH, DMF	MeOH, MeCN

obtained by using CCC and using semipreparative HPLC has been studied.[54] For the same enrichment of the desired product, from purity of 25% to 95%, the hydrodynamic mode CCC gave 0.4 g/L final concentration, which is about three times higher than that of 0.15 g/L by preparative HPLC. In addition hydrodynamic mode CCC consequently consumes less solvent than preparative HPLC. Unlike preparative HPLC, CCC can handle the very dirty materials, and no preliminary purification of the crude is required.

10.7. PREPARATIVE THIN-LAYER CHROMATOGRAPHY (TLC)

Procedures of preparative TLC similar to those of analytical TLC have been routinely used in screenings of product purity in the chemistry lab. Preparative TLC can separate and isolate materials from 10 mg to more than 1 g. With respect to precision, accuracy, sensitivity, and recovery, preparative TLC appears to be equivalent to preparative HPLC.[14,57] Preparative TLC is faster and more convenient than column chromatography, and less expansive than preparative HPLC in terms of instrumentation. The supporting materials and mobile phase are similar to those used in HPLC; however, the solvent consumption in TLC is much lower. The disadvantages of preparative TLC are that the procedure is more time-consuming than preparative HPLC, and the same separation efficiency as preparative HPLC cannot easily be achieved.

10.8. SUPERCRITICAL-FLUID CHROMATOGRAPHY

The SFC technique is closely related to HPLC, using much the same kind of hardware but with compressed gas such as CO_2 as a major component in the mobile phase. Therefore the solvent volume of the purified fraction of the desired product is very small and easily removed, which increases the productivity significantly. More recently supercritical-fluid chromatography (SFC) has begun to show promise as a good technology for purification of the combinatorial library.[58] The technique and applications of SFC are reviewed by several authors.[59–64]

Coleman has described an HPLC system modified to make use of SFC by a fast gradient (7 minutes) with UV and ELSD for detection and quantification.[65] The product purity and recovery yield for small organic molecules reached higher than 99% and 95%, respectively.[58]

A major advantage of supercritical-fluid chromatography (SFC) is that it offers the advantage of liquid-like solubility, with the capability to use a nonselective gas-phase detector such as flame ionization detector. Other advantages of using supercritical fluids for extractions are that they are inexpensive, contaminant free, and less costly to dispose safely than organic solvents.

Because of its increased efficiency, preparative SFC is being used for separations that are difficult to effect by HPLC. But, to take advantage of the narrow peaks obtained in SFC, very little overloading can be done for these difficult separations. As a result the maximum amount of material obtained in a run is on the order of 100 mg in SFC compared with the 1-g amounts obtainable sometimes in HPLC.

Wang and co-workers have reported that a preparative SFC system can be interfaced with a single quadrupole mass spectrometer for mass-directed fraction collection.[66] Samples with no chromophore (Ginsenoside Rb, Ginsenoside Rc, and Ginsenoside Re) were isolated near homogeneity. A more sophisticated preparative SFC system was patented by Maiefski et al.[67] There are four parallel channels in this system, and there is a UV detector for each channel. Since the eluent can be also splitted into a mass spectrometer, this system is capable of both UV and MS directed purification.

10.9.   CONCLUSION AND OUTLOOK

Various purification methods were reviewed in this chapter. Both preparative HPLC and solid-phase exaction are commonly used to purify large numbers of compounds with high recovery and purity, as well as sufficient productivity. Solid-phase exaction method is especially suitable for parallel high-throughput purification of combinatorial libraries if there is enough knowledge about the properties of the desired products and possible impurities. Preparative HPLC is a good technique for purification of the sample mixtures with low reaction yield and unknown impurities. Countercurrent chromatography can offer the very large-scale purification with high recovery but very low throughput. Further developments in SFC technique should offer the potential of significant advances over conventional preparative HPLC for high-throughput purification.

REFERENCES

1. H. N. Weller, M. G. Young, S. J. Michalczyk, G. H. Reitnauer, R. S. Cooley, P. C. Rahn, D. J. Loyd, D. Fiore, S. J. Fischman, *Mol. Diversity 3*, 61 (1997).

2. L. Zeng, D. B. Kassel, *Anal. Chem.* **70**, 4380 (1998).

3. J. P. Kiplinger, R. O. Cole, S. Robinson, E. R. Roskamp, R. S. Ware, H. J. O'Connell, A. Braisford, J. Batt, *Rapid Commun. Mass Spectrom. 12*, 658 (1998).

4. L. Zeng, L. Burton, K. Yung, B. Shushan, D. B. Kassel, *J. Chromatogr. A794*, 3 (1998).

5. J. Chen, B. Liu, Y. Tzeng, *J. Chromatogr. A830*, 115 (1999).

6. T. Fornstedt, G. Guiochon, *Anal. Chem.* **73**, 609A (2001).

7. J. Zhu, C. Coscolluella, *J. Chromatogr. B741*, 55 (2000).

8. A. Kulik, H.-P. Fiedler, *J. Chromatogr. A812*, 117 (1998).

9. R. A. Tommasi, L. W. Whaley, H. R. Marepalli, *J. Comb. Chem. 2*, 447 (2000).

10. A. Felinger, G. Guiochon, *J. Chromatogr. A796*, 59 (1998).

11. B. Kim, A. Velayudhan, *J. Chromatogr. A796*, 195 (1998).

12. L. Schultz, C. D. Garr, L. M. Cameron, J. Bukowski, *Bioorg. Med. Chem. Lett. 8*, 2409 (1998).

13. H. N. Weller, *Mol. Diversity 4*, 47 (1999).

14. B. A. Bidlingmeyer, ed., *Preparative Liquid Chromatograph, J. Chromatogr. Series*, Vol. 38, Elsevier, New York (1987).

15. R. L. Fahrner, P. M. Lester, G. S. Blank, D. H. Reigsnyder, *J. Chromatogr. A827*, 37 (1998).

16. I. Hughes, D. Hunter, *Curr. Opin. in Chem. Bio. 5*, 243 (2001).

17. R. Cole, K. Laws, D. Hiller, J. Kiplinger, R. Ware, *Am. Lab. 7*, 15 (1998).

18. L. Zeng, X. Wang, D. Kassel, *Comb. Chem. High Through. Screen.* **1**, 101 (1998).

19. D. Kassel, R. Xu, T. Wang, A. Brailsford, B. Smith, *Proceedings of 49th ASMS Conference on Mass Spectrometry and Allied Topics* (2001).

20. V. de Biasi, N. Haskins, A. Organ, R. Bateman, K. Giles, S. Jarvis, *Rapid Commun. Mass Spectrom.* **13**, 1165 (1999).

21. P. Renold, E. Madero, T. Maetzke, *J. Chromatogr. A* **908**, 143 (2001).

22. C. E. Palamarev, V. R. Meyer, M. D. Palamareva, *J. Chromatogr. A* **848**, 1 (1999).

23. P. Jandera, L. Petranek, M. Kucerova, *J. Chromatogr. A* **791**, 1 (1997).

24. D. J. Roush, F. D. Antia, K. E. Goklen, *J. Chromatogr. A* **827**, 373 (1998).

25. A. E. Osawa, R. Sitrin, S. S. Lee, *J. Chromatogr. A* **831**, 217 (1999).

26. C. Johnson, B. Zhang, P. Fantauzzi, M. Hocker, K. Yager, *Tetrahedron* **54**, 4097 (1998).

27. S. Peng, C. Henson, M. Strojnowski, A. Golebiowski, S. Klopfenstein, *Anal. Chem.* **72**, 261 (2000).

28. A. Studer, S. Hadida, R. Ferritto, S.-Y. Kim, P. Jeger, P. Wipf, D. P. Curran, *Science* **275**, 823 (1997).

29. G. A. Junk, J. J. Richard, M. D. Grieser, D. Witiak, J. L. Witiak, M. D. Arguello, R. Vick, H. J. Svec, J. S. Fritz, G. V. Calder, *J. Chromatogr.* **99**, 745–762 (1974).

30. U. J. Nilsson, *J. Chromatogr. A* **885**, 305 (2000).

31. E. M. Thurman, M. S. Mills, *Solid-Phase Extraction, Principles and Practice*, Wiley, New York (1998).

32. V. Pichon, *J. Chromatogr. A* **885**, 195 (2000).

33. J. S. Fritz, M. Macka, *J. Chromatogr. A* **902**, 137 (2000).

34. S. Ulrich, *J. Chromatogr. A* **902**, 167 (2000).

35. C. W. Huck, G. K. Bonn, *J. Chromatogr. A* **885**, 51 (2000).

36. M.-C. Hennion, *J. Chromatogr. A* **885**, 73 (2000).

37. D. T. Rossi, N. Zhang, *J. Chromatogr. A* **885**, 97 (2000).

38. M. Bouzige, V. Pichon, M.-C. Hennion, *J. Chromatogr. A* **823**, 197 (1998).

39. H. Lord, J. Pawliszyn, *J. Chromatogr. A* **885**, 153 (2000).

40. M.-C. Hennion, *J. Chromatogr. A* **856**, 3 (1999).

41. C. F. Poole, A. D. Gunatilleka, R. Sethuraman, *J. Chromatogr. A* **885**, 17 (2000).

42. J. J. Sun, J. S. Fritz, *J. Chromatogr.* **590**, 197 (1992).

43. U. J. Nilsson, E. J.-L. Fournier, O. Hindsgaul, *Bioorg. Med. Chem.* **6**, 1563 (1998).

44. B. A. Adams, E. L. Holmes, *J. Soc. Chem. Ind.* **54** (1) (1936). British Patents 450308–9.

45. S. Cheng, D. D. Comer, J. P. Williams, P. L. Myers, D. L. Boger, *J. Am. Chem. Soc.* **118**, 2567 (1996).

46. R. M. Lawrence, S. A. Biller, O. M. Fryszman, M. A. Poss, *Synthesis* 553 (1997).

47. L. M. Gayo, M. J. Suto, *Tetrahedron Lett.* **38**, 513 (1997).

48. B. Bookser, S. Zhu, *J. Comb. Chem.* **3**, 205 (2001).

49. K. Hostettmann, M. Hostettmann, A. Marston, *Preparative Chromatography Techniques: Application in Natural Product Isolation*, Springer Verlag, Berlin (1986).

50. W. D. Conway, *Countercurrent Chromatography—Apparatus, Theory and Applications*, VCH, New York (1990).

51. A. Marston, K. Hostettmann, *J. Chromatogr.* A*658*, 315 (1994).

52. A. P. Foucault, L. Chevolot, *J. Chromatogr.* A*808*, 3 (1998).

53. F. Yang, T. Zhang, Q. Liu, G. Xu, Y. Zhang, S. Zhang, Y. Ito, *J. Chromatogr.* A*883*, 67 (2000).

54. M.-C. Menet, D. Thiebaut, *J. Chromatogr.* A*831*, 203 (1999).

55. W. Wang-Fan, E. Küsters, O. Lohse, C. Mak, Y. Wang, *J. Chromatogr.* A*864*, 69 (1999).

56. A. P. Foucault, ed., *Centrifugal Partition Chromatography*, Chromatographic Science Series, vol. 68, Marcel Dekker, New York (1994).

57. L. Lin, J. Zhang, P. Wang, Y. Wang, J. Chen, *J. Chromatogr.* A*815*, 3 (1998).

58. T. A. Berger, K. Fogleman, T. Staats, P. Bente, I. Crocket, W. Farrell, M. Osonubi, *J. Biochem. Biophys. Methods* *43*, 87 (2000).

59. M. Caude, D. Thiebaut, ed., *Practical Supercritical-Fluid Chromatography and Extraction*, Hardwood, Amsterdam, 397 (1999).

60. C. Berger, M. Perrut, *J. Chromatogr.* *505* (1), 37 (1990).

61. K. Anton, C. Berger, *Supercritical-Fluid Chromatography with Packed Columns: Techniques and Applications*, Marcel Dekker, New York, 403 (1998).

62. K. Anton, C. Berger, *Supercritical-Fluid Chromatography with Packed Columns: Techniques and Applications*, Marcel Dekker, New York, 429 (1998).

63. T. A. Berger, *J. High Resolut. Chromatogr.* *14*, 312 (1991).

64. T. A. Berger, *J. Chromatogr.* A*785*, 3 (1997).

65. K. Coleman, *Analysis*, *27*, 719 (1999).

66. T. Wang, I. Hardt, D. Kassel, L. Zeng, *Proceedings of 49th ASMS Conference on Mass Spectrometry and Allied Topics*, Chicago (2001).

67. R. Maiefski, D. Wendell, C. Ripka, D. Krakover, US Patent (2001) No. US *6*,309,541.

CHAPTER

11

HIGH-THROUGHPUT PURIFICATION: TRIAGE AND OPTIMIZATION

JILL HOCHLOWSKI

11.1. INTRODUCTION

11.1.1. Background

It is generally desirable to achieve purity above a certain level for compounds synthesized in any format prior to high-throughput screening or further biological assay. Postsynthesis purification achieves several assay objectives. A high level of impurity, of either inactive contaminants or related components, can cause inaccuracy in the determination of IC50 results or toxicity evaluations, and lead to incorrect structure-activity relationship (SAR) conclusions. Impurities in a reaction mixture can cause false positive and false negative HTS assay results through synergy or antagonism effects. Therefore it is established procedure at Abbott Laboratories to purify all synthetically generated structures via high-performance chromatography methods with an objective to screen and biologically evaluate compounds only at a purity level at or above 95%.

11.1.2. Nonchromatographic High-Throughput Purification Methods

Many pharmaceutical companies possess high-throughput purification facilities, whose methodology can take the form of either a preliminary reaction cleanup or a high-performance technology. One such former technique is liquid–liquid extraction (LLE) as has been employed by workers at Procter and Gamble[1] for the purification of their combinatorial libraries. This automated system partitions reaction mixtures between an aqueous and a water-immiscible pair of solvents in a microtiter format, and provides a typical 90% pure product with >85% recovery of desired material. Liquid–liquid extraction can also be accomplished by employing a solid

Analysis and Purification Methods in Combinatorial Chemistry, Edited by Bing Yan.
ISBN 0-471-26929-8   Copyright © 2004 by John Wiley & Sons, Inc.

support for retention of the aqueous component as a "stationary phase" to facilitate automation. A parallel format solid-supported liquid–liquid extraction (SLE) technique has been employed at Johnson Pharmaceutical Research Institute[2] for the rapid purification of compound libraries. Organic acid and organic base contaminants can be effectively removed by this technique, which involves supporting an aqueous buffer stationary phase absorbed on a diatomateous earth solid support, subsequent application of the reaction mixture, followed by organic mobile phase elution. This SLE technique was first reported[3] from Arris Pharmaceutical, in an automated high-throughput format, for the purification a library of N-alkylaminoheterocycles.

Solid-phase scavenger methods are employed with increasing frequency as a preliminary reaction cleanup step in combinatorial chemistry, and have recently become commercially available (Argonaut, Calbiochem-Novabiochem, Varian, Alltech). Lilly researchers first reported on this approach,[4] employing solid supported electrophiles and nucleophiles for reaction purification in acylation and alkylation reactions. Yield and purity values reported were 90–95% and 50–99%, respectively, for a library generated by reductive amination. Parke-Davis researchers[5] achieved the removal of known reaction product impurities by the application of custom synthesized polymer supported reagents, specifically polystyrene-divinylbenzene supported derivatives of methylisocyanate and *tris*(2-aminomethyl)amine for cleanup of by-products resulting from urea, thiourea, sulfonamide, amide, and pyrazole libraries.

Solid-phase extraction (SPE) is a technique similar to solid-phase scavenger cleanup, but differs in that the impurities removed are not covalently bound to the resin as is most frequently the case with scavenger techniques. SPE packing materials are commercially available or easily setup in various formats, including cartridges or microtiter plates, each of which is amenable to facile automation. Researchers at Metabasis have reported[6] a detailed SPE investigation into the applicability of 11 different ion exchange resins for the efficacy of recovering carboxylic acids from parallel format syntheses. A microtiter well batch-wise format was employed throughout, and demonstrated achievable purity levels up to 98% by this capture-release process, with recovery yields typically on the order of 50% to 80%.

The above-mentioned techniques all constitute methodologies that can routinely achieve only preliminary reaction cleanup or the removal of library specific components, although in certain cases giving quite good results as far as achievable yield and purity. By contrast, high-performance chromatographies, specifically high-performance liquid chromatography (HPLC) and supercritical fluid chromatography (SFC) are more generally amenable to the purification of a wide range of structural types and routinely provide compounds of 95% or better purity. These same high-performance technologies are easily and routinely automated. Many companies have such high-performance, high-throughput purification capabilities, in either HPLC or SFC and in either one of two instrumentation access formats: as a service function or as open access systems.

11.1.3. HPLC Purification Methods

High-throughput HPLC purification systems described to date all rely on a detection methodology to recognize the elution of a peak of interest and, based on this signal, initiate the collection of eluting materials. Detection methods employed include ultraviolet (UV) spectroscopy, evaporative light-scattering detection (ELSD), and mass spectrometry (MS). These HPLC systems are almost universally reversed phase systems, PR-LC having been found to be the technique most widely amenable to the purification of druglike compounds. The first such high-throughput system was described by Weller et al. at Bristol-Meyers Squibb Pharmaceuticals.[7] This system was an open-access preparative HPLC system based on UV-threshold triggered fraction collection, and it was capable of unattended operation. Fast flow rates, short columns, and rapid universal reverse-phase gradient elution methods allowed the purification of up to 200 samples a day at weights of up to 200 mg. Similar systems, as will be described in detail in a later section, have been in operation at Abbott Laboratories for several years.[8] These systems are operated in a "service" rather than in an open-access format by a dedicated staff of purification chemists, and they are capable of triggering fraction collection based on either a UV or ELSD detector. Park-Davis researchers reported[9] on a UV-triggered fraction collection system for the purification of combinatorial libraries that operated in either a reverse-phase or a normal-phase mode. Their scheme entailed "scouting" analytical HPLC/MS data accumulation, whose conditions were selected based on structural information and subsequently applied to the selection of final purification conditions. A parallel processing format, UV-triggered HPLC system available from Biotage, termed Parallex[R] was co-developed with MDS Panlabs.[10,11] The system is comprised of four injectors, four parallel HPLC columns that are developed with identical gradients and a custom designed parallel flow path UV detector for the eluant, which allows the fraction collection from each to be triggered by two simultaneous wavelengths. This system accommodates the purification of 48-well microtiter format reaction mixtures, each carried out at a one millimolar scale. GlaxoSmithKline researchers[12] evaluated the high-throughput capabilities of the Parallex[R] system for the purification of their combinatorial arrays generated via solid-phase syntheses. They described the system as "robust" and capable of purifying 200 samples in 10 hours, with a 93% success rate, excellent purity, and acceptable recovery.

The most recently described fraction collection trigger methodology for HPLC has been mass spectrometry. The advantage of MW-triggered fraction collection over UV or ELSD initiated collection to combinatorial libraries lies in the fact that the single desired component from each reaction mixture is the only fraction collected, thus alleviating the necessity of postpurification deconvolution of fractions as well as of sample validation. Collaborations between CombiChem & Sciex and between Pfizer, Gilson & Micromass were responsible for the success of this MW-triggered fraction collection technology. Kassel et al.[13] at CombiChem reported the analysis and purification of combinatorial libraries by reverse-phase HPLC on a parallel analytical/preparative column dual operations system termed "parallel Analyt/Prep LCMS." This system employs an ESI-MS as the on-line detector that, in the preparative mode, initiates the collection of the reaction component of interest. The system has the capability to analyze and purify 100 samples a day and is available to the synthetic chemists on an open-access basis. In a further modification by the same group,[14] throughput was increased by the parallel formatting of two each of the analytical and preparative HPLC columns, thus expanding the instrument capability to the processing of more than 200 samples a day per instrument. The Pfizer group[15] designed a system capable of processing samples from about 10 to 20 mg in either the UV-triggered or the MW-triggered fraction collection mode. Using short columns and alternating column regeneration cycles, the system is capable of purifying a set of 96 samples in 16 hours.

A recent report by Kassel et al.[16] describes a high-throughput MW-triggered HPLC purification system operated in a four-column parallel format. This is achieved by directing the flow from each of four HPLCs into a multiplexed ESI ion source, which the authors term a "MUX" system. The MUX ion source is comprised of separate electrospray needles around a rotating disc, allowing independent sampling from each individual sprayer by the MS. MW-triggered collection from each of the four HPLC is directed to each of four fraction collectors. This technology is capable of expansion to accommodate up to a total of eight HPLC systems. The authors describe the application of this system to the purification of sample sizes from 1 to 10 mg, wherein they achieve recovery values above 80% and typical 90% purity of materials after optimization of the fraction collector's valve-switching timing. Three MW-triggered fraction collection HPLC systems are commercially available at present from Agilent (Wilmington, DE), Waters/Micromass (Milford, MA), and ThermoFinnigan (Waltham, MA).

11.1.4. SFC Purification Methods

Supercritical fluid chromatography (SFC) is the second high-performance chromatographic technique that has been applied to the purification of combinatorial chemistry libraries. SFC and HPLC techniques are similar in that each achieves separation by the adsorption of crude materials onto a stationary phase column followed by the preferential elution of components by a mobile phase. SFC differs from HPLC in the nature of the mobile phase employed, which in SFC is comprised of high-pressure carbon dioxide and miscible co-solvents such as methanol. The advantage of carbon dioxide as a mobile phase lies in the alleviation of sample dry-down which, particularly in large sample set purifications, can be a rate-limiting step in the total process. Preparative scale SFC is a recent addition to the arsenal of purification tools available for library purification, and instrumentation is only now catching up to the application on a research scale.

Berger Instruments (Newark, DE) developed the first semipreparative scale high-throughput SFC instrument[17] to be employed for the purification of combinatorial chemistry libraries. This commercially available system is comprised of several modified or developed components to achieve chromatography where a compressible fluid is the primary mobile phase. The system is comprised of a pump modified to achieve accurate delivery of carbon dioxide, a high-pressure mixing column, a manual injector, and a UV detector modified with a high-pressure flow cell. Back-pressure regulation is employed to control the column outlet pressure downstream of the UV detector, and a Berger Instruments "separator" prevents aerosol formation as fluid-phase methanol/carbon dioxide expands to the gas phase and separates from the liquid organic modifier. Fractions are collected at elevated pressure into a "cassette" system comprised of four individual compartments, each with a glass collection tube insert, allowing efficient collection of up to four components per chromatogram.

The purification of combinatorial libraries on a Berger system is described by Farrell et al.[18] at Pfizer for their parallel solution-phase syntheses. The overall process employs as well analytical SFC in combination with mass spectrometry and nitrogen chemiluminescence detection off-line of the preparative-scale SFC systems. Pre-purification analytical SFC/MS/CLND allows the triage of samples for purification, and an in-house software package analyzes data for predicted quality based on an evaluation of UV and MS data for the potential of co-eluting peaks during purification. This same software package selects a collection time window for purification, which is necessary to limit the number of fractions per sample. This system accommodates the purification of samples up to 50 mg in weight. Postpurification analytical SFC/MS/NCD is used as well to validate purified samples.

Our group also has custom modified[19] a Berger Instruments preparative SFC for use in support of the purification[20,21] of libraries synthesized by the high-throughput organic synthesis group and have integrated this system into the existing preparative HPLC purification processes. Specifics of this instrument and applications to the purification of libraries will be detailed within the text of this chapter.

MW-triggered fraction collection in SFC has been reported by the Dupont group,[22] whose system is comprised of a Gilson (Middletown, WI) semipreparative SFC and autosampler and a PE Sciex (Foster City, CA) mass spectrometer with ESI source, operated in the positive ion mode. A simple foil seal is placed over the collection tubes to improve recovery into the fraction collection system, and provide a loose barrier that encourages retention of the methanol portion of the eluant and departure of the carbon dioxide. Two solvent makeup pumps were added to the system: one to provide additional flow to the fraction collector as carbon dioxide departs and a second to provide auxiliary flow of methanol containing formic acid additive to the mass spectrometer stream in order to dilute sample concentration as well as to improve ion signal and peak shape. Fraction collection is triggered when signal for a component of the desired molecular weight from each sample goes above a pre-set threshold value and is controlled through the mass spectrometer software package. This system is capable of purifying crude samples up to 50 mg in weight for chromatographically well-behaved samples, and it is used by DuPont scientists for chiral separations as well as general library purifications. Kassel from the DuPont group has also reported[23] preliminary efforts toward a MW-triggered preparative SFC system on the Berger platform.

Ontogen researchers[24,25] have custom designed a MW-triggered semipreparative SFC system for purification of their parallel synthesis libraries. This parallel four-channel system is capable of processing four microtiter format plates simultaneously. A protocol is used whereby UV detection is first employed to identify peaks from each sample as they elute, followed by diversion of a split stream from the eluant to a mass spectrometer for molecular weight determination. Each of the four SFC columns is monitored by a dedicated UV detector, but as peaks are identified by their respective detectors, all are diverted to a single time-of-flight mass spectrometer. Fraction collection is initiated by the mass spectrometer, which identifies the desired target peak and then triggers fraction collection into a "target plate." In addition to collection of the desired component, reaction by-products are diverted into an ancillary plate whose fraction collection is initiated by identification by the mass spectrometer of a MW other than that of desired product. Collection plates have a 2 mL deep-well microtiter format and are custom designed with "expansion chambers" to accommodate evaporation of carbon dioxide as it departs. A 31-second "timeout" at the flow rate of 12 mL/min is used to avoid collection above a desired volume per sample and to achieve collection of samples into single wells. In-house designed software tracks all compounds collected with respect to plate and well location.

11.1.5. Abbott Laboratories HTP Scope

The high-throughput purification (HTP) group at Abbott Laboratories is based strictly on high-performance chromatography techniques so as to yield purities at 95% and above. Further HTP is set up as a service function rather than as open-access instrumentation. The synthetic chemists drop off samples at a centralized location, the purification work is carried out by a staff of purification chemists dedicated to the task, and purified compounds are returned to the synthetic chemist along with structural validation based upon MW. A wide range of synthetic formats are submitted to this HTP service, including those from a high-throughput organic synthesis (HTOS) group (who synthesize, on a service basis, 48 member libraries), a combinatorial chemistry projects support (CCPS) group dedicated to specific therapeutic targets and synthesize highly variable-sized libraries (from ten to hundreds of members), as well as numerous medicinal chemists, who typically synthesize single or small numbers of compounds of a particular class at a time. Similarly the weights of materials synthesized by these three distinct groups of synthetic chemists vary as do their library sizes. Whereas the HTOS group generate strictly 10 to 50 mg member libraries, the CCPS libraries may vary from 10 to 100 mg and the medicinal chemists single samples from 50 to 300 mg. This high variance in library size and entity weight requires a highly flexible purification service. Additionally each of the three synthetic schemes presents specific purification challenges. HTOS libraries, based on standardized chemistries, tend to give a high variability in product yield. The CCPS libraries are often carried out on valuable "core" in late-stage development, and hence reliability in return of the purified material is of paramount importance. Compounds synthesized by the medicinal chemist are submitted to HTP in "sets" of one, two, or few reaction mixtures and hence present a challenge in terms of processing efficiency.

The scope of the present HTP service is capable of processing samples sized from 10 to 300 mg in weight, maintains for the client chemists a 72 hour turnaround, has the capacity to purify 25,000 samples a year, and operates with a 99% success rate. To maintain this broad charter of service, it is required for submission of samples to HTP that analytical proof be provided for the presence of desired component via an analytical HLPC/MS trace, and that the desired component of the crude reaction mixture be present in at least an estimated 5% yield. Additionally the purifications requested must be of the routine type, excluding "natural products mixtures," no proteins, large peptides, or highly insoluble (in DMSO or MeOH) materials.

11.2. PURIFICATION SYSTEMS AND ANALYTICAL SUPPORT

11.2.1. Purification

To accommodate the variant needs of the different synthetic chemistry groups, a series of high-performance chromatographic systems are employed, including UV-triggered HPLC, ELSD-triggered HPLC, MW-triggered HPLC, and UV-triggered SFC. Each sample set submitted to the HTP service is directed to the specific instrument and technique that best meets the purification requirements of the library, based on considerations of efficiency and compatibility of the chromatographic technique to the structures, weights, and purification complexity.

HPLC (UV/ELSD)

The purification system first set up for our HTP service is a series of UV/or/ELSD-triggered fraction collection HPLC instruments.[26] Each UV/ELSD-HPLC is comprised of a Waters (Milford, MA) PrepLC 4000 solvent delivery and control system, a Waters 996 Photo PDA detector, Waters 717 plus auto-sampler, an Alltech (Deerfield, IL) Varex III ELSD, and two Gilson (Middletown, WI) FC204 fraction collectors. The chromatography columns used are Waters Symmetry[R] columns in 7-μ particle size, radial compression format, employing different sizes: 10×100 mm, 25×100 mm, and 40×100 mm for the purification of samples sized from 10–20, 20–70, and 70–300 mg, respectively. The standard gradient elution methods used are 0–100% CH_3CN:0.1% aqueous TFA or ammonium acetate, 10–100% CH_3CN:0.1% aqueous TFA or ammonium acetate, 20–100% CH_3CN:0.1% aqueous TFA or ammonium acetate, and 50–100% CH_3CN:0.1% aqueous TFA or ammonium acetate. The fraction collection is triggered by PDA detection at 220, 240, or 254 nm or by ELSD, based on an evaluation of analytical LC/MS data.

To ensure return to the synthetic chemist of the desired product or products from the UV- or ELSD-triggered fraction collection, loop injection mass spectrometry is employed. The MS validation system is comprised of a Finnigan (Waltham, MA) LCQ MS, operated in either ESI or APCI probe, in positive or negative ion mode, a Gilson 215 liquid handler for loop injection delivery of samples to the instrument, and a Gilson 307 solvent delivery system for addition of MeOH/10 mM NH$_4$OH (7:3) into the mass spectrometer.

Custom software packages were written in-house in order to track sample submissions, track fractions collected, make selections of fractions for loop injection MS validation, label samples, and generate chromatographic and mass spectrometry reports for return to the synthetic chemist. Sample submission begins with an Intranet-based log-in system (as shown in Figure 11.1). The chemist provides laboratory notebook designation (for sample tracking purposes), structure, and MW (if available) for the desired component(s), weight of the crude material to be purified (for the selection of column size), and any comments regarding special handling or information relevant to purification. This log-in system has a manual sample mode in which individual compounds can be entered singly and is also compatible with our company's electronic notebook system so that all data for large library sets can be populated automatically from the electronic notebook.

Samples submitted via the electronic notebook appear on the worksta-

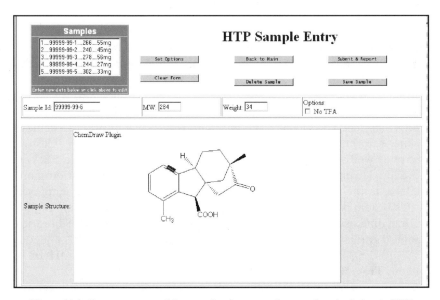

Figure 11.1. Screen capture of Intranet log-in system for sample submission to HTP.

tion for each of the purification chemists in HTP in a custom software package. Whenever possible, samples from multiple sources are grouped for purification. This can be done whenever the optimal wavelength for fraction collection trigger and the optimal gradient elution method are deemed identical for multiple sample sets. Doing so increases efficiency of the overall process. The software allows the assignment of purification parameters to a specific preparative scale HPLC. Upon completion of purification, data files and fraction collection racks are passed on for loop injection MS of fractions selected by the purification chemist. This system of UV-triggered fraction collection followed by MS validation has been the mainstay purification technique for HTP at Abbott for several years. It was adopted on the principle that several relatively inexpensive yet slow HPLCs feeding into a single more costly but fast MS give good value and allow for expansion. We have thus succeeded in accommodating initially two and presently four preparative UV/ELSD-triggered HPLCs.

At this point custom software allows importation of chromatograms and correlated mass spectrometry data to be reviewed by the purification chemist (as shown in Figure 11.2). Those fractions corresponding to the desired component(s) are selected for sample labeling and dry-down. The sample, along with a Purification Report (as shown in Figure 11.3), is returned to the submitting chemist.

HPLC (MS)

In addition to the UV or ELSD triggered HPLC instruments described above, an MW-triggered HPLC instrument is available in our HTP facility. This is commercially known as the Agilent (Wilmington, DE) 1100 series LC/MSD, which includes a PDA detector and mass spectrometer and a Gilson 215 liquid-handler for injection and fraction collection. An Antek (Houston, TX) 8060 nitrogen specific detector (Chemiluminescent Nitrogen Detection–CLND) was integrated into the system in order to achieve on-line determinations of sample weights during purification. Custom macros as well as a custom data browser were written for sample tracking and data viewing,[27] which includes CLND quantitation information made available in the final client report. The chromatography column employed on this system is a Phenomenex Luna[R] C8, 5-μ particle size column of 50 × 21.2 mm dimensions, used for the purification of 10 to 50 mg HTOS libraries. Standard gradient elution methods used are 0–100% MeOH:0.1% aqueous TFA, 10–100% MeOH:0.1% aqueous TFA, 20–100% MeOH:0.1% aqueous TFA, 50–100% MeOH:0.1% aqueous TFA, with methanol being used in lieu of acetonitrile so that solvent does not interfere with on-line CLND quantitation results.

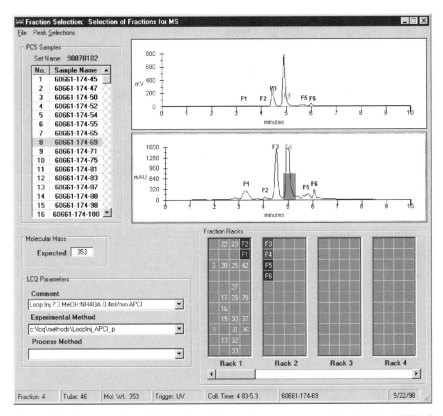

Figure 11.2. Custom software for sample tracking and mass spectrometric validation of HPLC fractions.

SFC (UV)

The supercritical fluid chromatography (SFC) instrumentation employed by our HTP service was built[19] from purchased and modified components. To a "manual" version of a Berger Instruments (Newark, DE) semipreparative SFC we integrated a Gilson 232 (Middleton, WI) autosampler and a Cavro (Sunnyvale, CA) pipetting instrument customized in-house to serve as a fraction collector. A custom designed fluid/gas outlet or "shoe" on the fraction collector enables collection of samples at atmospheric pressure, and a methanol wash system is incorporated into the fraction collection line to ensure high recovery and eliminate cross-contamination between fractions. Communication among the proprietary Berger SFC control software,

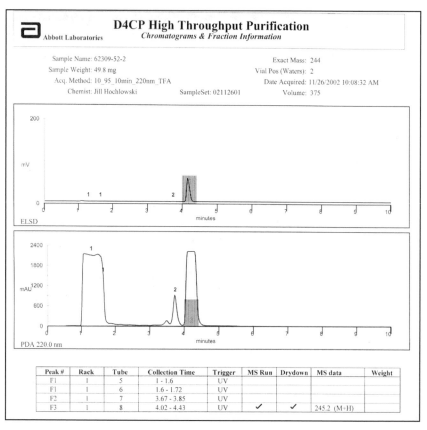

Figure 11.3. Purification Report returned to chemist, containing chromatogram with returned peaks highlighted, and MS results.

the Gilson autosampler scripts, and the custom fraction collector control software, each of which runs on a separate computer, is accomplished through a series of digital control lines that indicate the status of a particular subsystem to its two companions. In addition analog and digital input lines running from sensors on the Berger SFC to the fraction collection computer are used to track all fractions, to log their timing, and to record the overall chromatograms.

The chromatography columns used are Berger Instruments "Amino," "Cyano," and "Diol" in 6-μ particle size with a 21.2 × 150 mm format. The standard gradient elution methods employed are 5–50% methanol in carbon dioxide, 10–60% methanol in carbon dioxide, 20–60% methanol in carbon dioxide, with "hold" period of 2 minutes at the highest percentage of methanol at the end of each run. A tertiary modifier may be added to the methanol component of the solvent system, typically 0.5% triethylamine or 0.5% 1,4-dimethylpiperazine. Fraction collection on our semipreparative SFC is UV triggered from a Knauer UV detector with high-pressure flow cell purchased through Berger Instruments.

Postpurification validation is accomplished on the same instrumentation described above for loop injection MS. The software packages also described above for the UV-triggered, MS-validated preparative HPLC setup already in place were expanded to accommodate SFC data and samples. This software tracks mass spectrometry results and fraction collection data, correlating test tube location within a rack to chromatogram peaks as well as to MW. Thus samples purified on each of four HPLCs and one SFC by UV trigger can be validated on a single MS from identical footprint fraction collector racks and via a single software package.

11.2.2. Analytical Support

All samples submitted to the HTP service are required to be provided along with an analytical HPLC/MS report demonstrating that the desired reaction product has been generated. Additionally information from the analytical LC/MS/UV/ELSD reports are used by the purification chemist for the selection of optimal chromatographic method and fraction collection trigger wavelength. These reports are acquired on one of a series of open-access analytical HPLC/MS/UV/ELSD instruments that are set up at locations convenient to the synthetic chemists sited either within or near to individual synthesis laboratories. Two instrumentation formats are available to the chemist: injection vial and microtiter tray. The stations are operated as open-access systems by the synthetic chemists, and submission rules are enforced for the first format instruments, as sets of three samples only may be acquired during regular working hours and larger sets must be run overnight. However, we presently we have no analytical SFC capabilities in-house. The samples purified by SFC methods are selected based on known correlations of standardized chemistries, and the resultant functionalities are generated by the HTOS group.

Analytical LC/MS/UV/ELSD: Injection Vials

A series of instruments set up in "injection vial" format are each comprised of a Finnigan (San Jose, CA) Navigator with data control software, an

Agilent (Wilmington, DE) 1100 Series HPLC and PDA detector, and a Sedere (Alfortville, France) Sedex 75 detector. The columns used on the analytical instruments are Phenomenex Luna^R C8 columns in a 5-µ particle size and 2.1 × 50 mm format. Standard gradient elution systems used are 10–100% acetonitrile:0.1% aqueous TFA, at 1.5 mL per minute flow rate with positive ion APCI mode detection or 10–100% acetonitrile:10-mM ammonium acetate, at 1.5 mL per minute flow rate with negative ion APCI. These systems are mostly employed by medicinal chemists. Reports generated on this system are sent directly to the desktop of the submitting chemist via a custom program integrated to the company's e-mail system. Data are archived to the central server and are available to the synthetic chemist for further processing or future retrieval.

Analytical LC/MS/UV/ELSD: Microtiter Format

A microtiter injection format analytical HPLC/MS/UV/ELSD is available to the high-throughput organic synthesis (HTOS) group at Abbott so as to be compatible with their automated library syntheses format. This system is comprised of a Waters (Milford, MA) ZMD/Alliance LC/MS with APCI probe operated in the positive ion mode and a Sedere (Alfortville, France) Sedex 75 detector. Columns used on the analytical instruments are Phenomenex Luna^R C8 columns in a 5-µ particle size and 2.1 × 50 mm format. A standard gradient of 10–100% acetonitrile:0.1% aqueous TFA, at 1.5 mL per minute flow rate is used exclusively on this instrument. A software package—Masslynx Diversity Browser^R, commercially available from Waters—allows the HTOS chemist a global view of the success of library prior to submission to purification.

11.2.3. Purification Workflow

Samples are submitted for purification on a continual basis to the HTP service by a large number of chemists. These samples are generated in diverse formats, varying in weight of individual sample, library set size, yield, polarity and functionality, and in chromatographic behavior. For each sample or set submitted, however, the same general work flow is followed. This work flow begins with an Intranet-based sample log-in system, available to client chemists from their desktop. Data entered into the sample log-in system begins the automated process of sample tracking and provides information used in the selection of a purification method, such as sample weight and structure. Data are imported in different formats into the various purification systems available within HTP, but the system is designed such that output format and sample labeling are transparent to the submitting chemists. As described in the previous instrumentation section, samples are purified either by SFC or HPLC and by UV-, ELSD-, or MW-triggered fraction collection. In each case mass spectral validation data are returned to the synthetic chemist generated during either purification, as in the case of MW-triggered HPLC, or postpurification by loop injection MS, in the case of UV- or ELSD-triggered collection. After the acquisition of mass spectra, samples to be returned to the chemist are selected within interactive sample processing software packages, as described in the previous section. Labels are generated upon peak selection, and the appropriate fractions pulled, labeled, and dried-down on a series of centrifugal evaporation systems. These dried and labeled samples are returned to the submitting chemist along with a purification report and MS data, and an e-mail is sent to the chemist informing them that samples are ready for pickup. The process is accomplished within 72 hours of sample receipt. In the event that the submitting chemist subsequently requests the return of additional samples from the crude reaction mixture after evaluation of the purification report, all fractions collected from each run are retained for approximately one week.

11.2.4. Purification Triage

Of the systems just described, each may be more amenable to the purification of specific sample sets based on considerations such as sample sets size, sample weights, and structural characteristics. Additionally consideration is given to the availability of each instrument at the time at which samples are submitted and how the purification can be more efficiently and rapidly returned to the submitting chemist.

UV-triggered HPLC represents the highest purification capacity in HTP with four instruments available. This technique is therefore employed as the first of choice for all routine samples. These instruments are always used when chromatographic and spectroscopically amenable for the many small sample sets submitted in a given day. Further, when more than one component is requested for return, or particularly when the MW of some requested components are unknown, UV-triggered HPLC is employed. All samples in weight above 70 mg are purified on these systems as they are the only instruments presently set up in an automated column–switchable format that accommodates larger sized columns. The only requisite for use of the systems is that a good UV chromophore, else a good ELSD response, be observable based on the analytical LC/MS data. In order to most efficiently use the instruments available, several small sample sets from different submissions may be combined into a single purification set if the optimal gradient method and trigger wavelength are compatible.

MW-triggered HPLC purification is selected as the method of choice when large libraries are submitted with high variability in reaction yield. When no good UV chromophore is present in the reaction products, either ELSD-triggered fraction collection as described above or MW-triggered fraction collection must be employed. The requisite for fraction collection triggered by MW is the presence of a good ionizable group within the reaction product. Approximately one-half the face time is necessary on the part of HTP staff technicians for MW-triggered purification due to the elimination of a second MS validation step and sample sorting after UV-triggered purification. Therefore the single MW-triggered HPLC available to HTP is used at capacity at all times.

SFC is selected as the purification method of choice for several sample types. As SFC is a normal-phase chromatography technique, it is used for the purification of highly lipophilic compounds. When the submitting chemist requests that TFA not be used in purification, SFC may be used as an alternative to the less desirable ammonium acetate buffer alternative in HPLC. SFC can offer advantages in terms of increased resolution, hence crude samples with a high degree of complexity are preferentially purified by SFC. The requisite for purification by SFC is good solubility of material in an injection solvent compatible with carbon dioxide (typically methanol, dichloromethane, acetonitrile, and mixtures thereof) and, since this system is set up only in a UV-triggered mode, a good UV chromophore. In general, we have worked out methods specific for the purification of the standardized chemistries generated by the HTOS group,[21] so SFC is applied only to the purification of libraries from this client set. Since HTOS represents approximately one-half of the total number of samples submitted to HTP, this instrument is always used at capacity.

11.3. LIBRARY PURIFICATION EXAMPLES

The following examples (selected from typical structural types synthesized for the purpose of purification triage) illustrate application of each of the purification systems described above: UV-triggered HPLC, MW-triggered HPLC, and UV-triggered SFC. Upon submission of libraries to HTP, each of the three purification technologies is considered in terms of the relative merits and liabilities of the technique. The library characteristics evaluated include sample weights, sample set sizes, variability of yields within a library, the presence of good UV chromopore or good MS ionization character, and finally what samples are in process and those in queue as this affects instrumentation readily available at the time of submission. Each of the examples in the following sections depicts the purification of a single member from a set of similar samples submitted as a set.

11.3.1. UV-Triggered HPLC Purification Examples

UV-triggered HPLC represents the highest capacity instrumentation currently available within HTP. Therefore it is employed for the purification of most routine samples when a good UV chromophore is present. The ELSD-triggered format available on this same instrumentation is initiated only if necessary, as it is somewhat less reliable than is UV-triggered fraction collection. Owing to the large numbers of such sets submitted on a daily basis, all small sample sets (of one through six members) are purified by UV-HPLC if feasible.

Example 1 was a set of two reductive amination reaction samples submitted for purification. Each sample was approximately 200 mg and therefore was above the weight that can be processed in a single injection by our SFC instrumentation. As it is both more efficient and more desirable for the synthetic chemist that materials be purified and returned as a single fraction, HPLC was the method of choice for purification in this case. Each of the two samples had a good UV chromophore at both 220 and 254 nm and was therefore amenable to UV-triggered fraction collection. As a routine, high-weight, small sample set with good UV characteristics, UV-triggered HPLC was the method selected for processing. Figure 11.4A and 11.4B depict the structure and chromatography for one member of this sample set.

Example 2 was a 48-membered library of amine formation reaction products submitted for purification. Each sample was approximately 50 mg in weight, and all reactions proceeded with similar high yields. Each sample had a good UV chromophore at 220 nm, which was distinct from other reaction components as evidenced by the analytical HPLC/MS supplied upon submission. Therefore as routine library with a good UV product chromophore and similar reaction yields, HPLC, UV triggered at 220 nm was selected as the purification method for this library. Figure 11.5A and 11.5B depict the structure and chromatography, respectively, for a representative member of this library.

11.3.2. MW-Triggered HPLC Examples

A single MW-triggered HPLC instrument represents the total capabilities of this technology within HTP and is generally employed at capacity. Therefore this technique must often selected only when necessary, such as for the

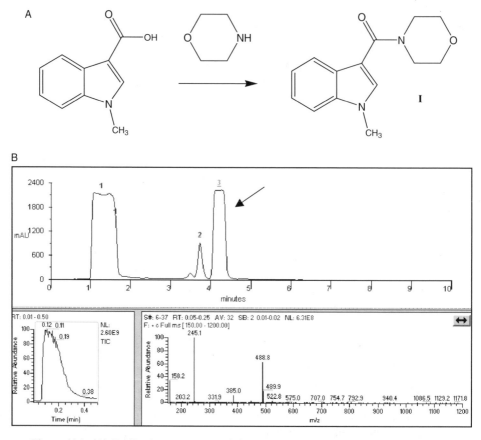

Figure 11.4. (*A*) Purification example 1. (*B*) Reaction and product depicted in a chromatogram of the UV-triggered HPLC purification. Chromatography conditions: C8 column (40 × 100 mm, 7-μ, radial compression, Symmetry[R], Waters), 10–95% aqueous 0.1% TFA/CH$_3$CN at 70 mL/min over 10 minutes. Fraction collection triggered at 220 mm. Arrow indicates the desired reaction product.

purification of samples with a poor UV chromophore or where low yield or highly variable yield in a library set can be more reliably processed by MW-triggered fraction collection.

Example 3 was a 48-membered library of amide formation reaction products submitted for purification. Each reaction had been set up to provide approximately 50 mg of product, but the individual reactions subsequently proceeded with highly variable yields. Although a UV chromophore at 254 nm showed a good extinction coefficient, the variability in yield presented the product as a peak of highly variable intensity. Purification by either UV-triggered HPLC or SFC would have required that the

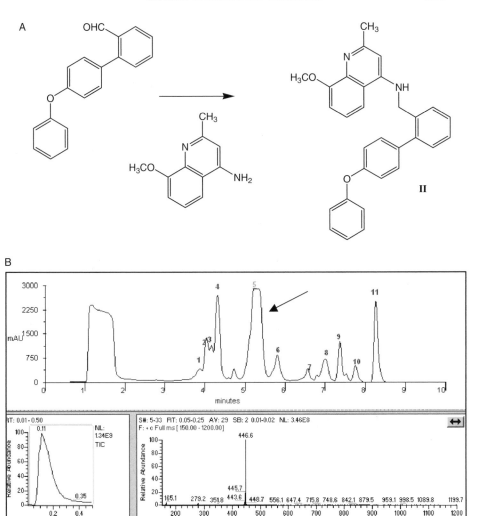

Figure 11.5. (*A*) Purification example 2. (*B*) Reaction and product depicted in a chromatogram of the UV-triggered HPLC purification. Chromatography conditions: C8 column (25 × 100 mm, 7-μ, radial compression, Symmetry^R, Waters), 10–95% aqueous 0.1% TFA/CH$_3$CN at 70 mL/min over 10 minutes. Fraction collection triggered at 220 mm. Arrow indicates the desired reaction product.

samples to be split into separate sets with different UV-triggered threshold settings. Since the desired products from all members the library gave good ionization in positive ion mass ESI spectrometry, MW-triggered HPLC was selected as the purification method for this library. Figure 11.6*A* and 11.6*B*

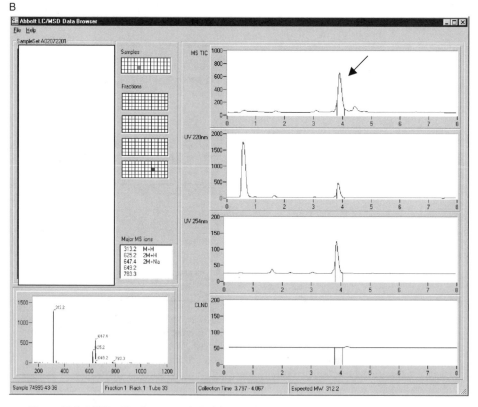

Figure 11.6. (*A*) Purification example 3. (*B*) Reaction and product depicted in chromatogram of the MW-triggered HPLC purification. Chromatography conditions: C8 column (21.2 × 50 mm, 5-μ, Luna[R] Combi-HTS, Phenominex), 10–100% aqueous 0.1% TFA/MeOH (0.5 minute hold, 5.5 minute gradient, 1 minute hold) at 40 mL/min over 8 minutes. Fraction collection triggered by positive ion ESI-MS. Arrow indicates desired reaction product.

depict the structure and chromatography, respectively, for a representative member of this library.

Example 4 was a 48-membered library, where each reaction was set up to generate 20 mg in crude weight. As in example 3, the individual reactions proceeded with highly variable yields. Although a good UV intensity could be observed at each of 220, 240, and 254 nm, the variability in yield presented the product as a peak of highly variable intensity as well as percentage of crude reaction mixture. So again, purification by either UV-triggered HPLC or SFC would have required that the samples to be split into separate sets with different UV-triggered threshold settings and, optimally, different wavelengths. Since the desired products from all library members gave good ionization in positive ion mass ESI spectrometry, MW-triggered HPLC was selected as the purification method for this library. Figure 11.7A and 11.7B depicts the structure and chromatography, respectively, for a representative member of this library.

11.3.3. UV-Triggered SFC Examples

Although not a requisite for the selection of SFC as the purification technique of choice, highly lipophilic samples are preferable so as to avoid solubility problems often encountered in reverse-phase HPLC processing. It should be noted, however, that highly polar materials can be purified by SFC so long as materials can be dissolved at a level of 50 mg/mL of methanol, along with co-solvents such as acetonitrile or dichloromethane. Further, when a synthetic chemist specifically requests that samples be returned in their non-TFA salt form, SFC is selected as it can be frequently developed in methanol/carbon dioxide or, alternatively, volatile small amine modifiers such as triethylamine. Finally, it is generally possible to achieve better resolution with an SFC rather than the HPLC method of our present setup, so in instances where a difficult separation of desired product is anticipated based on LC/MS analysis, SFC may be preferentially chosen over HPLC.

Example 5 was a 48-membered library submitted for purification, each sample being approximately 20 mg in crude weight. Each sample had good UV intensity at 220 nm, while exhibiting poor ionization in mass spectrometry as evidenced by analytical LC/MS. As the purification of amide libraries by SFC is well worked out owing to the large numbers of these libraries generated by our HTOS group, this method was readily selected for purification. Figure 11.8A and 11.8B depict the structure and chromatography, respectively, for a representative member of this library.

Example 6 was a 48-membered library submitted for purification, each sample being approximately 20 mg in crude weight and each proceeding

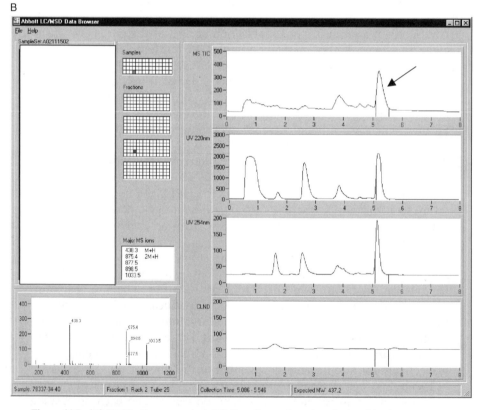

Figure 11.7. (*A*) Purification example 4. (*B*) Reaction and product depicted in chromatogram of the MW-triggered HPLC purification. Chromatography conditions: C8 column (21.2 × 50 mm, 5-μ, Luna[R] Combi-HTS, Phenominex), 10–100% aqueous 0.1% TFA/MeOH (0.5 minute hold, 5.5 minute gradient, 1 minute hold) at 40 mL/min over 8 minutes. Fraction collection triggered by positive ion ESI-MS. Arrow indicates desired reaction product.

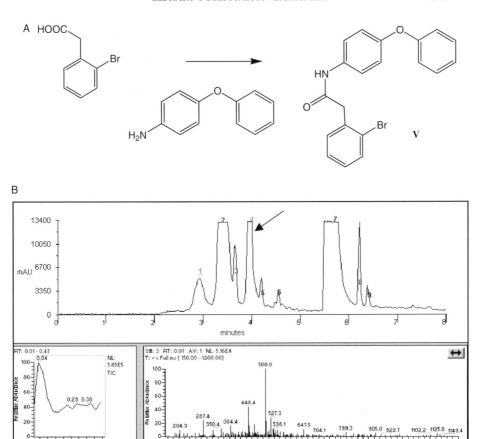

Figure 11.8. (*A*) Purification example 5. (*B*) Reaction and product depicted in chromatogram of the UV-triggered SFC purification. Chromatography conditions: Diol column (21.2 × 150 mm, Berger Instruments), 0.5% Et_3N/MeOH in carbon dioxide, 5–60% over 6 minutes and hold at 60% for 2 minutes. Fraction collection was triggered by UV at 254 nm. Arrow indicates desired reaction product.

with similar high yields. Each sample had good UV intensity at 254 nm, while exhibiting poor ionization in mass spectrometry as evidenced by analytical LC/MS. Further it was observed that the reaction product and amine core were not well resolved by HPLC for many monomers within the library. Therefore UV-triggered SFC was selected as the purification method of choice for this library. Figure 11.9*A* and 11.9*B* depict the structure and chromatography, respectively, for a representative member of this library.

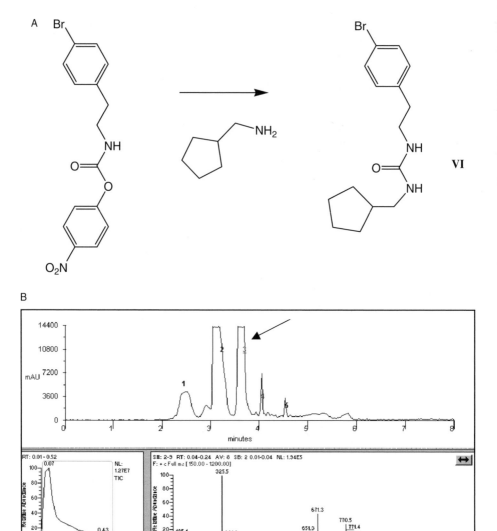

Figure 11.9. (*A*) Purification example 6. (*B*) Reaction and product depicted in chromatogram of the UV-triggered SFC purification. Chromatography conditions: Diol column (21.2 × 150 mm, Berger Instruments), 0.5% Et$_3$N/MeOH in carbon dioxide, 5–60% over 6 minutes and hold at 60% for 2 minutes. Fraction collection was triggered by UV at 220 nm. Arrow indicates desired reaction product.

REFERENCES

1. X. Peng, C. Hensen, M. Strojnowski, A. Golebiowski, S. Klopfenstein, *Anal. Chem.* **72**, 261–266 (2000).

2. J. Breitenbucher, K. Arienti, K. McClure, *J. Comb. Chem.* **3**, 528–533 (2001).

3. C. Johnson, B. Zhang, P. Fantauzzi, M. Hocker, K. Yager, *Tetrahedron*, **54**, 4097–4106 (1998).

4. S. Kaldor, M. Siegel, J. Fritz, B. Dressman, P. Hahn, *Tet. Lett.* **37**, 7193–7196 (1996).

5. J. Booth, J. J. Hodges, *Am. Chem. Soc.* **119**, 4882–4886 (1997).

6. B. Bookser, S. J. Zhu, *Comb. Chem.* **3**, 205–215 (2001).

7. H. N. Weller, M. G. Young, J. D. Michalczyk, G. H. Reitnauer, R. S. Cooley, P. C. Rahn, D. J. Loyd, D. Fiore, J. S. Fischman, *Mol. Divers.* **3**, 61–70 (1997).

8. M. Routburg, R. Swenson, B. Schmitt, A. Washington, S. Mueller, J. Hochlowski, G. Maslana, B. Minn, K. Matuszak, P. Searle, J. Pan, *Implementation of an Automated Purification/Verification System*, International Symposium on Laboratory Automation and Robotics, Boston, October (1996).

9. C. E. Kibby, *LAR 9*, 309–321 (1997).

10. P. Coffey, *Lab Automation News 2*, 7–13 (1997).

11. L. Schultz, C. D. Garr, L. M. Cameron, J. Bukowski, *Bioorg. Med. Chem. Lett.* **8**, 2409–2414 (1998).

12. C. Edwards, D. Hunter, *J. Comb. Chem.* (2002). *ASAP article*, A–F.

13. L. Zeng, L. Burton, K. Yung, B. Shushan, D. B. Kassel, *J. Chromatog. A* **794**, 3–13 (1998).

14. L. Zeng, D. B. Kassel, *Anal. Chem.* **70**, 4380–4388 (1998).

15. J. P. Kiplinger, R. O. Cole, S, Robinson, E. J. Roskamp, R. S. Ware, H. J. O'Connell, A. Brailsford, J. Batt, *Rapid Commun. Mass Spectrom.* **12**, 658–664 (1998).

16. R. Xu, T. Wang, J. Isbell, Z. Cai, C. Sykes, A. Brailsford, D. Kassel, *Anal. Chem.* **74**, 3055–3062 (2002).

17. T. A. Berger, K. Fogleman, T. Staats, P. Bente, I. Crocket, W. Farrell, M. Osonubi, *J. Biochem. Biophys. Methods 43*, 87–111 (2000).

18. W. Farrell, M. Ventura, C. Aurigemma, P. Tran, K. Fiori, X. Xiong, R. Lopez, M. Osbonubi, *Analytical and Semi-Preparative SFC for Combinatorial Chemistry*, Supercritical Fluid Chromatography, Extraction, and Processing, Myrtle Beach, SC, August (2001).

19. J. Olson, J. Pan, J. Hochlowski, P. Searle, D. Blanchard, *JALA*, *7* (4), 69–74 (2002).

20. J. Hochlowski, P. Searle, G. Gunawardana, T. Sowin, J. Pan, J. Olson, J. Trumbull, *Development and Application of a Preparative Scale Supercritical Fluid Chromatography System for High Throughput Purification*, Prep 2001, Washington, DC, May (2001).

21. J. Hochlowski, J. Olson, J. Pan, D. Sauer, P. Searle, T. Sowin, *J. Liq. Chromatogr.* **26** (3), 327–348 (2003).

22. T. Wang, M. Barber, I. Hardt, D. B. Kassel, *Rapid Commun. Mass Spectrom.* **15**, 2067–2075 (2001).

23. D. Kassel, *Possibilities and Potential Pitfalls of SFC/MS as a Complementary Tool to HPLC/MS for High Throughput Drug Discovery*, Supercritical Fluid Chromatography, Extraction, and Processing, Myrtle Beach, SC, August (2001).

24. W. C. Ripka, G. Barker, J. Krakover, *DDT.* **6** (9), 471–477 (2001).

25. R. Maiefski, D. Wendell, W. D. Ripka, J. D. Krakover, *Apparatus and Method for Multiple Channel High Throughput Purification*, PCT Int. Appl. WO 0026662.

26. P. Searle, *An automated preparative HPLC-MS system for the rapid purification of combinatorial libraries*, Strategic Institute Conference: High-Throughput Compound Characterization, Dallas, TX, March (1998).

27. P. Searle, J. Hochlowski, *High Throughput Purification of Parallel Synthesis Samples; An Integrated Preparative-LC/MS System with Quantitation by Chemiluminescent Nitrogen Detection*, HPLC 2002, Montreal, Canada, June (2002).

CHAPTER

12

PARALLEL HPLC IN HIGH-THROUGHPUT ANALYSIS AND PURIFICATION

RALF GOD and HOLGER GUMM

12.1. INTRODUCTION

In the pharmaceutical industry high-throughput screening (HTS) has become a standard procedure in drug discovery and lead optimization. For this reason the preparation and purification of a huge number of test samples for HTS has become an indispensable prerequisite in the drug development process. Automated synthesis of compound libraries and fractionation of natural product extracts are the two widely used approaches to get access to a huge number of substances covering the supply of the high-throughput screening systems. However, it is of major importance to provide these test samples with a purity of higher than 80% to gain reliable and reproducible test results. This challenging demand has created a new discipline in high-throughput purification (HTP).

Over the last three decades high-performance liquid chromatography (HPLC) has been proved to be a versatile and very robust method in non-destructive sample analysis and purification. HPLC has become one of the most important chromatography methods in research and industry. Since the beginning of this evolution the improvement of stationary phases, standardization of equipment and consumables, as well as development of sophisticated detection systems and coupling techniques has been a major task for scientists and equipment manufacturers. All these basic efforts have lead to a very high standard. Robustness and reliability are not a concern any more. Nowadays HPLC has come close to its technological limits. However, the increasing demand for higher throughput in HPLC triggered by the needs of HTS has created a new and challenging bottleneck. Solving this topic is the basic idea in the development of modern HPLC systems of a second generation.

Analysis and Purification Methods in Combinatorial Chemistry, Edited by Bing Yan.
ISBN 0-471-26929-8 Copyright © 2004 by John Wiley & Sons, Inc.

12.2.   HOW TO SPEED UP THROUGHPUT IN HPLC?

It is a common agreement that a high degree of automation in HPLC will help to optimize throughput, sample logistics, and up-time of the instrumentation. But automation is not a real restriction in this development, as computer and controller systems to deal with this task are available for reasonable cost. Software with graphical user interface (GUI) supporting the operator to get easily an overview of the running system is state of the art.

However, when looking at HPLC separation speed itself, there is a physical boundary that must be overcome for the desired high resolution of the separated chromatographic peaks. In a simplified way for a gradient separation this can be illustrated by (12.1), which is applicable for analytical separations but can approximately be used for preparative HPLC as well:

$$k' \sim t_{\mathrm{g}} \cdot \frac{F}{V_0}. \tag{12.1}$$

The capacity factor k', and therefore the resolution of the chromatographic peaks, is proportional to the gradient time t_{g} and the flow rate F at a given interparticle volume, which is the void volume V_0 of the column. For a given resolution the gradient time t_{g}, though, can only be shortened by increasing the flow rate F and decreasing the void volume V_0 of the column. Both measures increase the backpressure of the column dramatically, and therefore shorter columns with a lower void volume, namely with smaller particles in the stationary phase, have been used for speeding up gradient separations.[1,2]

Another way to speed up HPLC that does not interfere with the existing gradient separation method is by parallel operation of several HPLC columns. The development in this direction started a couple of years ago[3,4] when fast gradient separation was first combined with mass spectrometry-based detection. Parallel column operation was achieved by a single pumping system and a splitter tee that transferred the gradient flow onto two HPLC columns.[4] The effluent of the two columns was simultaneously sprayed into a modified ion spray interface of a quadrupole mass spectrometer.[4] From the overlay chromatogram both desired and previously known compounds were identified after their molecular ions were filtered from the total ion current (TIC). In this first system, however, it was difficult to enhance the parallelization, and the detection system created a bottleneck.[4]

12.3. PARALLEL HPLC

Today several parallel HPLC systems are available commercially and also adapted to an individual use in laboratories.[4–6] The extent of parallelization varies greatly, as there is no clear definition for what is a "parallel HPLC" system.

From an operator's point of view, it is quite obvious that the parallelization should lead to easier handling, enhance the sample throughput and ease the workload. By using this definition, an easy concept would be hardware in simultaneous operation involving several HPLC systems by way of one computer, with a user friendly software interface for instrument control, data acquisition, and sample tracking.

From an economical point of view, a parallel HPLC system should more clearly indicate a distinct reduction in the number of building blocks and much higher efficiency compared to known stand-alone devices. These are objectives that can only be achieved by a parallel or multiplex use of the individual components, including the gradient pump, autosampler, detection system, and fraction collector. All the equipment is easily be controlled by one personal computer. The concept suggests more compact dimensions as well as lower cost of ownership compared to other customary systems. The following paragraphs describe a parallel HPLC system that was designed to satisfy this economical conception but mainly to meet the operator's needs.

When only one pump is used for parallel operation of several HPLC columns, it is almost impossible to get the same flow conditions in the individual branches, even if the columns are taken from the same manufacturing lot. This is because of the variation in backpressure of the individual columns, and it becomes even more predominant in a gradient run and can dramatically change over the lifetime of a column.

For this reason the basic building block of a parallel HPLC device is the flow control unit (see Figure. 12.1). The flow control can be plugged to any high-pressure isocratic or gradient pumping system, and it provides steady streams of eluent to the multichannel injection system connected to the individual separation columns. After separation the eluent streams pass through the detection cells of a multiplex photodiode array detector (PDA). This detection system triggers the fraction collector, which collects the separated samples in microtiterplates.

12.4. FLOW CONTROL UNIT

The active flow control unit[7] consists of a flow measurement device, flow adjustment valves, and a feedback control unit. Flow measurement is

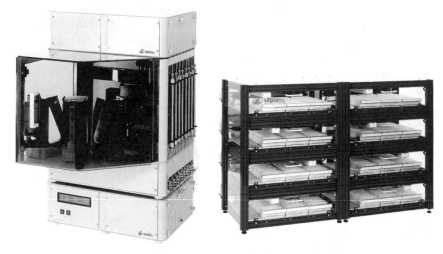

Figure 12.1. Parallel HPLC device (*left*) with flow control unit (*bottom*), autosampler, injection valves, separation columns (*middle*), multiplex photodiode array detector (*top*), and fraction collector (*right*) for parallel fraction collection into 8 × 12 = 96 microtiterplates.

obtained by metering the pressure drop Δp in a defined restrictor capillary with high-precision pressure sensors. According to Hagen-Poiseulle's law for laminar flow in capillary tubes, the measured pressure drop is proportional to the flow in each line. Flow adjustment is achieved by specific valves consisting of an orifice and an electromagnetically driven plate that covers the exit slit of the orifice with an adjustable force.

To see more clearly how this flow control unit works, let us take a simplified model of fluid resistance and consider it in terms of Ohm's law. Equation (12.2) is very similar to resistance of an electrical circuit, and it is applicable to fluidic resistance R in this system with laminar flow conditions. The pressure drop Δp is caused by the restrictor at a specific flow rate F:

$$R = \frac{\Delta p}{F}. \tag{12.2}$$

A looking at the branched system (see Figure 12.2) consisting of identical fluidic restrictors $R_1, R_2, \ldots, R_n$ used as measurement capillaries in a combination with valves for flow adjustment shows the flow rates $F_1, F_2, \ldots, F_n$ in a set of HPLC columns to be the same in all branches if the individual pressure drops $\Delta p_1, \Delta p_2, \ldots, \Delta p_n$ are the same. These pressure drops Δp_i

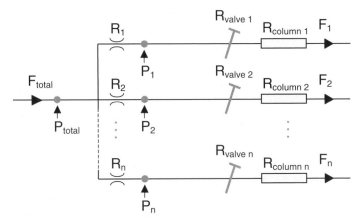

Figure 12.2. Arrangement for an active flow control unit in parallel HPLC. The partial flows F_1 to F_n resulting from the total flow F_{total} can be measured by metering the pressure drop along the restrictor capillaries R_1 to R_n. The flows are adjustable by the valves in an individual branch. $R_{valve\ 1}$ to $R_{valve\ n}$ can be compared to potentiometers in an electrical circuit. The specific settings of the valves lead to an equal flow in all the separation columns with different fluidic resistance $R_{column\ 1}$ to $R_{column\ n}$ and thus different backpressure.

$(i = 1, 2, \ldots, n)$ can be measured according to (12.3) as the difference of the total system pressure P_{total} and the specific pressure P_i after a measurement capillary:

$$\Delta p_i = P_{total} - P_i. \tag{12.3}$$

In principle, these values could be directly used as regulating variable in a feedback system to control the valves for flow adjustment in the different branches. However, from (21.2) and Hagen-Poiseulle's law, it follows that

$$\Delta p = R \cdot F = \frac{(8 \cdot \eta \cdot l)}{(\pi \cdot r^4)} \cdot F \qquad (kg \cdot m^{-1} \cdot s^{-2}), \tag{12.4}$$

where l and r are the constant length and the constant inner radius of the measurement capillary and η is the dynamic viscosity of the solvent. Relation (12.4) clearly shows that fluidic resistance R and thus pressure drop Δp is dependant on viscosity η for a given flow F. This means that pressure drop Δp representing the regulating variable will change when the composition of the solvent is changed by the HPLC gradient.

Figure 12.3 shows the course of pressure drop Δp when running a gradient from 100% water to 100% methanol for 8 minutes. Due to the fact

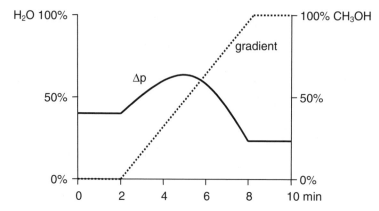

Figure 12.3. Typical course of pressure drop Δp during a water/methanol gradient.

that the viscosity of water is higher than the viscosity of methanol, the pressure drop Δp is higher when starting the gradient than in the end. Maximum viscosity, and thus maximum Δp, is reached approximately in the middle of the gradient when solvent composition is nearly 50% methanol and 50% water. For a flow control unit that has also to deal with gradient flow, it is important to have a regulating variable that is independent from fluidic parameters such as dynamic viscosity:

$$Q_i = \frac{P_{total}}{\Delta p_i} = \frac{P_{total}}{P_{total} - P_i}. \tag{12.5}$$

In (12.5) the quotient Q_i resulting from the system pressure P_{total} divided by pressure drop Δp_i has no dimension and thus is not influenced by any fluidic parameters; that is, it remains constant during a gradient run. Because of its direct proportionality to the fluidic resistance of a specific branch, it serves as a perfect regulating variable in the PI-control circuit, which is now also applicable for gradient HPLC systems.

With this insight it is now easy to complete the feedback control circuit for a flow control unit in a parallel HPLC. As Figure 12.4 shows, the instantaneous value x_n can be measured and compared to a given setpoint w_n for identical flow in all the branches. The deviation e_n of the setpoint w_n and actual value x_n are calculated, and the controller transmits the controller output y_n to the valves to adjust $R_{valve\ 1}$ through $R_{valve\ n}$ to the appropriate resistance.

This active flow control unit is plugged to an isocratic or gradient pumping system. It divides the total pump flow F_{total} into exactly equal split

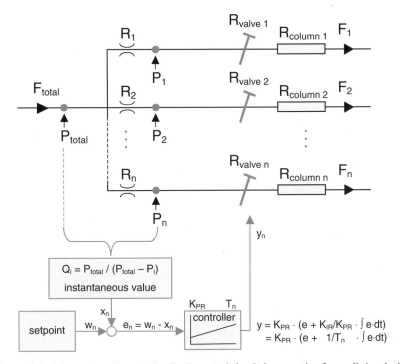

Figure 12.4. Schematic outline of a feedback control circuit for an active flow splitting device.

flows F_1 through F_n by using a sophisticated feedback control circuit. In a simplified way it can be seen as a black box that is plugged to both the pump and the injection system. The operator's only two commands to this box are "flow control on" and "off." A typical setup is with four or eight separation columns within the range of 0.25 to 1.5 mL/min, 1.5 to 6 mL/min, and 6 to 20 mL/min per branch. Especially in the setup with eight columns the flow control unit is significantly less expensive and space requiring than individual HPLC pumping systems. The positive effect of levelling out retention times when switching on the flow control unit can be seen in Figure 12.5.

12.5. AUTOSAMPLER FOR PARALLEL INJECTION

In high-throughput purification (HTP) the 96-well microtiterplate is the common housing for sample handling. Mainly deep-well or the shallow-well plates with a sample capacity of about 1 mL or 50 μL per well are used. Thus

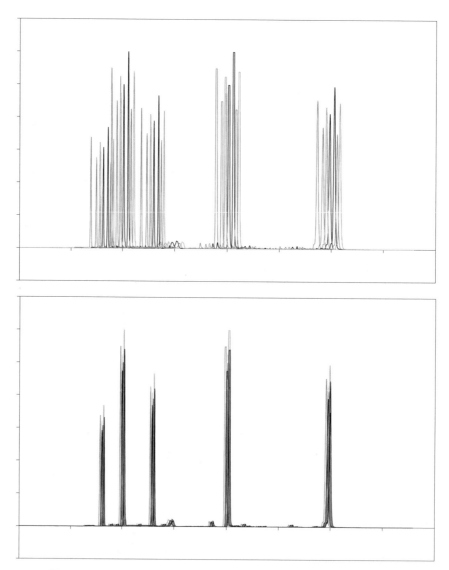

Figure 12.5. Parallel HPLC runs without (*top*) and with (*below*) flow control.

the autosampler must be able to deal with these microtiterplates and to inject samples from the plates either in a parallel mode or independently in the so-called sample-picking mode.

The handling of microtiterplates is solved by a robotic arm (see Figure

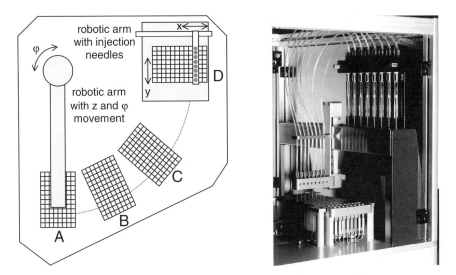

Figure 12.6. Autosampler for microtiterplate handling and parallel sample injection.

12.6, left), which stacks the plates onto each other in three magazines A through C, whereby two are usually filled with new microtiterplates and one is empty to store the already used plates. Position D is used to present a microtiterplate to another robotic arm with eight independently operating injection needles. This arm can be moved in x and the table with the fixed plate in y direction. By this mechanism it is possible to fill the sample loops of the injection valves in parallel with one row of eight samples or to pick the samples independently by a specific needle from a selected well. The injection needles are operated by a syringe pump (see Figure 12.6, right). A washing station is fixed in position D to clean the needles after each injection. The injection valves with sample loops can be seen in Figure 12.1 on the right side of the instrument, and the valves are directly plugged to the flow control unit and the separation columns. During the injection process these valves are switched simultaneously into the eluent flow, and the samples in the filled sample loops are flushed to the columns.

12.6. MULTIPLEX PHOTODIODE ARRAY AND MASS SPECTROMETRIC DETECTION

Like the flow control unit the detection system is an important component of parallel HPLC. Photodiode array and mass spectrometric detection provide the characteristic data for almost any compound and therefore are

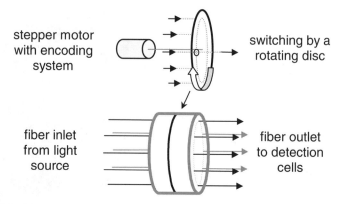

stepper motor
with encoding
system

switching by a
rotating disc

fiber inlet
from light
source

fiber outlet
to detection
cells

Figure 12.7. Schematic drawing of the optical switch used in multiplex photodiode array (PDA) detector.

the two preferred detection systems in analytical and preparative HPLC. However, the high cost of the photodiode array detector (PDA) or the mass spectrometric detector with an HPLC interface is a major drawback. In parallel HPLC reasons of economy can prohibit one from installing several of these detectors in parallel. Thus a solution for parallelizing these detection systems was inevitable. In both detection systems the analyzer part is the most expensive part and so could not be duplicated. For this reason the most feasible way to go was the development of a multiplex solution.

The multiplex design for a PDA is achieved by way of optical fiber guides. The light of a deuterium UV light source is split into individual beams by separating the fiber bundle into several individual channels connected to an optical multiplex switch. The principle of this switch is shown in Figure 12.7. A rotating disc with a small hole is stepped through the channels to single open channel for a short moment while the others are closed. The optical fiber guides on the outlet of the multiplex switch are connected to the detection cells to provide UV light to the optical inlet of the cell while a channel is active. The transmitted light is collected by another fiber guide at the outlet of the detection cell and is transferred onto the entrance slit of a photodiode array. The correlation of spectra and detection cells is guaranteed by an encoding system that is installed on the stepper motor axis of the optical switch. The rotation frequency of the rotating disc is approximately 1 Hz. This means that per channel every second a complete PDA spectrum is recorded. Compared to modern single channel devices, the recording frequency of an multiplex eight-channel PDA is about 10 times lower and thus not sufficient for trace analysis, but in sensitivity it perfectly covers all of the applications in high-throughput analysis and purification.

A similar mass spectrometric detection system is available commercially and is realized by a multiplexing electrospray inlet in combination with a time-of-flight (TOF) mass spectrometer that has a sufficiently high data acquisition rate.[5] Like the above-described optical switch, the effluent of the HPLC channels is nebulized in individual probe tips and sprayed toward a rotating cylinder with a slit. By stepping this slit very rapidly from one probe tip to another, again a data acquisition rate of approximately one spectrum per channel can be reached in 1 second. Unlike the multiplex PDA detector, this multiplex mass spectrometric detector is normally connected to a HPLC system by a flow splitter.[5]

12.7. OVERALL SYSTEM

The parallel HPLC system discussed above, with its specific characteristics in microtiterplate handling, active flow control, simultaneous sample injection, and multiplex PDA detection, was designed for industrial use where a large number of samples have to be analyzed or purified under identical chromatographic conditions. It is a second-generation HPLC system that exploits the available technology in automation, detection, and control and combines a high level of experience in HPLC separations accumulated over three decades. A schematic of the overall system is given in Figure 12.8. Compared with eight conventional HPLC systems, the result in space and cost savings is approximately 80% and 30%, respectively. The user-friendly software has significantly simplified the logistics of the sample and maintenance of the instrument. In a basic version the autosampler can handle up to 10 deep-well or 30 shallow-well microtiterplates. This is equivalent to injection of 960 or 2880 samples, respectively. Because of parallel operation these samples can be processed in 120 or 360 HPLC runs in an unattended mode, which is typically the sample load for two days. Sample collection is triggered by the multiplex PDA detector, and the fraction collector can collect up to 9216 fractions, which corresponds to $8 \times 12 = 96$ deep-well microtiterplates.

Compared to other high-throughput systems in this parallel HPLC device, the use of a flow control unit with only one pumping system is a decisive factor. This setup guarantees stabilized flow in an extensible number of simultaneously operated separation columns and thus, in principle, can be combined with any other sophisticated column switching and detection system. A specific version of this instrument was adapted to preparative fractionation and purification of natural product extracts[8] by

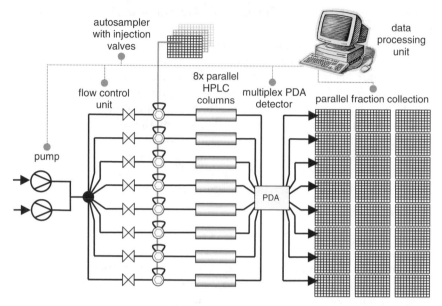

Figure 12.8. Parallel HPLC device with one gradient pumping system, flow control unit, and multiplex photodiode array (PDA) detector.

integration of solid-phase extraction[9] (SPE) columns for sample injection and for the final sample workup (see Figure 12.9). In this device SPE columns coupled[10] with HPLC separation columns allowed injection of mixtures with a high polarity range dissolved in dimethyl sulfoxide (DMSO) as a potent and unique solvent. Eluting fractions undergo an on-line and automated workup on another set of SPE columns and thus can be transferred without buffer and in a small amount of pure organic solvent into the deep-well microtiterplates of the fraction collector.

Parallel HPLC is one of the most promising approaches for speeding up sample analysis and purification. The fact that most large companies[6] are concentrating their research effort in this area shows clearly the importance of continuing advancements of throughput in HPLC. We expect the next generation of industrial HPLC equipment to be characterized by further improvements in handling for the operator and a distinct reduction in the number of building blocks involving features like flow control[7] and the multiplex detection principle. Within the next few years the line of development should resolve the bottleneck in high-throughput sample analysis and purification.

Figure 12.9. Parallel HPLC device for fully automated high-throughput fractionation and purification of natural product extracts on a large scale.

REFERENCES

1. H. N. Weller, M. G. Young, S. J. Michalczyk, G. H. Reitnauer, R. S. Cooley, P. C. Rahn, D. J. Loyd, D. Fiore, S. J. Fischman, *Molecul. Div.* **3**, 61–70 (1997).

2. W. K. Goetzinger, J. N. Kyranos, *Am. Lab.* **4**, 27–37 (1998).

3. L. Zeng, L. Burton, K. Yung, B. Shushan, D. B. Kassel, *J. Chromatogr. A* **794**, 3–13 (1998).

4. L. Zeng, D. B. Kassel, *Anal. Chem.* **70**, 4380–4388 (1998).

5. V. de Biasi, N. Haskins, A. Organ, R. Bateman, K. Giles, S. Jarvis, *Rapid Commun. Mass Spectrom.* **13**, 1165–1168 (1999).

6. some manufacturers providing parallel HPLC equipment: Waters Corp., Milford, MA 01757, USA, *www.waters.com*. Micromass UK Ltd., Manchester, M22 5PP, UK, *www.micromass.co.uk*. Merck KGaA, D-64293 Darmstadt, Germany, *www.merck.de/chromatography*. SEPIAtec GmbH, D-12487 Berlin, Germany, *www.sepiatec.com*. Biotage Inc., Charlottesville, VA 22906, USA, *www.biotage.com*.

7. German Pat. DE 199 57 489 A1 (May 31, 2000), L. Mueller-Kuhrt, R. God, H. Gumm, J. Binkele (to SEPIAtec GmbH).

8. R. God, H. Gumm, H. Stump, *Am. Lab.* **12**, 20 (2001).

9. E. M. Thurman, M. S. Mills, *Solid-Phase Extraction: Principles and Practice. Chemical Analysis*, vol. 147, Wiley, New York (1998).

10. R. God, H. Gumm, C. Heuer, M. Juschka, *GIT Lab. J.* **3**, 188–191 (1999).

PART

IV

ANALYSIS FOR COMPOUND STABILITY AND DRUGABILITY

CHAPTER

13

ORGANIC COMPOUND STABILITY IN LARGE, DIVERSE PHARMACEUTICAL SCREENING COLLECTIONS

KENNETH L. MORAND and XUEHENG CHENG

13.1. INTRODUCTION

The chemical, biological, and analytical activities that comprised early pharmaceutical drug discovery have been most aptly summed by Richard Archer (*The Automation Partnership*) in his commentary on "the drug discovery factory" as the inevitable result in the evolution of high-throughput parallel processing.[1] The introduction of high-throughput technology platforms, such as those discussed in this book, has included combinatorial chemistry and high-throughput screening. Both have had a profound effect on drug discovery and led to the advancement of novel biological targets and compound generation and optimization. In as much as it is necessary to instill quality control measures and process management in a manufacturing or production operation, the use of high-speed drug discovery tools in combinatorial chemistry has necessitated the development of a vast array of analytical technologies to support the characterization or qualification of large numbers of new chemical entities generated by robotic and automated chemistry platforms. Further the interdependency of these technologies in the acceleration of the drug discovery process has lead to an increased awareness of the importance of a systems organizational approach.

Modern drug discovery relies heavily on high-throughput biological screening (HTS),[2,3] as the goal is to identify a well-characterized organic compound or a "lead" that may be used to initiate a medicinal chemistry program. Major pharmaceutical companies generally maintain a large repository of compounds for lead discovery and optimization efforts. These compounds are synthesized internally through combinatorial or medicinal chemistry, purchased from commercial sources, or obtained from natural

Analysis and Purification Methods in Combinatorial Chemistry, Edited by Bing Yan.
ISBN 0-471-26929-8 Copyright © 2004 by John Wiley & Sons, Inc.

products. The size and the diversity of the repository compound collection are generally deemed important for the success of the initial lead discovery efforts. Thus a significant effort is devoted to increasing the numbers and the structural diversity of the collection available for HTS.

Unfortunately, often the compound screening effort is conducted with little thought to the origin, quality, and stability of the organic compounds being used in the screening process. In fact, in many instances compound synthesis and acquisition, or more generally compound supply and compound storage, are separated from the screening activity. While the widespread use of combinatorial chemistry has resulted in an awareness of the importance of the quality and purity of the library compounds measured at the time of synthesis,[4,5] there have been very few studies on the purity and stability of large, diverse screening collections. We will discuss the organization of these functions in an organic compound process cycle, which necessarily includes not only compound supply and screening but, additionally, compound storage. A map for this compound-processing cycle, as well as the primary analytical needs for each function, is illustrated in Figure 13.1.

A recent issue of the *Drug and Market Development* report addressed the concern of compound screening and its interdependency with compound supply and storage, stating, "... it may no longer be sufficient to provide increased throughput for screening while doing nothing to affect downstream bottlenecks in later-stage screening. Alternatively, it may no longer be sufficient to provide high-throughput screening solutions that fail to effectively interface with compound storage and retrieval systems."[6] More simply stated, in the management of compounds, storage is a critical factor between synthesis and screening efforts (Figure 13.2), as the com-

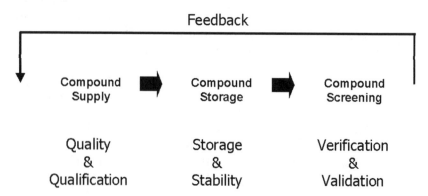

Figure 13.1. Organic compound process cycle for synthesis, storage, and screening of new chemical entities in early drug discovery programs.

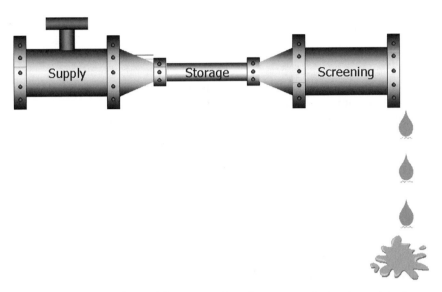

Figure 13.2. Interdependency of the compound supply, storage, and screening and can have a profound effect on the output of the organic compound process cycle. Compound stability, as well as robotic and data integration, within the compound supply function can significantly diminish high-throughput and high-speed biological assay platforms.

pound supply may be diminished by inappropriate or poor design, integration, and the storage of compounds in the corporate repository.

In particular, the quality of the repository compounds is important to drug discovery efforts. Poor sample quality can create many problems in biological assays. Impurities and decomposition products may result in false positives or a lower than expected concentration of the target compound. They may cause difficulties in determining accurate IC50 and Kd values or even preclude the possibility of finding an important drug lead in a primary screening. In addition unstable compounds, if not handled properly, can cause assay inconsistency if time-dependent changes occur in the concentration of target compounds and impurities. It is not uncommon for a large pharmaceutical company to have a repository of several hundred thousand to a million or more compounds. Thus, if even 10% to 15% of the collected compounds is poor quality, hundreds of thousands of dollars a year will be wasted in primary screening costs. A poor quality compound can create even more waste when it is subsequently tested in follow-up or functional assays.

As is common practice, repository compounds are stored either in the solid state or in solution at nominally mM concentrations. It is generally believed that storage in the dry state affords a longer shelf life compared with storage in solution. However, the use of solvent storage is generally necessary in the automated dispensing of large numbers or small amounts of samples. It is relatively easy to automate the transfer of liquids with available high-speed pipetting equipment, and many companies routinely store portions of some or all of their collections in a solvent for this purpose. Dimethyl sulfoxide (DMSO) is the most common solvent used because of its good solvating ability, its relative chemical inertness, and high boiling and freezing points. The samples are stored either at room temperature or in a frozen state and are thawed as needed.

Unfortunately, prolonged storage of organic compounds in solution can result in significant sample degradation and a general decline in the analytical qualification of the screening collection due to the exposure of the compound supply to a variety of atmospheric conditions. As an obvious deterrent to compound degradation, many pharmaceutical organizations now store their screening collections as frozen DMSO solutions in order to mitigate or retard compound degradation. A common approach is to store the sample solutions at temperatures ranging from as −20°C to 4°C, (−80°C for biologics), and low relative humidity, such as 5% to 20%. However, even under stable storage conditions, such as −20°C, and low relative humidity, compounds are exposed to a variety of atmospheric changes that can degrade the sample prior to use in a biological screening assay. For example, compounds stored at low temperature must be thawed prior to usage. Therefore they may be subjected to multiple freeze–thaw cycles throughout their shelf life. The repeated freeze–thaw cycles will precipitate or degrade the compound over time, and will eventually compromise its biological screening results. Likewise DMSO solutions stored at room temperature in order to lessen the deleterious effects of repeated freeze–thaw cycling are susceptible to accelerated degradation due to exposure to changing temperatures or sample precipitation resulting from the absorption of water by the DMSO solvent. It is apparent that the upfront analytical and synthetic efforts to generate high-quality samples for the compound supply can be readily lost because of poor storage conditions. In addition to the storage form and the sample handling conditions, the many other factors that can affect repository compound stability include physiochemical properties, humidity, atmosphere and oxygen content, storage container materials, salt form, or the presence of impurities. Information about the relative effects and actual magnitudes of these factors is desirable in order to select the optimal format and conditions for repository compound storage and handling.

Until recently there had been a paucity of information published on the proper design of compound storage facilities or their proper interface to the compound supply and screening functions in the overall compound processing cycle. More important, there has been little data published on the best storage conditions or on the stability profiles of large, diverse pharmaceutical compound screening collections. While the breadth of compound classes, functionality, and physiochemical properties associated with such a diverse set of compounds makes setting any one standard an arduous process, the proper design and use of the storage facility along with well-established expiration limits or expiry dates for compound retention is essential to extending the life and value of the compound supply and to ensuring the robustness and reliability of the biological screening data. Recently there has been increased activity in this area resulting in several publications[7–12] and more than a dozen conference reports in the last three years[13–27] compared to pre-2000 period.[28–30] Some of the interest here may be due to HTS maturing to become an integral part of most drug discovery programs, and some may be due to the recognition that the efficiency of discovering a novel drug is largely dependant on the quality and qualification of the biological screening compound collection (i.e., size, chemical diversity, purity, and stability).

Some of the issues associated with long-term soluble compound storage are addressed in Figure 13.3. While sparse and far from complete, we will summarize major findings from the authors' laboratories, as well as those presented at recent combinatorial and high-throughput screening conferences. This will include a discussion of the factors influencing compound

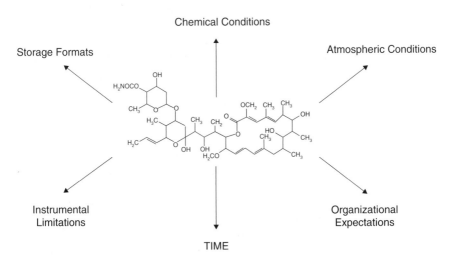

Figure 13.3. Environmental factors that can lead to compound degradation.

stability, such as atmosphere, humidity, temperature, container materials, storage format, and storage time. We will conclude by outlining future directions in this area. However, because of the relatively new interest in the stability of large, diverse compound screening collections, the reader is cautioned that much of the work published in this chapter is yet unpublished in peer review journals. Therefore the conclusions drawn on the part of the authors are preliminary and should be validated in subsequent experiments and laboratories. Finally, while studies on the stability of individual drugs in specific formulations at preclinical and clinical stages have been documented extensively as required by the Food and Drug Administration (FDA) for drug approval,[31-34] these studies are not directly applicable to repository compounds and will not be included in this review. It is suggested that the reader consult these references for more information.

13.2. COMPOUND STABILITY UNDER DIFFERENT STORAGE CONDITIONS

Compound loss related to the liquid storage of organic libraries for biological screening may result from a variety of processes such as precipitation, absorption, and decomposition, as well as the structural characteristics of compounds that determine reactivity, solubility, and absorption. However, the presence of other components in the sample such as salts, impurities, or excess acids and bases can also result in the increased reactivity or accelerated degradation for the compound of interest, while environmental factors, such as temperature, humidity, atmosphere and light and operational and processes parameters (e.g., storage time, storage format of solid state vs. DMSO solution), storage container type, sealing, and processes such as repeated freeze/thaw may also affect compound stability.

In this chapter we address many of the confounding factors that may compromise compound integrity following long-term solution storage. The chapter is divided to show the various environmental, container, and process conditions that influence compound behavior in solution (i.e., DMSO). The data presented were obtained from a variety of sources and include findings from a number of recent conference proceedings as well as work conducted in the authors' laboratories. However, the results we discuss should not be considered conclusive or a comprehensive as our access to corporate repository collections, storage systems and processing procedures was limited. We hope, however, that the results we present will provide useful guidance to readers developing, maintaining, or screening large diverse collections of pharmaceutical compounds.

13.2.1. Environmental Conditions—Temperature, Humidity, Oxygen

Temperature

During long-term storage, repository compounds are affected by a variety of the environmental factors. First and foremost, temperature can affect compound stability by accelerating or depressing the rate of degradation reaction when stored in solution. Most degradation processes have a positive activation energy[35,36] and will be accelerated as the solution temperature is increased. This effect may be modeled by an Ahrrenius type of relationship to predict degradation at various temperatures.[32,35,36] Darvas et al.[7,8] used kinetic modeling to develop a systematic method for assessing the stability of compounds generated using combinatorial chemistry and high-throughput organic synthesis. By their approach, Stabex™, compound stability is measured experimentally at several elevated temperatures, such as from 115°C to 190°C, for a selected set of compounds from a chemical library in order to determine an average range for the Arrhenius parameters. The experimental Arrhenius parameters were subsequently used to predict the expiry rate or shelf life of the library at selected storage temperatures. The prediction model was verified using experimental measurements at various intermediate temperatures, such as 75°C and 100°C. Darvas and co-workers observed that the apparent shelf life for the majority of the compounds studied experimentally at an intermediate temperature agreed well with the predictions based on kinetic parameters obtained at elevated temperatures. In one case 8 of 10 compounds exhibited good agreement with <15% difference between predicted and experimental stability profile.[8] For further verification, the authors have since applied the Stabex™ method in the study of a small repository of 3000 compounds obtained from approximately 300 chemical libraries, with the results projecting 96% of the compounds to be stable for two years or longer and 80% of the compounds to be stable for five years or longer at 25°C in the solid state.[8] We expect in the future to evaluate the accuracy of the model as data are obtained for compound retained for longer storage times.

For storage in a liquid DMSO format, several recent studies have shown the effect of temperature on the compound stability. Heaton et al. observed in a study of 530 selected compounds degradation rates of approximately 8% following four and a half months of storage at room temperature as compared to only 4% degradation after six months of storage at 4°C. Significant degradation was noted that was greater than 50% from the compound purity determined initially by NMR characterization at the start of the study. The samples were stored in sealed NMR tubes to minimize other confounding environmental effects.

A follow-up study of approximately 300 compounds stored in lidded, 96-well microtiter plates using LC/MS quantitation showed that storage at −20°C was more stable than at 4°C, which in turn is more stable than compound storage under room temperature conditions.[24] More recently Cheng et al. showed for a set of 644 structurally diverse compounds that the rate of change and decrease in concentration can be much higher at 40°C than at 22°C.[9,19] Compounds selected for this study were stable at room temperature for approximately half a year and only 15 weeks at 40°C. Interestingly inspection of the percentage of compounds that retained 80% or more of their initial concentration showed that the difference between the degradation rates at the two temperatures (22°C vs. 40°C) was close to 2 (Figure 13.4). This difference in concentration change is considerably smaller than would be expected based on Ahrrenius kinetics when including reasonable activation energies for thermal degradation.[35]

As a rule of thumb, it is generally assumed that there is a two- to four-fold increase in degradation rate for every 10°C increase in temperature.[32,36]

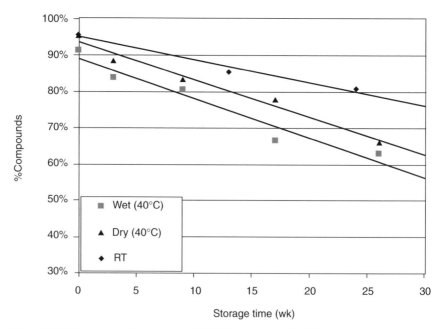

Figure 13.4. Percentage of compounds with 80% or more of the initial concentration remaining versus storage time at two different temperatures. The wet DMSO samples contained 5% water (v/v) and the dry and room temperature (RT) samples are in water-free DMSO.

However, when we examined the 52 most unstable compounds, we found that for samples stored at 40°C versus 22°C, those exhibiting observable degradation peaks in the UV chromatogram following storage to have much higher decreases than a factor of 2 (Figure 13.5). A similar analysis of 45 unstable compounds without new UV peaks after storage showed a smaller difference in the rate of compound loss between 22°C and 40°C (Figure 13.6). It is possible that the change in sample concentration for the latter set of unstable compounds resulted from some other mechanism than thermal degradation, such as precipitation or absorption, and that these processes did not abide by the same Arrhenius kinetics as thermal degradation. Nevertheless, the our observations indicate a limitation to projecting the data obtained at elevated temperatures in predicting compound stability at lower temperatures, especially for large and diverse compound collections. As compound loss processes may have weak temperature dependence or different mechanisms of compound loss at different temperatures, recognition of such confounding factors will be important in the design and utility of large repository storage systems.

Pure DMSO has a melting point of 18.4°C at normal atmospheric pressure. DMSO-solubilized compounds are often stored at or below 4°C in a frozen form to prolong their shelf life beyond a few months (e.g., 6–9

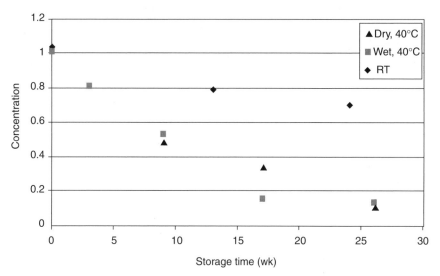

Figure 13.5. Average concentration for 52 unstable compounds for which new UV peaks were observed after storage.

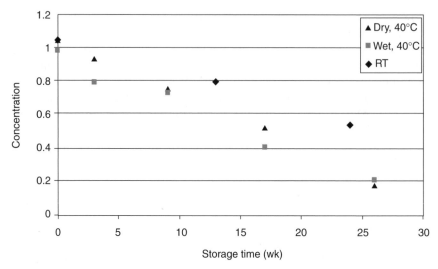

Figure 13.6. Average concentration for 45 unstable compounds for which no new UV peaks were observed after storage.

months), which is appropriate for room temperature storage. However, it is uncertain as to how different the stability would be for compounds stored in a frozen form at much lower temperatures than just below the DMSO freezing point (e.g., 4°C vs. −20°C vs. −80°C). Previous studies have indicated that compound storage at −20°C is preferable to 4°C, resulting in a more stable storage environment for the sample as the DMSO solution is presumably frozen. Heaton et al. studied 297 compounds with lidded plate seals and observed approximately 12% degradation at −20°C compared with 23% degradation at 4°C after six months of storage.[24] However, there is uncertainty about the sample state at 4°C as the melting temperature of the DMSO solution will depend greatly on the concentration and physiochemical properties of the analytes, as well as level of impurities and water in the solution. In fact 4°C may be relatively close to the melting point of the DMSO solutions for many compounds depending on the analyte concentrations. Nevertheless, the preponderance of the available data indicates that storage as a truly frozen DMSO solution is generally much more stable than in liquid format and offers satisfactory retention of compounds for up to five years when humidity and temperature are well controlled to minimize water uptake. In a subsequent study Heaton et al. observed only 3.5%

compound degradation by LC/MS analysis for 342 compounds stored as DMSO solutions at −20°C for two years with proper sealing.[24] Cheng et al. observed that compounds stored in DMSO at −80°C for approximately four years exhibited only 13% notable degradation in an HPLC analysis.

Aside from the absolute temperature, an equally important, yet unaddressed factor influencing the rate of compound loss is the rate at which the solubilized samples are frozen. The freezing rate has not been addressed in any of the aforementioned studies and the authors are not aware of work that has been presented or published on this issue or on subsequent compound integrity and precipitation rates. However, it should be noted that research on compound storage in combinatorial chemistry and high-throughput biological screening has only recently become important to the pharmaceutical industry, and therefore most published work has focused on predicting compound behavior within a framework of the existing storage systems and parameters. No doubt, as future studies focus more broadly on the theoretical aspects of compound integrity, the issue of freezing rates will be appropriately addressed.

Humidity

The presence of water can affect compound stability in several different ways. Water can participate or catalyze degradation reactions in solid or DMSO solution format. Additionally, in solid-state storage, water absorption can alter the morphology of the sample and the mobility of the molecules in the sample and thereby affect the rate of degradation or compound loss. In liquid storage, the presence of water can change the solubility of compounds in DMSO and, depending on the nature and concentration of the analyte in solution, cause compound precipitation.

Savchuk et al. recently showed the effect of relative humidity (RH) on the stability of 2212 compounds selected from a collection of combinatorial library compounds stored as frozen DMSO solutions at 4°C. Their results indicated that after only one year of storage there was significant degradation in 2% of the compounds stored at 20% RH, while there was a marked increase in the level and number of compounds exhibiting significant degradation in compounds maintained at a similar storage temperature but at higher relative humidity, namely an increase from 2% to 20% degradation at 40% RH and an increase to 24% degradation at 60% RH.[26]

Additionally, Cheng et al. have demonstrated in an accelerated study conducted at 40°C that the presence of 5% water in the DMSO samples has a clear detrimental effect on compound stability compared with water-free DMSO samples (Figure 13.4).[9,19] The authors suggest that compounds from the study, which are less stable in wet DMSO, may be related to those compound classes that are hydrolytically more labile under aqueous conditions, or alternatively, those compounds that are of lower solubility in wet DMSO. Interestingly no significant new peaks or signal were observed in the UV chromatogram for the compounds identified as less stable in wet DMSO samples at the zero time point when the 5% water was initially added to the compound solution. This indicated that sample precipitation is a more likely and important factor than compound degradation in the compound loss observed between wet and dry DMSO samples in this study.

Oxygen

Similar to the relative humidity of the storage environment, atmospheric oxygen can have a deleterious effect on the long-term storage of compounds, as oxygen in the atmosphere can react under certain conditions with compounds in the screening collection to form oxidation products. Oxidative degradation of pharmaceutical compounds is not uncommon, and there are numerous instances reported in the literature for individual compounds, most commonly in stability profiling of drug formulation.[37] Recently Yan et al. reported examples of oxidative degradation of compounds derived from a combinatorial library containing a thiophene substituent.[21] Cheng et al. examined the effect of compound storage in high relative humidity and "synthetic air" environments with varying levels of oxygen content.[9,19]

The latter study, Cheng et al., was conducted at an elevated temperature of 40°C under several different conditions to determine the relative importance of water and oxygen on compound stability.[9,19] The experimental design was such that each factor could be investigated independently or in relation with one another. Samples were stored in atmospheric control chambers filled with pure nitrogen or synthetic "air" (20% oxygen, 80% nitrogen) and dissolved in either DMSO or DMSO/H_2O (95:5 v/v). The determination of the percentage volume of water in DMSO was made based on the level of water absorption observed for the typical lifetime of HTS samples. Table 13.1 shows the relative concentration change for the samples under each set of atmospheric conditions at the 6-month time point. It is apparent from these data that the difference in concentration change between the oxygen-free and oxygenated samples are much smaller than the differences between the dry and wet samples and may suggest that humidity is a more important factor in compound stability.

Table 13.1. Concentration for 644 Compounds after Treatment at 40°C for 6 Months under 4 Different Conditions

Condition	DN	DO	WN	WO
Average	0.74	0.74	0.71	0.71
Median	0.86	0.87	0.84	0.82
80% remaining[a]	57%	60%	53%	53%

Note: DN = dry samples under nitrogen atmosphere; WN = wet samples under nitrogen atmosphere; DO = dry samples under a mixture of 20% oxygen and 80% nitrogen; and WO = wet samples under a mixture of 20% oxygen and 80% nitrogen.

[a] Percentage of compounds with 80% or more of initial concentration remaining.

13.2.2. Storage Formats—Solid, Dry Film, and Liquid or Frozen DMSO Solution

In practice, repository compounds are often stored in any number of formats: solid, dry film, liquid, or frozen solution, depending on the specific needs of the screening and repository functions. It is not uncommon for an organization to store each compound in dry powder form for long-term storage or archival, as a frozen DMSO solution or dry film for intermediate-term storage of one to five years, and as a room temperature solution for short-term storage and direct access for biological screening. Further there may be one archived collection in the most stable format (typically solid state in low humidity conditions at or below room temperature) and additional copies for central dispensing or for specific usage applications. Nonarchived collections are generally formatted to facilitate high-speed liquid dispensing, and therefore the long-term or multiple-year shelf life of the sample is sacrificed in favor of simple and rapid dispensing protocols relative to the laborious sample handling procedures necessary for samples stored in either a solid or frozen state format. Similarly the use of a dry film format has been especially prominent for the intermediate-term storage of combinatorial chemistry libraries for which the total amount of synthesized compound is at best only a few milligrams and the compounds are dried by solvent evaporation following HPLC separation. Finally we should note that this discussion on compound storage formats is primarily limited to those formats that are directly accessible for soluble or solution-based high-throughput biological screening. There are several alternative storage formats, such as bead-based storage or ChemCard technologies, that may be equally suitable for maintaining compound stability and sample integrity. However, these technologies are not uniformly used throughout the drug discovery industry and therefore are not included in this discussion.

A few recent studies have considered compound stability and integrity for each of the aforementioned storage formats. For example, Kozikowski et al.[20] evaluated the stability of approximately 1500 compounds maintained in a dry film format and stored for various lengths of time from one to four years. DMSO solutions of the compounds were analyzed by flow-injection analysis mass spectrometry (FIA-MS) prior to evaporation of the DMSO and dry film storage. Among the 1194 compounds that were detected in the initial analysis, 1055 (88%) were observed to be stable and present once reconstituted in DMSO following two years of storage in a dry film format.[20] Additionally Yan et al. conducted a stability study on 1% of the compounds in their collection of more than 300,000 compounds stored in a dry film format at −78°C. Similar to the results presented by Kozikowski, the authors reported that most of the compounds demonstrated good stability after two years of storage as a dry film.[21]

Darvas et al. investigated the stability of 3000 compounds selected from approximately 300 diverse combinatorial libraries by LC/MS analysis and observed that 96% of the compounds had a projected shelf life of two years or longer with less than 10% compound loss, while 80% of the collection had a shelf life of greater than five years when stored at 25°C in solid state storage. These stability studies were conducted, however, at 75°C for eight days, and the shelf life at 25°C was extrapolated from the data at 75°C and kinetic parameters obtained at elevated temperature for selected compounds in the test set.[7,8] Similarly Savchuk et al. conducted a stability study of 2212 selected compounds stored as frozen DMSO solutions at 4°C and 20% relative humidity. They observed 98% retention of the compounds following a single year of storage.[26]

These data confirm that storage in a solid state, either as a dry film or as a dry powder, will generally allow one to maintain the integrity of a large and diverse compound collection for a number of years (5 years or longer) when stored at or below room temperature with proper humidity control. Compound storage as frozen DMSO solutions is also suitable as long as the temperature is well below the melting temperature of the sample solution (see earlier discussions on temperature effect). On the other hand, compounds stored in DMSO at room temperature are much less stable than those stored in solid-state formats, and therefore current compound storage systems must be designed to minimize the exposure and length of time compounds are stored or processed at ambient atmospheric conditions prior to biological screening. There have been numerous studies designed to evaluate the shelf life of diverse compound collections stored as room temperature DMSO solutions.

In a recent study Kozikowski et al.[11] evaluated the stability of a broad diversity of compound classes representative of a typical pharmaceutical screening collection over a time period of one year in order to develop a predictive model representative of the stability behavior of their entire repository. The approximately 7000 compounds included in the study were a chemically diverse as a set and representative of the diversity of the chemical repository at Procter and Gamble Pharmaceuticals as determined by principal component analysis and comparison against the P&GP repository collection; that is, a total of 67% of variance in the original chemical descriptor space is accounted for by the first three principal components. Samples were prepared as 10 mM solutions in DMSO and stored at ambient atmospheric conditions in 96-well microtiter plates with fitted covers. The large sample number included in the study dictated that a high-throughput analytical method such as FIA-MS be used instead of quantitative method such as high pressure liquid chromatography–mass spectrometry (HPLC-MS). Therefore this study defines compound fate as sample purity or concentration were not determined. Samples were analyzed at three time points: at the $t = 0$ or the starting point of the study, and $t = 12$ months or the end point of the study, and at an intermediate time between 1 and 11 months. As expected, sample loss occurred with prolonged storage. After one month of storage at room temperature the probability of observing the compound was 95.3%, the probability of observing the compound after six months was approximately 80%, and after one year in DMSO the probability had been reduced to 52% (Figure 13.7).

More generally, by these results and subsequent independent studies by other research groups, the shelf life—defined as the time after which 20% or more of the compounds show substantial degradation (e.g., >50% degradation versus $t = 0$; Figure 13.6)—for large and diverse screening collections stored as room temperature DMSO solutions is between 6 to 12 months.[15,16] Further these studies indicate that 50% of the compounds will undergo substantial degradation after four years of storage in DMSO at room temperature. Nevertheless, it should be recognized that results may vary for the various compound classes, sample concentrations, and other noncontrolled environmental conditions.[15,19,20]

13.2.3. Sample Container Material—Plastic or Glass

Besides environmental conditions, sample storage container materials can affect compound stability and recovery because of the potential for the

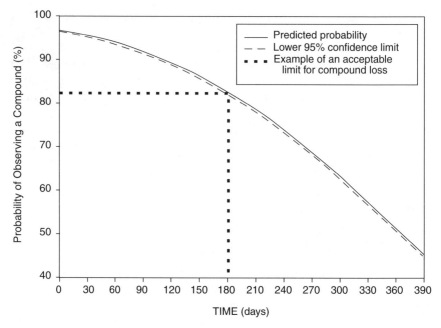

Figure 13.7. Relationship between the probability of observing a compound as a function of time in storage for a series of pharmaceutically relevant compounds stored as 10mM DSMO solutions at ambient atmospheric conditions.

compound to be absorbed into the container or adsorbed onto the surfaces of the container. Further the porosity of the container can result in the permeation of oxygen and water from the surrounding environment, and in some instances the surface of the containers may catalyze compound degradation. For example, residual silanol groups can be present on the surface of glass containers, so plastic containers are a more inert and appropriate storage material for long-term storage. However, many plastic containers contain plasticizers and other additives that can be extracted by DMSO, so they present yet another manner to contaminate the sample solution or promote degradation. In either case the outcome will be poor or nonvalidated biological screening results. Cheng et al. recently studied the effect on compound stability for a library of compounds stored as DMSO solutions in plastic and glass containers for five months.[9,19] A polypropylene microtube, which exhibits a low level of extractable material based on the manufacturer's specifications, was selected as the plastic storage container. Earlier studies reporting on the stability of marketed

Table 13.2. Concentration for Glass versus Plastic Containers and Available Data
for Long-Term Room Temperature Storage Samples

Experiment	Glass	Plastic	RT	RT
Storage time (month)	5	5	3	6
Average	0.93	0.95	0.91	0.88
Median	0.98	0.99	0.98	0.96
80% remaining[a]	87%	90%	85%	80%

[a] Percentage of compounds with 80% or more of initial concentration remaining.

drugs in glass as opposed to plastic syringes indicated no detectable difference between the two types of containers.[38] In the present study the authors observed little change in the concentration of 644 compounds stored in either glass or polypropylene containers, Table 13.2, even following five months of storage at room temperature. ^{1}H-NMR measurements, however, indicated the presence of extractable materials from the polypropylene container at the level of 20 µM (total hydrogen atom integration) when 500 µL DMSO was stored in the container for five months. Table 13.2 shows the results for each of the container materials, as well as a comparison with the results for the long-term room temperature stability study. Polypropylene containers are generally easier to work with in automated systems, and therefore they are appropriate for storing repository compounds in DMSO. Nevertheless, these results are for a limited number of storage container material types and therefore should not be considered predictive for other storage containers. We expect further studies to provide a much broader profiling of the available container formats and materials.

However, DMSO extractable compounds in the container and sealing materials can directly affect biological screening results, leading to false positive biological activity. Yaskanin et al.[22] considered the effect of DMSO extractables derived various plate sealing materials on target inhibition in an enzyme assay. Microtiter plates loaded with DMSO were sealed using various commercially available sealing technologies. The sealed plates were stored at either 22°C or 40°C in both an upright and inverted configuration to promote extraction of the sealing materials. Following storage for 48 hours, a qualitative LC/MS analysis and enzymatic assay were performed on the DMSO solutions to determine the presence of potential extractable compounds and their effect on enzyme inhibition. Control DMSO samples were stored in glass vials and in unsealed plates, and they did not show an appreciable level of enzyme inhibition nor were peaks observed by LC/MS analysis. For the seven seals tested in this study, enzyme inhibition levels ranged from as low as 7% to 92% for plates stored at 22°C and increased to a range from 15% to 96% for plates stored at 40°C. Interestingly extractable compounds were only identified by LC/MS analysis for two of the seven seals tested in this study, although all seven sealing materials showed some level of enzyme inhibition.

13.2.4. Freeze–Thaw Cycles

Depending on the storage format, repeated freeze–thaw cycles may be required for compound dispensing, and sample stability under such conditions is of general importance. Water absorption and compound precipitation are important practical issues when freezing and thawing samples. Kozikowski et al. examined the effect of freeze–thaw cycling on stability for compounds stored as 20 mM DMSO solutions. Approximately 320 compounds selected from a number of commercial vendors were included in the freeze–thaw study. The set of compound had average initial purity of 96%. Compounds were stored at 4°C under argon in pressurized canisters to simulate a low humidity environment and subjected to 25 freeze–thaw cycles while being exposed to ambient atmospheric conditions after each thaw cycle to simulate the time and manner by which compound plates are exposed to the atmosphere during typical liquid handling processes. The compounds were analyzed by HPLC-MS with evaporative light-scattering detection following every fifth freeze–thaw cycle to quantitate the change and amount of material remaining in solution. Samples were neither agitated or vortexed to facilitate the dissolution of any precipitated material in the plates. Control plates were stored either at room temperature under argon or at 4°C under argon without freeze–thaw cycling, and were evaluated only at the midpoint and end of the study. Interestingly the percentage of compound remaining at the end of the study was greatest for the frozen and never-thawed controls, and was least for the compounds exposed to the multiple freeze–thaw cycles (83.1% vs. 55.8%, respectively) as seen in Figure 13.8. As expected, the percentage of compound remaining decreased for all three storage methods. Additional peaks were not observed in any of the HPLC chromatograms, indicating the absence of soluble degradation products. However, solid precipitate was observed in many of the solutions at the end of the study.[17]

In contrast Heaton et al. observed no significant compound loss following 25 freeze–thaw cycles at 2 mM in DMSO.[14,24] Additionally Cheng et al. conducted a freeze–thaw study using 10 mM DMSO solutions for 644 com-

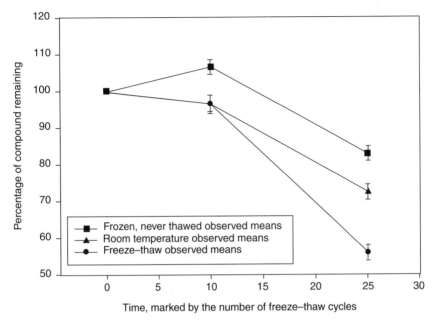

Figure 13.8. Relationship between the percentage of compound remaining and the number of freeze–thaw cycles for a series of pharmaceutically relevant compounds prepared as 20 mM dimethyl sulfoxide (DMSO) solutions and stored at 4°C under argon in pressurized canisters to simulate a low humidity environment.

pounds.[9,19] In this study compounds were frozen at −15°C and were thawed under a dry nitrogen atmosphere to prevent water absorption. Two methods of compound dissolution were selected following each freeze–thaw cycle: agitation with a vortexing apparatus (AG) and mixing with repeated aspiration and dispensing using a liquid handling robot ("suck-and-spit," or SS). Figure 13.9 shows the results of this freeze–thaw study. Overall only a slight decrease in the compound concentration was observed following the 11 freeze–thaw cycles. These results are quite similar to the results presented by Heaton et al. at a sample concentration of 2 mM in DMSO,[14,24] but markedly different from the results of Kozikowski et al. for the study conducted at 20 mM in DMSO.[17]

In the study by Cheng et al. there was no apparent difference between the freeze–thaw results and results from added dissolution, such as the agitation and the repeated aspiration/dispensing methods. This is understandable as the compound loss overall from freeze–thaw cycling is minimal as is compound precipitation. Visual inspection of the glass vial's bottom indi-

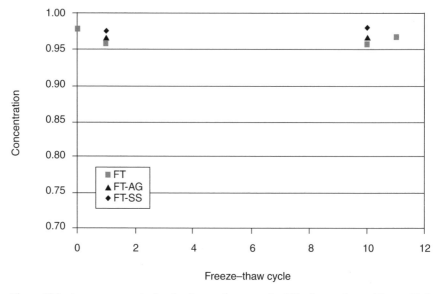

Figure 13.9. Average concentration for freeze–thaw samples. FT = freeze–thaw with no added re-dissolvation; FT-AG = freeze–thaw followed by re-dissolvation with agitation; FT-AG = freeze–thaw followed by re-dissolvation with repeated aspiration and dispensing.

cated that precipitation of 2% of the compounds occurred after one freeze–thaw cycle, 3% after 10 cycles, and 3% after the 11 cycles. In general, there was good correlation between the visual observation of compound precipitation and the decrease in concentration, suggesting that precipitation is an important factor in compound loss during freeze–thaw cycling. As shown in Figure 13.10, only 11% of the compounds that exhibited no observed precipitation had a final concentration of less than 80% of initial following 11 freeze–thaw cycles versus 55% of the compounds for which precipitation was observed. Additionally no significant new peaks were observed in the UV chromatograms for these samples after 11 cycles of freeze–thaw, indicating that degradation is not important for compound loss in the freeze–thaw process.[9,19]

It is possible that the differences reflected in these three studies is based solely on the disparity in the concentration of samples used in each experiment, as it would be expected that the higher concentration solutions would likely result in a higher level of compound precipitation following a freeze–thaw cycle. However, in the Kozikowski study the compound solutions were thawed under argon and then exposed to ambient atmospheric conditions for about 2 hours to simulate the time that compound plates are

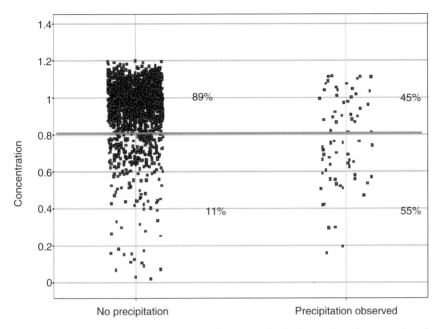

Figure 13.10. Correlation between compound concentration in freeze–thaw (average of tenth and eleventh cycles) and observed compound precipitation during freeze–thaw.

exposed to the atmosphere during liquid handling. Accordingly the DMSO solutions under these conditions would have absorbed larger amounts of water, leading to higher levels of sample precipitation at 20 mM. Finally, and perhaps most important, differences in the thaw and freezing rate were not noted for any of the three studies, although the freezing rate has a significant impact on compound precipitation. The differences in study designs and results for each of these studies only underscore the extreme difficulty in comparing results across various laboratories for even a single element of the compound storage process. As additional issues are addressed and presented in peer-review literature, differences similar to those presented in these studies should be resolved to benefit the entire industry.

13.2.5. Compound Structure, Physical/Chemical Properties, and Salt Form

Factors intrinsic to the nature of the repository compound samples such as structure, physical/chemical properties, salt form, the presence of other components (stereoisomers, impurities), and pH values of the samples have a variety of effects on compound stability.

Savchuk et al. has reported on a number of examples of compound degradation following solid-state storage and suggested that hydrolysis was a major pathway for compound degradation in their compound collection.[26] However, the collection may be especially sensitive to water absorption during storage and sample or liquid handling procedures. Also Yan et al. discussed several instances of compound degradation due to hydrolysis and oxidation mechanisms,[21] while Darvas et al. demonstrated the stability of a combinatorial library containing a labile core structure with an electro-rich pyrrole moiety could be improved by using electron-withdrawing or bulky substitution.[8] Cheng et al. observed several elimination reactions, such as loss of 17 Da (water), 17 Da (ammonia), 36 Da (HCl), and 28 Da (CO).[19] In contrast, very few direct oxidation products were observed (e.g., +16, +32 Da) in their study, Table 13.3, and these results correlate well with the observation that the water content in the sample has a more significant effect on compound stability than the atmospheric oxygen content. Figure 13.11 gives an example of the degradation products observed for the compounds in this study. Additionally many compounds showed a significant decrease in concentration, with no new components identified in the UV or MS chromatograms. These results suggest that the observed compound loss was due to precipitation. As the unstable compounds show a broad base-line in the UV chromatogram with no clearly distinguishable new peaks, this is probably due to polymerization. Figure 13.12 shows two examples where compounds were lost with no readily identified new UV peaks after storage.

Table 13.3. Degradation Products for 644 Compounds Stored at 40°C or Room Temperature Conditions

Mass Change	Possible Transformation
−16 Da	Loss of O atom (reduction)
−17 Da	Loss of NH3
−18 Da	Loss of H2O
−28 Da	Loss of CO
−33 Da	Loss of HONH2
−36 Da	Loss of HCl
+12 Da	Addition of C atom (reaction with DMSO?)
+16 Da	Addition of O atom (oxidation)

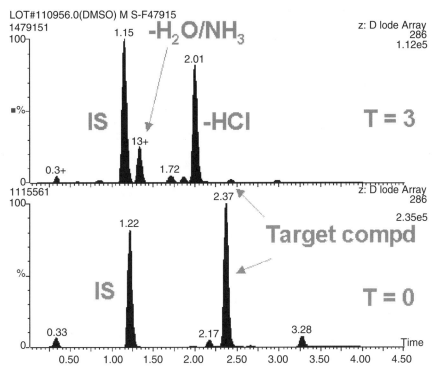

Figure 13.11. Observed degradation products for a compound after storage at 40°C for 18 weeks at 10 mM in DMSO.

Clearly, HPLC has become a preferred choice for the purification of combinatorial chemistry libraries. Trifluoroacteic acid (TFA) is frequently used as an additive to the HPLC mobile phase during gradient elution to improve the quality of the compound separation for compounds with ionizable side chains. Following HPLC-based purification the collected sample fractions are dried by solvent evaporation along with the residual, nonvolatile TFA. Recently Yan et al. reported a study where compound degradation was accelerated for compounds purified and stored as TFA salts.[21] Hochlowski et al. conducted a similar study using the TFA salts for appriximately 50 compounds and also observed that the TFA salt increases degradation in several types of structural fragments.[10] Newer compound purification methods, such as supercritical fluid chromatography, that do not use TFA should eliminate these issues, and we look forward to the continued developments in these techniques.

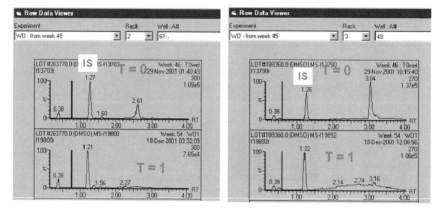

Figure 13.12. Examples of unstable compounds for which no readily identified products of degradation were observed. The two compounds were stored for three weeks at 40°C in 10 mM DMSO.

13.3. SUMMARY

Large, diverse compound libraries are used in pharmaceutical research for drug discovery and development. Historically the efforts have been largely to increase the size and the chemical diversity of repository compound collections in order to increase the efficiency of HTS. Benchmarking the effects of prolonged storage on the integrity of such collections is important. This involves determining the retention time and expiration limits for compound storage to ensure a high-quality screening collection. There is growing interest to examine the purity and chemical stability of large diverse compound collections as poor sample quality and unstable compounds can interfere with subsequent processes in new drug discovery. There are many environmental conditions that may ultimately lead to compound degradation: chemical, atmospheric, or simply the formats used for storage of the compound supply. As a community of combinatorial and analytical chemists a large portion of the effort to date has been on the development of applications and technologies that contribute to the analytical qualification of chemical libraries. While this effort is important, it must be balanced against the need to develop similar applications and technologies for maintaining and monitoring the long-term integrity of the collection during storage.

The data presented in this chapter were accumulated from the literature and the authors' laboratories. They provide for us some insight into the effects of storage format, environmental conditions, and sample handling procedures on compound stability. Structural features, which lend to compound degradation, can now be identified and unstable compounds eliminated or phased out from existing compound collections. We find additionally that the chemical stability of a compound should be included in the list of properties that make it suitable for inclusion in the repository compound collection and in the selection criteria for compound synthesis or purchase. Because of the profound effect of atmospheric humidity on compound loss, a limited number of freeze–thaw cycles should be performed on frozen DMSO solutions prior to the compound stock being depleted or discarded. We recommend that other options, such as pre-aliquotted solid-state or frozen DMSO solutions, be considered as there is no need for repeated freeze–thaw cycling and both forms of storage have been proved to provide an acceptable retention and life cycle.

However, in the future more studies must be made on the effects of salt forms (e.g., TFA), impurity profilings, freezing and thawing rates, and computational models for predicting compound class and library stability. As we have briefly discussed in this chapter, TFA salts are present in large quantities in many compound collections because of the widespread use of TFA modifiers in HPLC-based purification applications and as a cleavage agent for solid-phase synthesis in combinatorial chemistry. Recently there has been limited observation of the long-term deleterious effects of TFA salts on the stability of compounds stored as DMSO solutions due to water uptake and the potential for acid-catalyzed degradation. As a result compound purification methods that do not use TFA are preferred to the existing HPLC methods that rely heavily on the use of TFA as mobile phase additive. Additionally we discussed QSAR methods that optimize the storage conditions for highly labile compounds, as they can be applied to stability data sets to generate compound class specific structure-property relationships that correlate to compound degradation.

The need to properly manage compound storage and stability is being widely recognized by the pharmaceutical industry. At the time this chapter was being prepared, the authors had just completed a half-day workshop on repository compound stability studies at the 2002 IBC Drug Discovery Technology World Congress.[19,21–26,39] Not surprisingly, the meeting attracted individuals from a host of industries to include the pharmaceutical and biotechnology industries as well as representatives from compound suppliers, brokerage houses, combinatorial chemistry and biological screening companies, and storage system and lab automation manufactures. More important, the meeting attracted scientists with a broad range of expertise in analytical and medicinal chemistry, computational chemists, and high-throughput biological screening, which supports our contention on the importance of the issue of compound degradation and loss. Clearly, the design of robust storage systems is not something that can be easily or quickly solved by any single organization. In as much as this chapter was a collaborative effort to summarize the efforts of a broad group of people and organizations, we hope that a collaborative effort of the pharmaceutical industry, compound suppliers, and storage system and consumables manufactures will address such issues and contribute to improvements in the productivity and acceleration of the drug discovery process.

ACKNOWLEDGMENTS

The authors would like to thank Drs. Jill Hochlowski, David Burns, Carl Beckner, and Alex Buko; and Robert Schmitt, Hua Tang, Darlene Hepp, Stan Kantor, and Fanghua Zhu from Abbott Laboratories, as well as Drs. Barbara Kozikowski, Barbara Kuzmak, Sandra Nelson, and David Stanton; and Tom Burt, Kathy Gibboney, Nathan Hall, Debra Tirey, and Lisa Williams from Procter and Gamble Pharmaceuticals for contributing to the work and for helpful discussions in the preparation of this chapter.

REFERENCES

1. R. Archer, *Nature Biotechnol.*, *17*, 834 (1999).

2. R. P. Hertzberg, A. J. Pope, *Curr. Opin. Chem. Biol.* *4*, 445–451 (2000).

3. J. J. Burbaum, N. H. Sigal, *Curr. Opin. Chem. Biol.* *1*, 72–78 (1997).

4. B. D. Dulery, J. Verne-Mismer, E. Wolf, C. Kugel, L. V. Hijfte, *J. Chromat. B725*, 39–47 (1999).

5. D. A. Yurek, D. L. Branch, M.-S. Kuo, *J. Comb. Chem.* *4*, 138–148 (2002).

6. K. Rubenstein, C. Coty, *Drug and Market Development Report* 1–12 (May 2001).

7. F. Darvas, T. Karancsi, P. Slegel, G. Dorman, *Gen. Eng. News* *7*, 30–31 (2000).

8. F. Darvas, G. Dorman, T. Karansci, T. Nagy, I. Bagyi, in K. Nicolaou, R. Hanko, W. Hartwig, eds., *Handbook of Combinatorial Chemistry*, Wiley-VCH, New York, pp. 806–828 (2002).

9. X. Cheng, J. Hochlowski, H. Tang, D. Hepp, C. Beckner, S. Kantor, R. Schmitt, *J. Biomolecul. Screen.*, *8* (3), 292–304 (2003).

10. J. Hochlowski, X. Cheng, D. Sauer, S. Djuric, *J. Comb. Chem.*, *5* (4), 345–349 (2003).

11. B. A. Kozikowski, T. M. Burt, D. A. Tirey, L. E. Williams, B. R. Kuzmak, D. T. Stanton, K. L. Morand, S. L. Nelson, *J. Biomolecul. Screen.*, *8* (2), 205–209 (2003).

12. B. A. Kozikowski, T. M. Burt, D. A. Tirey, L. E. Williams, B. R. Kuzmak, D. T. Stanton, K. L. Morand, S. L. Nelson, *J. Biomolecul. Screen.*, *8* (2), 210–215 (2003).

13. K. E. Milgram, M. J. Greig, *Proceedings of 48th ASMS Conference on Mass Spectrometry and Allied Topics*, Long Beach, CA (2000).

14. Z. Heaton, S. Holland, R. Hughes, R. Lifely, L. Robb, *Society for Biomolecular Screening Conference*, Baltimore, MD (2001).

15. S. Nelson, *Society for Biomolecular Screening Conference*, Baltimore, MD (2001).

16. L. E. Williams, B. A. Kozikowski, T. Burt, K. Morand, *Society for Biomolecular Screening Conference*, Baltimore, MD (2001).

17. B. A. Kozikowski, L. E. Williams, D. Tirey, B. Kuzmak, K. Morand, *Society for Biomolecular Screening Conference*, Baltimore, MD (2001).

18. J. A. Kelley, C. C. Lai, J. J. Barchi, M. C. Nicklaus, J. H. Voigt, L. Anderson, N. M. Malinowski, N. Neamati, *Proceedings of 50th ASMS Conference on Mass Spectrometry and Allied Topics*, Orlando, FL (2002).

19. X. Cheng, *Post DDT2002 session on Repository Compound Stability Studies*, Boston (2002).

20. B. A. Kozikowski, *Post DDT2002 session on Repository Compound Stability Studies*, Boston (2002).

21. B. Yan, *Post DDT2002 session on Repository Compound Stability Studies*, Boston (2002).

22. D. Yaskanin, *Post DDT2002 session on Repository Compound Stability Studies*, Boston (2002).

23. J. Warhurst, *Post DDT2002 session on Repository Compound Stability Studies*, Boston (2002).

24. Z. Heaton, *Post DDT2002 session on Repository Compound Stability Studies*, Boston (2002).

25. M. Turmel, *Post DDT2002 session on Repository Compound Stability Studies*, Boston (2002).

26. N. Savchuk, *Post DDT2002 session on Repository Compound Stability Studies*, Boston (2002).

27. K. Morand, *Cambridge Health Institute 7th Annual Conference on High-Throughput Organic Synthesis*, February 13–15 San Diego, CA (2002).

28. M. J. Ortner, *CHI conference on Studies on the Impact of Solvent and Storage Conditions on the Stability of Different Chemical Groups within High Throughput Screening Libraries*, UK (1995).

29. F. Darvas, *SBS 5th Annual Conference and Exhibition*, Edinburgh, UK, September 13–16 (1999).

30. M. Valenciano, *Proceedings of the SBS CSS Discussion Group on Substances, SBS 5th Annual Conference and Exhibition*, Edinburgh, UK, p. 127 (1999).

31. DHHS, *Guideline for Submitting Documentation for the Stability of Human Drugs and Biologics, Food and Drug Administration*, February (1990).

32. B. Kommanaboyina, C. T. Roodes, *Drug. Dev. Ind. Pharm.* **25**, 857–868 (1999).

33. R. J. Lantz Jr., *Drug Div. Ind. Pharm.* **10**, 1425–1432 (1984).

34. D. C. Monkhouse, *Drug Div. Ind. Pharm.* **10**, 1373–1412 (1984).

35. L. Kennon, *J. Pharm. Sci.* **53**, 815–818 (1964).

36. G. Anderson, M. Scott, *Clin. Chem.* **37**, 398–402 (1991).

37. Y. Wu, *Biomed. Chrom.* **14**, 384–396 (2000).

38. W. F. Stanaszek, I.-H. Pan, *Proc. Okla. Acad. Sci.* **58**, 102–105 (1978).

39. Workshop on Repository Compound Stability Studies, *IBC USA Conferences at the 2002 Drug Discovery Technology World Congress*, Boston, August 8 (2002).

CHAPTER

14

QUARTZ CRYSTAL MICROBALANCE IN BIOMOLECULAR RECOGNITION

MING-CHUNG TSENG, I-NAN CHANG, and YEN-HO CHU

14.1. INTRODUCTION

Biomolecular recognition is fundamental to biochemical and biomedical research. These include functional genomics and proteomics, combinatorial chemistry, disease diagnosis, new drug discovery, and clinical immunology. Most of the biomolecular recognition involve specific interactions between biomolecules, such as antibody-antigen, receptor-ligand, and enzyme-substrate.[1-3] Study of biomolecular recognition is essential in order to understand the mechanisms involved. Until now many methods used to determine binding interactions are immuno-based assays, including enzyme-linked immunosorbent assay (ELISA), fluoroimmuno assay (FIA), and radioimmuno assay (RIA). Although these methods carry good specific selectivity and sensitivity to evaluate the biomolecule recognition, they are still problematic. For example, all assays require chemical labels. In addition ELISA is time-consuming and RIA has the waste disposal problem. FIA must use an expensive fluorometer for measurements. Therefore it always has needs to develop rapid, in-time, label-free, and effective analytical methods to evaluate and screen specific bindings between biomolecules.

Recent advances in analytical and microengineering technologies have made the impact on the new development for studies of biomolecular recognition. One of the current interests is focusing on biosensors. The transducers for these sensors include surface plasmon resonance (SPR),[4] optics (OP),[5,6] surface acoustic waves (SAW),[7] quartz crystal microbalance (QCM),[8,9] and electrochemistry (EC).[10-14] In comparison with conventional immunoassays, most of these sensors are rapid and real-time measurements, and some are more sensitive. Among them, SPR-, QCM-, and SAW-based technologies have the advantages of being label-free, monitored on-line,

Analysis and Purification Methods in Combinatorial Chemistry, Edited by Bing Yan.
ISBN 0-471-26929-8   Copyright © 2004 by John Wiley & Sons, Inc.

351

and available commercially. Because of the current technical difficulties of fabrication and component packaging, the SAW technique has lacked immediate commercialization.

As a mass sensor, QCM directly measures the changes in mass density occurring on the sensor surface.[15,16] The SPR transducer detects the dielectric constant differences on the sensor surfaces.[4,16,17] Both techniques employ the evanescent waves on the sensor surface.[16] That is to say, both techniques are limited by distance in their detection from sensor surfaces. The typical values for QCM range from 250 to 400 nm in the 5 to 10 MHz region;[18–20] in the case of SPR, the values are 200 to 300 nm in visible region of spectrum.[4,17] Beyond these distances the analyte can only be detected if the noise from bulk solution is completely eliminated. Both QCM and SPR are well suited for flow-injected analysis systems; in the other words, the techniques require no washing steps when applied to biochemical systems. Interestingly the principles of operation for QCM and SPR are different, but they come to the same results in evaluating the binding interaction between biomolecules.[16] In this chapter we discuss these principles and applications of QCM in some detail.

14.1.1. Basic Theory of QCM Technique

In 1880 Jacques and Pierre Curie discovered that a piezoelectric property exists in crystals such as quartz, rochelle salt ($NaKC_4H_4O_6 \cdot 4H_2O$), and tourmaline. This is a property that produces a small amount of voltage as pressure is applied to the crystal. The phenomenon was then termed the "piezoelectric" effect.[21] The prefix "piezo" was borrowed from Greek *piezein*, which means pressure. The Curies also demonstrated that as these crystals were supplied with certain voltages by connecting them to an electrical circuit, the crystals produced a mechanical stress, a so-called reverse piezoelectric effect. It is now known that the crystal does not possess central symmetry; that is, the projection of any ion through the center is to an ion of opposite polarity. As mechanical stress is applied to the crystal, the temporal electric dipole frozen and perpendicular to the direction of stress, and produces a temporal electric field.[22] Consequently, if the piezoelectric crystal is connected to an oscillation circuit, the piezoelectric crystal will vibrate at a certain frequency. Among 21 crystallographic systems lacking a central symmetry, 20 of them are piezoelectric. Among the piezoelectric crystals, quartz is most widely used in industrial applications, mostly because of its low cost, inert chemical properties, and good mechanical properties.

In 1957 Sauerbrey quantified the relationship between the vibrating frequency and deposition mass in the piezoelectric quartz crystal. The relationship was described as[23–25]

$$\Delta F = -2F_R^{\ 2} \frac{\Delta M}{(\rho_Q \mu_Q)^{1/2} A} = C F_R^{\ 2} \frac{\Delta M}{A}, \tag{14.1}$$

where ΔF is the measured frequency shift, F_R is the frequency of the quartz crystal prior to mass deposition, ΔM is the mass deposition on quartz crystal, $\rho_Q = 2.65\,\mathrm{g\,cm^{-3}}$ is the density of the quartz, $\mu_Q = 2.95 \times 10^{11}\,\mathrm{g\,cm^{-1}s^{-1}}$ is the shear modulus of the quartz crystal, and A is the piezoelectricity active area on quartz crystal, $C = -2.26 \times 10^{-6}\,\mathrm{cm^2/(Hz \times g)}$. From (14.1) it is obvious that the frequency of quartz is decreased as the mass loaded on quartz surface is increased. Therefore the quartz microbalance (QCM) can be used to measure the mass deposited on a surface, and this is important in the semiconductor industry. In 1972 Lewis found that QCM was only capable of weighing up to about 10% of the weight of the quartz crystal.[26] This confirmed that QCM has its own limitation in detection, as was previously suggested. Also the equation is suitable only when the deposited material has elastic properties.[26] Since then QCM has been used to detect air pollutants, such as carbon dioxide, sulfur dioxide, volatile organic compounds, and organic sulphides.[27–29]

What about the detection behavior of QCM if applied in solutions; that is, does the deposited material have viscoelastic properties? In 1981 Nomura and Iijima first reported the QCM measurement in liquid medium.[30] Since then much effort has been devoted to measuring QCM in solution. It appears that the frequency of quartz changes with the density, viscosity, conductivity, and dielectric constants of the solution studied.[24,25,31–33] In addition the roughness of deposition material[34] and the nature of the electrode[35–37] used on the quartz's surface can affect the frequency of QCM. The Sauerbrey expression in (14.1) is therefore modified as[24,25,38–40]

$$\Delta F = \frac{C F^2 \Delta M}{A} + C F^{3/2} (\Delta\eta_L \Delta\rho_L)^{1/2}, \tag{14.2}$$

where $\Delta\eta_L$ and $\Delta\rho_L$ are the change of the viscosity and the density of the solution, respectively. In (14.2) we see that the frequency shift of the QCM in solution is related not only to the mass loading on the QCM but also to the solution properties. By using continuous flow as the means for QCM experiments, we can keep the physical properties of the running buffer solution constant (i.e., negligible $\Delta\eta_L$ and $\Delta\rho_L$). The second term in (14.2) can then be neglected. Thus (14.2) becomes (14.1), and the frequency shift is only dependent on the mass loading of QCM. In 1983 Guilbault immobilized the biomolecule on QCM to detect formaldehyde in air. It was the first time that QCM was used as a biosensor.[29] Since then various attempts have succeeded in developing QCM for biosensor applications.[15,41] Recently the QCM biosensor has begun to play an important role in investigations of the interactions among biomolecules,[42] including those of affinity and kinetic measurements. Compared to the commercialized SPR technique, the QCM has the advantages of simple instrumentation design and very low cost. In this chapter we use the QCM biosensor to investigate the biomolecular recognitions, quantitative binding constant measurements, and high-throughput applications.

14.2. MOLECULAR RECOGNITION

The study of specific interactions between ligands and proteins is important in today's biochemical research. Most of the interactions are not covalently bound, and exist between substrates and enzymes, antigens and antibodies, hormones, drugs, and receptors, nucleotides and proteins, and proteins and small organic molecules.

14.2.1. Materials

1. Research grade reagents and distilled deionized water for all experiments.
2. Apparatus consisting of a QCM system, which is available commercially (ANT Tech. *www.anttech.com.tw*) and includes a frequency meter, an oscillator, a port connected to a computer, a software to collect and display on a computer screen, a flow-injected analysis system.
3. AT-cut piezoelectric crystals (9 MHz, ANT Tech.) with vapor deposition of Cr metal first, and Au metal later on each side.
4. An auto-sampling system.
5. A crystal cleaning solution: 1 N NaOH and 1 N HCl. A clean crystal surface is important for successful QCM experiments.

14.2.2. Real-Time Molecular Recognition by QCM

Aryl Sulfonamides as Carbonic Anhydrase Inhibitors

It is well documented in the literature that the sulfonamide anion (SO_2NH^-) specifically binds to the carbonic anhydrase and inhibits its activity. This sulfonamide anion forms a complex with zinc cation at the active site of carbonic anhydrase.[43] Taylor proposed a two-step mechanism for this interaction of aromatic sulfonamide with carbonic anhydrase.[44,45] In the first step, the aryl sulfonamide molecule enters the active site of the enzyme by the hydrophobic interaction; in the second step, the sulfonamide associates with the zinc ion of the carbonic anhydrase. Various methods have been employed to investigate the thermodynamics and kinetics of the two steps; these include affinity capillary electrophoresis (ACE),[46,47] fluorescence,[44,48] surface plasma resonance (SPR)[49], and the carbon dioxide hydration inhibition method.[50] The dissociation constant of this binding interaction is on the order of 10^{-6} to 10^{-9} M. For the QCM experiments, we prepared a sulfonamide derivative having a free thiol group. This thiol is the basis of sulfonamide immobilization on the gold surface of the quartz crystal. The self-assembled monolayer (SAM) established the sulfur–gold chemisorption bonding interaction and was ready for specific binding with the carbonic anhydrase (Figure 14.1).

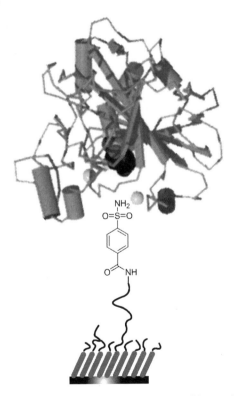

Figure 14.1. Binding of carbonic anhydrase to a sulfonamide-terminated self-assembled monolayer (SAM) on QCM.

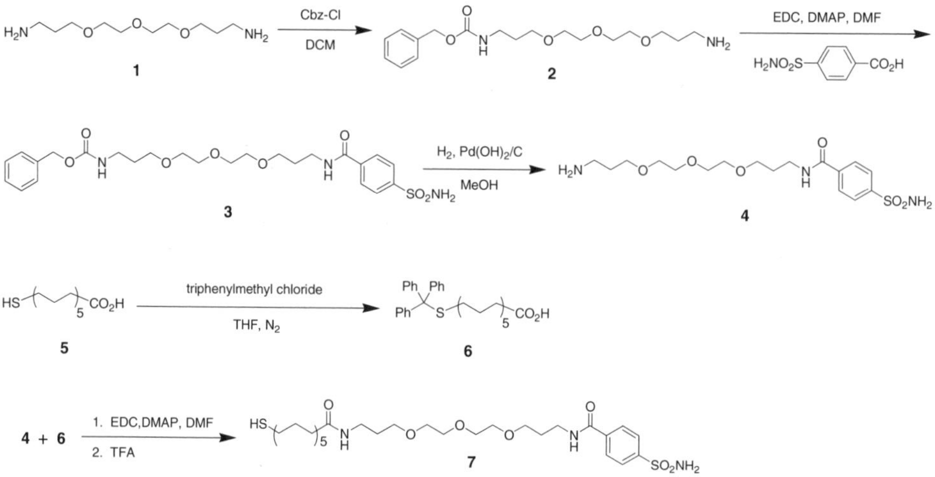

Figure 14.2. Synthesis of the sulfonamide inhibitor **7**.

The experimental procedure for the synthesis of the sulfonamide compound **7** has been successfully carried out, as shown in Figure 14.2. To prepare the inhibitor **7**, the diamine **1** was first monoprotected using Cbz–Cl in dichloromethane to yield compound **2**. With EDC used as the coupling reagent in dimethyformamide, compound **2** was reacted with 4-carboxy-benzenesulfonamide to result in compound **3**. The sulfonamide compound **4** was readily obtained from **3** using a standard deprotection scheme: H_2 and $Pd(OH)_2$ in MeOH. The overall isolated yield of compound **4** from the starting diamine **1** was 71%. Finally the amide bond formation reaction of sulfonamide **4** with a thiol-protected acide **6** led to inhibitor **7**, which had a good reaction yield of 59%. The SAM of sulfonamide **7** was readily formed by immobilizing on gold electrode surface of the quartz in QCM.

Monitor the Specific Interaction between Carbonic Anhydrase and Its Sulfonamide Inhibitor by QCM

Figure 14.3 shows the binding of bovine carbonic anhydrase (BCA) to an immobilized sulfonamid **7**. The top curve in Figure 14.3 clearly indicates that with BCA (1 nM) as the sample for injection, QCM correctly measured the binding of BCA to its inhibitor **7** at pH 7.2. In another feasible binding experiment (the middle curve from Figure 14.3), no direct mass deposition (i.e., binding) was detected when a soluble sulfonamide inhibitor (1 μM) was present in the BCA sample mixture at pH 7.2. This result implies that

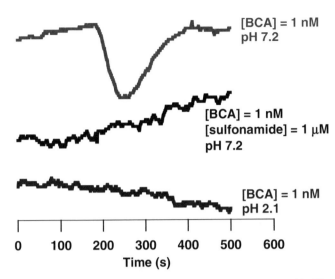

Figure 14.3. Specific molecular recognition of the immobilized sulfonamide **7** by QCM with bovine carbonic anhydrase (BCA) used under three different experimental conditions.

the soluble inhibitor has competed with the **7** for the BCA. Also this non-binding demonstrates again that the binding of the top curve is real and that any nonspecific interaction in the experimental condition was not detected. As shown in the bottom curve of Figure 14.3, the BCA was unable to recognize its inhibitor when the experiment was carried out outside the enzyme's optimal pH range. It should be noted that all three of the QCM experiments presented in the figure were performed using the same SAM surface.

The results of Figure 14.3 demonstrate that QCM can readily determine specific binding interactions among biomolecules, and this makes it potentially applicable to a wide range of biochemical systems. Real-time and reliable binding results may be obtained by QCM if carefully controlled experiments are planned and carried out in advance.

14.2.3. Binding Constant Measurement by QCM

Using Vancomycin–Peptide Binding Interaction as a Model System

As a natural glycopeptide antibiotic from *Nocardia orientails*, vancomycin is an important antibiotic to fight against drug-resistant Gram-positive bacterial infections. It shares a common heptapeptide structure with other antibiotics. The side chains of this heptapeptide cross-link with one another to form a stable structure as the drug active site. This active site of vancomycin is responsible for the binding with peptides whose C-termini are D-Ala-D-Ala, resulting in preventing the cross-linking of bacterial cell wall and ultimately leading to the death of bacteria.[50] The affinities between vancomycin and D-Ala-D-Ala containing peptides are in the range of 10^5 to 10^6 M^{-1}. Common methods used to quantitatively measure the vancomycin–peptide interactions have been the ACE,[51-55] nuclear magnetic resonance (NMR),[56-65] and the SPR.[66] We applied QCM biosensor system to verify the affinity between the vancomycin and the D-Ala-D-Ala containing peptides.

Synthesis of Peptidyl Ligands

The D-Ala-D-Ala containing peptide and control peptides tethered with a thiol functional group can be readily prepared by standard solid-phase peptide synthesis. As shown in Figure 14.4, for the study of their binding to vancomycin three compounds were prepared for the QCM experiments: ligand A, the D-Ala-D-Ala containing peptide; ligand B, a L-Ala-L-Ala control peptide that does not associate with vancomycin; ligand C, a SAM

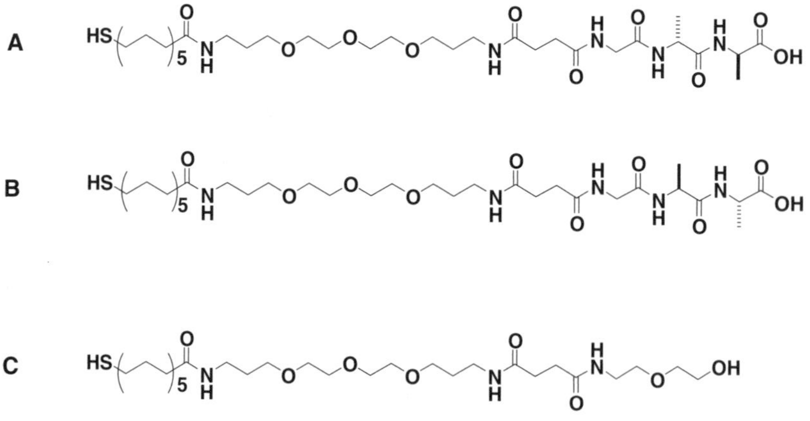

Figure 14.4. Ligand A: the D-Ala–D-Ala containing peptide tethered with an end thiol group that is capable of recognizing vancomycin. Ligand B: the L-Ala–L-Ala peptide having a thiol group for SAM formation on the gold electrode. Ligand B was used as a control molecule, since vancomycin only binds with Ala-Ala containing peptides of the D-form but not the L-form. Ligand C: a hydrophilic SAM component for the formation of 1:1 mixed monolayer with ligand A or B on QCM surfaces.

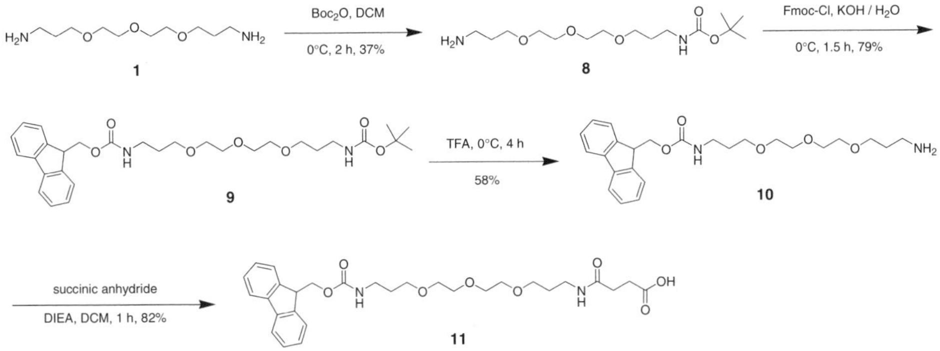

Figure 14.5. Preparation of the Fmoc-protected linker used for the synthesis of ligands A, B, and C.

component for the formation of 1:1 mixed monolayer with ligand A or B on QCM surfaces.

Since oligoethylene glycol is known to prevent the adsorption of biomolecules such as proteins, a triethylene glycol-like compound was employed as the linker to incorporate into the ligands A, B, and C. The preparation of the linker using standard Fmoc chemistry on solid support is illustrated in Figure 14.5.

For ligands A and B, the D(L)-Ala-2-chlorotrityl resin was first submerged in dimethylformaldehyde (DMF) where it swelled. The resin was then coupled with excess Fmoc D(L)-Ala in DMF for 2 hours using the standard Fmoc chemistry. The resulting resin mixture was filtered, washed in sequence by DMF, dichloromethane (DCM), and isopronpanol. Piperidine (20% in DMF) was employed to deprotect the Fmoc group for 15 minutes; the resin was subsequently washed again by DMF, dichloromethane, and isopronpanol. The coupling procedure was repeated using Fmoc-Gly, Fmoc-Gly, compound **11**, and compound **7** to yield the resin-bound product whose structure is shown in Figure 14.6. The final ligand could be readily isolated using the straightforward cleavage by trifluoroacetic acid (TFA, 50% in DCM).

Immobilization of Peptide Ligands on QCM Sensor Surfaces

Ligands A, B, and C have similar chemical structures. They all carry an end thiol functional group (for attaching onto a gold surface of a QCM device), an alkane chain (to form the well-packed SAM[67]), and a triethylene glycol linker (to prevent nonspecific adsorption[68]). The resulting 1:1 mixed SAMs

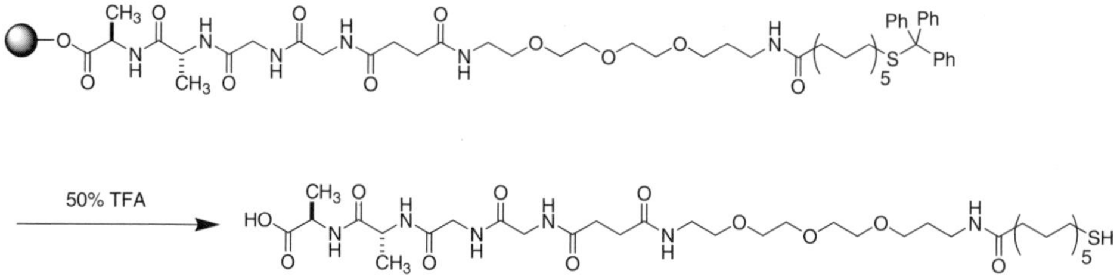

Figure 14.6. Synthesis of solid-phase peptide with a thio-end group.

Figure 14.7. A simulated molecular architecture of the A:C mixed SAM sensor surface.

(A:C and B:C) are believed to have the characteristics of self-assembly, monolayering, and minimal nonspecific binding. While the B:C surface was used as the control, the A:C surface could bind vancomycin and gave signals in QCM. Figure 14.7 simulates the molecular architecture of surfaces.

Binding Measurements of Vancomycin to the Immobilized Peptides on QCM Sensor Surfaces

Three QCM experiments based on vancomycin–peptide interactions were conducted. First, vancomycin was used as the sample to inject to the A:C mixed monolayer surface and PBS was used as the running buffer (pH 7.3). The second experiment was carried out under a condition similar to the first experiment, but the value of the pH was changed to 2.1. The third experiment used the B:C surface and PBS running buffer (pH 7.3) for vancomycin–peptide binding study. Since second and third QCM experiments were employed as the control for the first measurement, as expected, vancomycin could only be correctly associated with the D-Ala-D-Ala containing peptide on the A:C surface at the physiological pH (Figure 14.8).

Using QCM, we could not only qualitatively determine the specificity of individual binding interaction but also quantitatively measure their binding constants. Figure 14.9 shows a representative QCM frequency shift as a

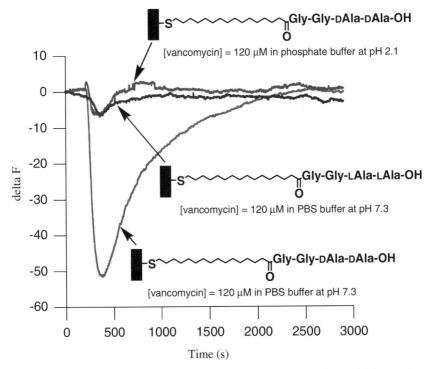

Figure 14.8. Use of QCM to determine binding specificity of vancomycin–peptide interactions.

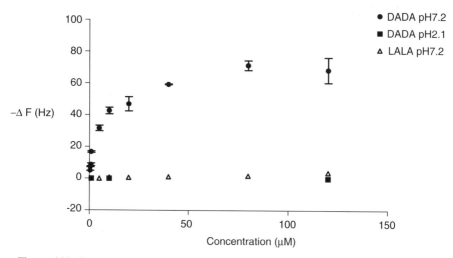

Figure 14.9. Frequency shift increases in QCM measurements along with increases of vancomycin concentration in the binding interaction study.

function of the injected vancomycin concentration under equilibrium conditions in the vancomycin–peptide binding system. It was found that on the A:C surface at pH 7.3, the frequency shift increased with each increase of vancomycin concentration. In the A:C mixed monolayer on the QCM sensor surface at pH2.1, the frequency changed insignificantly under the various vancomycin concentrations. This result demonstrated that in vancomycin–peptide interactions the binding must take place at physiological pHs. As expected for the B:C sensor surface at pH 7.3, the shifts in QCM frequency provided no evidence of difference for the various vancomycin concentrations, implying that vancomycin does not bind with the L-Ala-L-Ala containing peptide. All of the above-mentioned results clearly show that given the QCM conditions of our experiments, the nonspecific interactions between vancomycin and immobilized peptides are minimal, if not entirely avoided. The frequency shifts of the QCM, when detected experimentally, seem to result from specific binding. The results also show that without optimal conditions for binding (i.e., pH 2.1, instead of pH 7.3), no specific recognition occurs between vancomycin and the peptides.

Quantitative Dissociation Constant Measurements

The measurement of dissociation constants greatly helps the understanding of the mechanisms involved in biomolecular recognition. The dissocia-

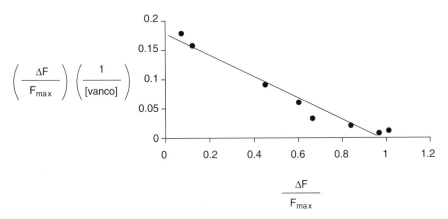

Figure 14.10. Scatchard analysis of the binding interaction between vancomycin and a D-Ala-D-Ala containing peptide (ligand A) measured by QCM.

tion constant K_d between vancomycin and a D-Ala–D-Ala containing peptide can be calculated using the equation (the Scatchard equation)

$$L + \textbf{Vanco} \underset{K_d}{\overset{K_a}{\rightleftharpoons}} [\textbf{L} \cdot \textbf{Vanco}]$$

$$\frac{\Delta F}{F_{max}} \frac{1}{[\textbf{Vanco}]} = \frac{1}{K_d}\left(1 - \frac{\Delta F}{F_{max}}\right). \tag{14.3}$$

The slope of the curve in the plot of $\Delta F/(F_{max} \times [\text{Vanco}])$ versus $\Delta F/F_{max}$ gives the binding constant of a biochemical system.[69] From (14.3), K_d can be readily obtained. The binding data presented in Figure 14.9 can be used for Scatchard analysis to calculate the corresponding dissociation constant. Figure 14.10 shows the result of such Scatchard analysis. The K_d for vancomycin binding with the peptide was 5.5 µM, which is correlated well with the value reported in the literature.

14.2.4. QCM for High-Throughput Screening of Compound Libraries

Combinatorial chemistry has been a useful tool for new drug development. Many researchers in the past have applied and utilized this technology for massive screening of the active compound from compound libraries in a relatively short time in order to find potential drug candidates. The same approach has been used to find the best substrates of enzymes. In successful combinatorial chemistry research, the development of high-throughput screening (HTS) assay methods is critical to the outcome. To facilitate our combinatorial chemistry program in discovering effective cell cycle inhibitors, we have integrated the QCM with an autosampling system to carry out so-called HTS-by-QCM experiments, for example, in the search for an active compound from a demethoxyfumitremorgin C library.

The unique feature of a QCM system is that the sensor chip is very sensitive to the mass change on the SAM gold surface. Since an equimolar concentration exists for every compound in a library, in the binding experiments by QCM the compound of high affinity should readily bind to the sensor's surface, effecting thus a good QCM response signal. Molecules that did not bind produce basal signals. This way, the magnitude of the frequency shifts can be directly related to the affinities of molecules bound to the QCM sensor chip, provided that similar molecular weight was met for every compound. By this assay method, high-affinity compounds can be readily identified from the library.

With the vancomycin–peptide binding system as a conceptual proof, an immobilized vancomycin was used to screen for candidate ligands from a peptide library that bind tightly with its receptor. Our preliminary result is shown in Figure 14.11, which clearly indicates that only the peptides with high affinity associated with the vancomycin surface and produced QCM responses. We are currently conducting other binding systems to demonstrate the applicability of QCM to library screening in combinatorial chemistry.

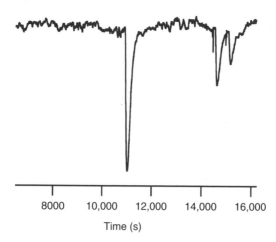

Figure 14.11. High-throughput screening of a small peptide library for an immobilized vancomycin sensor chip.

14.3. CONCLUSION

In this chapter we demonstrated the usefulness of QCM in biomolecules recognition, quantitative dissociation constant measurements, and high-throughput screening applications. In comparison with the popular SPR-based technology, QCM is much simpler and less expensive in its instrumental design and assembly. In addition QCM is highly sensitive to the mass change on the sensor surfaces.

Many more QCM applications are possible. Among these, we would include the analysis of protein–protein interactions often involved in signal transductions and DNA-drug interactions. The current limitation to QCM is that the area of binding surface is large (μL and nL for QCM and SPR, respectively). Therefore a complete solution has not yet been achieved, although much progress has been made to reduce the "reaction cell." As is the present trend for every analytical technology, a miniaturized QCM must be developed, and such a development could lead to QCM measurements that are both convenient and rapid. QCM's already apparent value in studies of molecular recognition, high throughput, and miniaturization will likely make accessible other unexplored areas in chemistry and biology.

ACKNOWLEDGMENTS

We gratefully acknowledge support of this work by a Grant-in-Aid and an instrumentation loan from ANT Technology (Taipei, Taiwan), and also in part by the National Science Council (Taiwan, R.O.C.). We thank our colleague Jerry Lo who participated in various aspects of this research for his critical comments and continual encouragement. M.C.T. is a recipient of the NSC Graduate Student Fellowship.

REFERENCES

1. J. F. Bach, ed., *Immunology*, Wiley, New York (1978).
2. J. M. Rini, U. Schulze-Gahmen, I. A. Wilson, *Science* **255**, 959 (1992).
3. A. L. Cox, et al. *Science* **264**, 716 (1994).
4. U. Jonsson, L. Fagerstam, S. Lofas, E. Stenberg, R. Karlsson, A. Frostell, F. Markey, F. Schindler, *Ann. Biol. Clin.* **51**, 19–26 (1993).
5. T. Vo-Dinh, K. Houck, D. L. Stokes, *Anal. Chem.* **66**, 3379 (1994).
6. X. Chen, X. E. Zhang, Y. Q. Chai, W. P. Hu, Z. P. Zhang, X. M. Zhang, A. E. G. Cass, *Biosens. Bioelecctron.* **13**, 451 (1998).

7. H. Su, K. M. R. Kallury, M. Thompson, *Anal. Chem.* **66**, 769 (1994).

8. Y. Okahata, Y. Matsunobu, K. Ijiro, M. Mukae, A. Murakami, K. Makino, *J. Am. Chem. Soc.* **114**, 8299 (1992).

9. F. Caruso, D. N. Furlong, K. Niikura, Y. Okahata, *Colloids Syrf.* **B10**, 199 (1998).

10. K. Hashimoto, K. Ito, Y. Ishimori, *Anal. Chem.* **66**, 3830 (1994).

11. K. M. Millan, A. Saraullo, S. R. Mikkelsen, *Anal. Chem.* **66**, 2943 (1994).

12. S. Takenaka, K. Yamashita, M. Takagi, Y. Uto, H. Kondo, *Anal. Chem.* **72**, 1334 (2000).

13. K. M. Millan, S. R. Mikkelsen, *Anal. Chem.* **65**, 2317 (1993).

14. E. Palecek, M. Fojta, *Anal. Chem.* **73**, 75A (2001).

15. A. D'Amino, C. D. Ntale, E. Verona, in E. Kress-Rogers, ed., *Handbook of Biosensors and Electronic Noses: Medicine, Food, and the Environment*, CRC Press, Boca Raton, FL (1997).

16. C. Kößlinger, E. Uttenthaler, S. Drost, F. Aberl, H. Wolf, G. Brink, A. Stanglmaier, E. Sackmann, *Sensors Actuators* **B24–25**, 107–112 (1995).

17. B. Liedberg, K. Johansen, in K. R. Rogers, A. Mulchandani, eds., *Affinity Biosensor: Techniques and Protocols*, Humana Press, Totowa, NJ (1998), Chapter 3.

18. T. W. Schneider, S. J. Martin, *Anal. Chem.* **67**, 3324–3335 (1995).

19. K. K. Kanazawa, J. G. Gordon II. *Anal. Chim. Acta* **175**, 99–105 (1985).

20. J. Rickert, A. Brecht, W. Göpel, *Biosen. Bioelectr.* **12**, 567–575 (1997).

21. P. Curie, J. Curie, *C. R. Acad. Sci.* **91**, 294 (1880).

22. J. Janata, *Principles of Chemical Sensors*, Plenum Press, New York (1989).

23. G. Sauerbrey, *Z. Phys.* **155**, 206 (1959).

24. D. A. Buttry, M. D. Ward, *Chem. Rev.* **92**, 1355–1379 (1992).

25. G. C. Dunham, N. H. Benson, D. Petelenz, J. Janata, *Anal. Chem.* **67**, 267–272 (1995).

26. C. S. Lu, O. J. Lewis, *Appl. Phys.* **43**, 4385 (1972).

27. A. Snow, H. Wohltjen, *Anal. Chem.* **56**, 1411 (1984).

28. D. Ballantine, H. Wohltjen, *Anal. Chem.* **61**, 705 (1989).

29. G. G. Guilbault, *Anal. Chem.* **55**, 1682–1684 (1983).

30. T. Nomura, M. Iijima, *Anal. Chim Acta.* **131**, 237 (1981).

31. Z. A. Shana, H. Zong, F. Josse, D. C. Jeutter, *J. Electroanal. Chem.* **379**, 21 (1994).

32. T. Zhou, L. Nie, S. Yao, *J. Electroanal. Chem.* **293**, 1 (1990).

33. S.-Z. Yao, T.-A. Zhou, *Anal. Chim. Acta.* **212**, 61 (1988).

34. M. A. M. Noel, P. A. Topart, *Anal. Chem.* **66**, 486 (1994).

35. M. Urbakh, L. Daikhin, *Lamgmuir* **10**, 2836 (1994).

36. S. J. Martin, G. C. Frye, A. Ricco, *J. Anal. Chem.* **65**, 2910 (1993).

37. M. Yang, M. Thompson, *Lamgmuir* **9**, 1990 (1993).

38. K. K. Kanazawa, J. G. Gordon, II. *Anal. Chim. Acta.* **175**, 99 (1985).

39. J. E. Roederer, G. J. Bastiaans, *Anal. Chem.* **55**, 2333 (1983).

40. M. Thompson, C. L. Arthur, G. K. Dhali, *Anal. Chem.* **58**, 1206 (1986).

41. M. Mascini, M. Minunni, G. G. Guilbault, R. Carter, in K. R. Rogers, A. Mulchandani, eds., *Affinity Biosensor: Techniques and Protocols*, Humana Press: Totowa, NJ (1998) Chapter 4.

42. Y. Okahata, M. Kawase, K. Niikura, F. Ohtake, H. Furusawa, Y. Ebara, *Anal. Chem.* **70**, 1288–1296 (1998).

43. S. J. Dodgson, R. E. Tashian, G. Gros, N. D. Cartwr, *The Carbonic Anhydrases: Cellular Physiology and Molecular Genetics*, Plenum Press, New York (1991).

44. P. W. Taylor, R. W. King, A. S. V. Burgen, *Biochem.* **9**, 2638 (1970).

45. R. W. King, A. S. V. Burgen, *Proc. R. Soc. London Ser. B* **193**, 107 (1976).

46. I. J. Colton, J. D. Carbeck, J. Rao, G. M. Whitesides, *Electrophoresis* **19**, 367 (1998).

47. L. Z. Avila, Y.-H. Chu, E. C. Blossey, G. M. Whitesides, *J. Med. Chem.* **36**, 126 (1993).

48. G. B. Sigal, G. M. Whitesides, *Bioorg. Med. Chem. Lett.* **6**, 559 (1996).

49. M. Mrksich, J. R. Grunwell, G. M. Whitesides, *J. Am. Chem. Soc.* **117**, 12009 (1995).

50. M. L. Cohen, *Science* **257**, 1050 (1992).

51. Y.-H. Chu, L. Z. Avila, J. Gao, G. M. Whiteside, *Acc. Chem. Res.* **28**, 461 (1995).

52. J. Gao, F. A. Gomez, R. Haerter, G. M. Whitesides, *Proc. Natl. Acad. Sci. USA* **91**, 12027 (1994).

53. M. Nieto, H. R. Perkins, *Biochem. J.* **123**, 773 (1971).

54. M. Nieto, H. R. Perkins, *Biochem. J.* **123**, 789 (1971).

55. P. H. Popienick, R. F. Pratt, *Anal. Biochem.* **165**, 108 (1987).

56. J. P. Brown, L. Terenius, J. Feeney, A. S. Burgen, *Mol. Parmacol.* **11**, 126 (1975).

57. O. Convert, A. Bongini, J. Feeney, *J. Chem. Soc. Perkin Trans.* **1**, 1262 (1980).

58. A. Bongini, J. Feeney, M. P. Williamson, D. H. Williams, *J. Chem. Soc. Perkin Trans.* **2**, 201 (1981).

59. D. H. Williams, M. P. Williamson, D. W. Butcher, S. J. Hammond, *J. Am. Chem. Soc.* **105**, 1332 (1983).

60. W. G. Prowse, A. D. Kline, M. A. Skelton, R. J. Loncharich, *Biochem.* **34**, 9632 (1995).

61. S. W. Fesik, T. J. O'Donnell, R. T. J. Gampe, E. T. Olegniczak, *J. Am. Chem. Soc.* **108**, 3165 (1986).

62. G. E. Hawkes, H. Molinari, S. Singh, L. Y. Lian, *J. Magn. Reson.* **74**, 188 (1987).

63. J. C. J. Barna, D. H. Williams, *Annu. Rev. Microbiol.* **38**, 339 (1984).

64. H. Molinari, A. Pastore, L. Lian, G. E. Hawkes, K. Sales, *Biochem.* **29**, 2271 (1990).

65. C. M. Pierce, D. H. Williams, *J. Chem. Soc. Perkin Trans.* **2**, 153 (1995).

66. J. Rao, L. Yan, B. Xu, G. M. Whitesides, *J. Am. Chem. Soc.* **121**, 2629 (1999).
67. A. Ulman, *An Introduction to Ultrathin Organic Films from Langmuir-Blodgett to Self-Assembly*, Academic Press, San Diego, CA (1991).
68. J. M. Harris, ed., *Poly(Ethylene Glycol) Chemistry: Biotechnical and Biomedical Applications*, Plenum Press, New York (1992).
69. C. R. Cantor, P. R. Schimmel, *Biophysical Chemistry Part III: The Behavior of Biological Macromolecules*, Freeman, New York (1980).

CHAPTER

15

HIGH-THROUGHPUT PHYSICOCHEMICAL PROFILING: POTENTIAL AND LIMITATIONS

BERNARD FALLER

15.1. INTRODUCTION

Physicochemical profiling has gained considerable importance in the last years as most companies realized that inappropriate physicochemical properties could lead to compound withdrawal later in development. The basic physicochemical parameters of interest for "drugability" prediction are solubility and permeability, the two components of the Biopharmaceutical Classification Scheme.[1] However, these two fundamental parameters are in turn influenced by other physicochemical parameters worth considering, particularly in the lead optimization phase. For example, permeability is influenced by lipophilicity (induces membrane retention) and pH (ionizable compounds), solubility is influenced by pH (ionizable compounds), and dissolution rate is linked to particle size, polymorphism, and wettability.

Initially, technologies available to determine physicochemical properties were not suited for combinatorial chemistry, mainly because the throughput was too low, the compound requirements were too high, and only very pure compounds (>95%) could be characterized. In the last five years a number of new technologies have been developed to address the points above and to cope with the increasing pressure for compound profiling at an early stage of drug discovery in the pharmaceutical industry. Today, physicochemical profiling of combinatorial chemistry libraries is becoming possible for a few key molecular descriptors, and a more complete panel is likely to become accessible by the end of the decade. High-throughput physicochemical profiling is still a relatively young discipline, with roots starting in the early to mid-1990s. The difficulty in setting up high-throughput physicochemical profiling assays is that one cannot rely on traditional generic readouts used in HTS like fluorescence or scintillation counting, simply because one does not look at the interaction between a

Analysis and Purification Methods in Combinatorial Chemistry, Edited by Bing Yan.
ISBN 0-471-26929-8   Copyright © 2004 by John Wiley & Sons, Inc.

collection of compounds and a molecular target that can itself be used to generate a signal (i.e., through labeling) but rather at the behavior of the compound itself in a changing environment (i.e., pH, polarity, lipophilicity, viscosity, and electrostatic fields). Typical detection systems used in physicochemical profiling assays are UV absorbance, electrochemical sensors, surface tension, and mass spectrometry. So far not all of the parameters mentioned above can be measured at a high throughput. The primary focus in establishing high-throughput assays is to substantially increase the number of determinations; usually from a handful to about 50 to 100 determinations a day while keeping reasonable accuracy, and reduce sample consumption as well as cost per analysis. There are two basic strategies to move an assay to high throughput: in the simplest approach one tries to directly transpose the assay in a different format (i.e., from vial to microtiterplate) and add labautomation to it (i.e., kinetic solubility with plate nephelometer). Such a straightforward approach is, however, not always possible, either because the technology cannot be easily transposed to the microtiterplate format (i.e., dissolution rate) or because the detection device lacks adequate speed (i.e., potentiometric titration). In this case one needs to work out another assay concept to access the desired parameter (i.e., $\log P$ based on RP-HPLC retention times or artificial membranes).

In this chapter we want to review the main technologies available for physicochemical profiling in the pharmaceutical industry, with focus on strengths and limitations of the currently available approaches, and then to give practical hints how to use the different technologies to derive experimental parameters of sufficient quality.

15.2. MOLECULAR PROPERTIES RELEVANT FOR ADME PROFILING

Biopharmaceutical profiling assays are used to predict the behavior of drug candidates against the number of hurdles a compound needs to overcome to become a drug, like absorption, distribution, metabolism, and elimination. The main absorption pathways in the GIT epithelium are shown in Figure 15.1, while Table 15.1 lists the parameters used to address the different aspects of drug disposition. As oral delivery has become a must due to patient compliance and marketing reasons, parameters governing GIT absorption are usually of primary interest. Solubility and permeability are the two main physicochemical parameters relevant for oral absorption. Compounds can be classified in four categories according to the BCS classification scheme (Table 15.2) proposed by Gordon Amidon.[1] The technologies currently used to access physicochemical parameters of interest in ADME profiling are listed in Table 15.3.

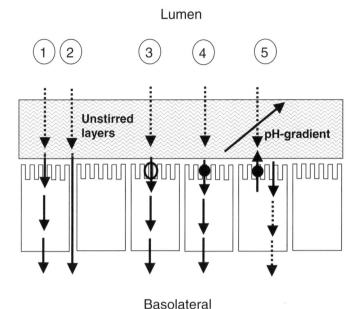

Figure 15.1. Absorption pathways. The first common step for all pathways in GIT absorption is the diffusion throught the unstirred layers. (1) Transcellular (most common pathway for drugs), (2) paracellular (small, water soluble compounds), (3) facilitated transport, (4) active tranport (saturable, requires ATP), and (5) efflux pumps (i.e., *pGp*. MRP).

Table 15.1. Biophamaceutical Classification Scheme[1]

	High Solubility	Low Solubility

High permeability	**Class I**
	Dissolution rate limits
	  absorption
	Carbamazepine
	Ketoprofen
	Propranolol
		Class II
		Solubility limits
		  absorption
		Chlorpromazine
		Diclofenac
		Nortriptyline

Low permeability	**Class III**
	Permeability limits
	  absorption
	Famotidine
	Nadolol
	Valsartan
		Class IV
		Significant problems for
		  oral delivery expected
		Ciclosporine
		Furosemide
		Terfenadine

Table 15.2. Parameters Used to Address Biopharmaceutical Properties Influencing Drug Disposition

Biopharmaceutical Property	Parameter
Oral absorption	Solubility
	Permeability
	Ionization constant (pKa)
Blood–brain penetration	Permability
	Efflux (pGp)
	Active transports
	Lipophylicity
Distribution excretion	Lipophilicity (log P)
	Molecular size
Stability	Metaboic stability in liver microsomes
	Chemical stability
Drug–drug interaction (safety)	P450 inhibition

Table 15.3. Methods to Measure Parameters of Interest in Drug Disposition Prediction

Property	Method
Partition/distribution coefficients (log P and D)	**RP-HPLC log k**
	Shake-flask/HPLC
	Dual-phase potentiometric titration
	Octanol coated columns
	Liquid artificial membranes
Permeability	Caco-2 monolayers
	MCDK monolayers
	PAMPA
	Hexadecane liquid membrane
	HPLC-IAM log k
Ionization constants	Potentiometric titration
	Spectrophotometric titration
	Fast-gradient spectrophotometric titration
	Capillary electrophoresis
Solubility	**Nephelometric titration**
	Turbidimetry
	Dissolution template titration
	Shake-flask/HPLC
	Flow cytometry titration
	UV shake-plate

Note: High-throughput methods are indicated in bold.

15.3. SOLUBILITY DETERMINATIONS

Solubility plays an essential role in drug disposition as passive drug transport across a biological membrane is the combined product of both permeability and solubility (Fick's first law). Poor solubility has been identified as the cause of numerous drug development failures.[2] It is one of the components of the BCS[1] and, if not properly taken into account, leads to erroneous results in a number of assays[3] (i.e., Caco-2 as well as artificial model membrane permeability, binding affinity determinations). In the ADME field, solubility is particularly important for immediate release BCS class II drugs, for which absorption is limited by solubility (thermodynamic barrier) or dissolution rate (kinetic barrier).[1,4] Although it may appear relatively simple at first glance, the experimental determination of solubility is often difficult due to the complexity of the solution chemistry involved. A number of factors like supersaturation, dissolution rate, aggregation reactions, micelle formation, effect of ions and buffers, presence of impurities, and ionic strength come into play, and if not properly taken into account, they lead to erroneous results. The measurement of aqueous solubility can be divided into five components[5]

- Temperature control
- Purification of solute and water
- Equilibration kinetics
- Separation of phases
- Analysis of the saturated solution

The difficulty to get reliable results usually increases as solubility decreases. Incorrect solubility values can result from improper execution of any of the steps above.

15.3.1. Equilibration Kinetics

Since dissolution is a time-dependent process, the contact between the powder and water is important. The kinetics depends on the dissolution rate, which is itself influenced by the type of agitation and the surface of contact between the solute and water (dictated by particle size). Dissolution kinetics is usually inversely correlated with solubility, although there is considerable variation in this relationship. The approach to equilibrium is usually asymptotic, which means that the solute initially enters in solution quite rapidly, and as it approaches equilibrium, the rate of dissolution decreases. To check that the equilibrium has been reached, one must take samples at various time points up to a constant concentration is obtained.

In practice, this is would be too labor intensive for research compounds, so one usually sonicates the suspension (to break down aggregates and maximize the surface of contact) and shake overnight. Dissolution kinetics can also be strongly influenced by the wettability of the solid. In this case kinetics can be extremely slow, even if the equilibrium solubility is not particularly low. One can then try to add a wetting agent or start by pre-dissolving the compound, for example by addition of acid or base.

15.3.2.   Separation of Phases

Separation of phases is a major issue in the saturation shake-flask method. Two techniques can be used to separate the solute from the supernatant: (1) filtration and (2) decantation (usually accelerated by centrifugation). Neither techniques is totally satisfactory. If filtration is used, the pore size and the nature of the filter material are important. The main problem is that the filter can remove solute from solution through sorption; therefore filtration is not recommended for hydrophobic molecules ($\log P > 3$–4). The other option to separate the phases is to centrifuge. Here care must be taken to avoid re-suspended material being carried over (would then lead to overestimation of solubility). Also, if the solute is less dense than water, it will float on the surface, and withdrawal of the aqueous solution will be more complicated. To summarize, the phase separation through filtration can lead to underestimation of solubility because of sorption, while phase separation through centrifugation may lead to overestimation of solubility due to carry over.

A new method that avoids the difficulties related to phase separation has been recently proposed to measure the solubility of ionizable compounds.[6,7] The method is based on potentiometric titration, and the separation between precipitate and solute is obtained through the selectivity of the pH-electrode. To summarize, the principle of the method is that only the dissolved substance participates to the protonation-deprotonation equilibrium.

15.3.3.   Equilibrium Solubility

Dissolution Template Titration

In our laboratory these experiments are carried out with a p-SOL3 solubility analyzer (pION Inc., Woburn, MA). The titration are performed under argon at 25.0°C in solutions containing 0.15 M KCl. Prior to any titration the electrode is standardized by titrating a strong acid (HCl) with a strong base (KOH).[8] This method requires that the ionization constants have been carefully determined previously. The system is seeded with an initial estimate of solubility, obtained through the semi-empirical equation below:

$$\log S_w = 1.17 - 1.38 \log P_{o/w}, \qquad (15.1)$$

where $\log P_{o/w}$ is either obtained from experimental determination (see Section 15.6) or calculated. Equation (15.1) works only when $\log P$ or $CLOGP > 2$. Another option is to use the value generated from high-throughput assays like nephelometric titration.[9,10] To avoid supersaturation one titrates from the low soluble to the high soluble end (i.e., from acid to alkaline for an acid). In the dissolution titration of a saturated solution, it is necessary to allow the system to return to a steady state following each titrant addition. Using the pKa and the estimated $S_{w(n)}$, the instrument simulates the entire titration curve before starting the assay. The pH domain embracing precipitation is estimated from the simulation and the data collection speed is set accordingly using modified version of the Noyes-Withney theory[11] modified by Serajuddin and Jarowsky[12] to take into account the diffusion of the different ionic species in the unstirred layer associated to the particles. This kind of potentiometric titration is called dissolution or Noyes-Withney template titration. About 20% of the data, nearest the point of lowest solubility, is allotted about 80% of the time. The amount of sample in the titration is calculated to cause precipitation during some portion of the titration. One experiment typically requires between 0.2 and 10 mg of sample, depending on the expected solubility. However, care must be taken to ensure that the amount of sample used does not exceed 1000 times the predicted intrinsic solubility, to avoid chloride or potassium salt precipitation of the compound due to the presence of the 0.15 M KCl (background electrolyte). For compounds with predicted solubility <0.5 g/L, the solid is first dissolved by the addition of strong acid (for a base) or strong base (for an acid) and then quickly re-precipitated and allowed to gestate for 10 to 60 minutes (depending on the estimated $S_{w(n)}$) before the data collection starts. The lower the expected solubility, the longer the assay time. Typically 2 to 12 hours are required for the entire titration. The intrinsic solubility and the solubility pH-profile are calculated from the difference between the titration in the presence of an excess of substance (causing precipitation) and the titration curve where no precipitation occurs (from which the pKa can be calculated). It is possible to derive the solubility pH profile from the solubility of the neutral species if the pKa is known.[13,14] At the end of the titration it is advisable to check the chemical stability of the molecule by comparison with a fresh standard using an appropriate analytical method (RP-HPLC/UV with a C8 or C18 column).

Saturated Shake-Flask

The first step is to establish a suitable analytical method to quantify the analyte. This is usually done by RP-HPLC with a C8 or C18 column. In some cases this step can be quite labor intensive, especially when none of the generic separation methods available in the laboratory is suitable for the substance of interest.

The second step is to build a calibration curve using the analytical method established previously. Dilutions of the standard are prepared in pure MeOH or MeOH/water mixtures. The calibration curve is obtained by plotting the peak surface area according to concentration. To minimize carryover, the standards are injected from the lowest to the highest concentration. In the presence of diastereoisomers, double peaks are sometimes recorded. In this case the sum of the two peaks is taken into account.

In our laboratory, saturation shake-flask solubility is routinely performed at pH 6.8 and pH 1.0 (for acids only) 0.5 mL of buffer is added to about 2 mg of sample in a small borosilicate vial (2 mL) and sonicated for 3 minutes using a sonicating bath. The suspension is then incubated at 25.0°C for 20 hours in a thermostated water bath at a shaking speed of 200 cycles/min. The phases are separated by decantation/centrifugation. The supernatant is carefully collected and the solute quantified using the analytical method defined previously. Before each injection, a blank is performed to avoid phantom peaks due to carryover or impurities present in the system. After the centrifugation it is advisable to re-check the pH of the medium.

15.3.4. Kinetic Solubility Assays

Nephelometric Titration

Technologies currently available to measure thermodynamic equilibrium solubility require relatively large amounts of materials and are time-consuming as well as labor intensive. Thus they are not feasible for the high-throughput screening of aqueous solubility in early drug discovery where a large number of compounds need to be characterized with minimal sample requirement. An alternative approach to estimate water solubility involves *in silico* prediction. It, however, remains a difficult problem mainly because water solubility is the result of a complex interplay between hydrogen-bond acceptor and donor properties, conformational effects, and crystal packing energy—effects that cannot be described by a simple summation of contributing groups. For these reasons standard deviations of computed water solubility compared to the experimental value (thermodynamic solubility) are still relatively large (usually 1–2 orders of magnitude). In addition some *in silico* predications may require other physicochemical parameters such as partition coefficient (log P) or melting point of the test compounds that in general are not available in the early stage of drug discovery. The concept of turbidimetric (also called kinetic) solubility was introduced a few years ago by Lipinski[2] at Pfizer, and it allows large number of samples to be analyzed. Although kinetic solubility cannot replace true thermodynamic equilibrium solubility, it serves as a useful guideline in early drug discovery. The principle of the assay described in the present report is similar to the original setup described by Lipinski except that microtiterplate nephelometry is used instead of turbidimetry coupled with a nephelometric cell. The potentials and limitations of the nephelometric kinetic solubility assay are outlined in Section 15.3.6.

Assay Principle

Detection of light scatter has been recognized for a long time as a way to measure solubility.[9] Today turdidimetric cells have been replaced by microtiterplate nephelometers.[10] Compounds are first solubilized in DMSO. Small aliquots of the DMSO stock solution are dispensed in the test media, and the formation of particles is checked by nephelometry throughout the titration. Solubility is determined as the last concentration before two consecutive readings higher than the background S/N are recorded. Correlation between thermodynamic equilibrium solubility and kinetic solubility is shown in Figure 15.2. As DMSO induces a slight increase of the solubility, particularly for solubility values lower than 0.01 mg/mL, it is advisable (1) to keep DMSO at a reasonably low concentration (<5%) and (2) to keep it constant throughout the titration

Flow Cytometry Titration

A variation based on flow cytometry light scattering is being developed by BD Gentest. The flow cytometry technology has a number of potential advantages over the conventional neplelometric plate readers or turbidimetric devices. In particular, there is a hydrodynamic focusing of the sample which enables the characterization of individual scatter events. Scatter intensity profiles are generated for each sample and contain information about particle size (0.2–20 μm range) and relative intensity of the individual traces. That could potentially help (1) increase sensitivity and (2) reduce interferences by impurities or low soluble contaminants. Also due to hydrodynamic focusing, the sample never touches the walls of the flow cell, so adsorption is minimized. 25 μl of sample is enough for a measurement at one concentration (vs. 100–300 μl in 96-well plates and 80 μl in 384 plates).

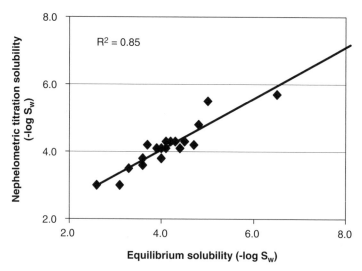

Figure 15.2. Correlation between kinetic solubility (nephelometric titration) and equilibrium solubility (pH-metric for ionizable compounds or saturated shake-flask for neutral compounds). The solubility values correspond to the solubility of the neutral species. The pH of the buffer was adjusted accordingly in the nephelometric titration experiments. Solubility values are expressed in negative log units of S_w in mole/liter.

Mixing procedure and incubation time play a major role in kinetic solubility determinations. According to the instrument manufacturer, good results were obtained with amiodarone, which is known to cause adsorption problems in the polystyrene plates used in nephelometric titration (and thus lead to overestimated solubility). However, most the advantages listed above remain theoretical until solid validation data with molecules relevant for drug discovery become available.

15.3.5. Shake-Plate UV Assay

A method based on filtration in 96-well plates followed by UV analysis has been developed by pION Inc. and marketed under the μ-sol evolution instrument (*www.pion-inc.com*). Basically aliquots of a 10-mM solution in DMSO are mixed with buffer, incubated, and filtered. The UV spectrum of the filtrate is then compared with a reference standard to quantify the amount of substance in solution. Respective shapes of the filtrate and the reference spectra are also analyzed in an attempt to determine whether an impurity has interfered with the assay. For ionizable molecules algorithms were also derived to account for the presence of DMSO.

15.3.6. Potential and Limitations of the Methods Described Above

Dissolution-Template Potentiometric Titration

This method is particularly well suited for the analysis of ionizable molecules, for which solubility is influenced by pH (70–80% of generic drugs). It does not require the development of any analytical method but the ionization constants have to be determined relatively accurately prior to the analysis.

The main advantages of this approach are (1) there is no need to separate the solute from the precipitate, (2) the pH is automatically adjusted by the addition of KOH/HCl, therefore no buffer is needed, (3) the equilibrium kinetics are taken into account by the system using a modified version of the Noyes-Whitney equation, (4) the Noyes–Whitney equation gives the solubility pH-profile, and (5) it covers an exceptional dynamic range (from mg/mL down to ng/mL).

There are, however, a number of limitations inherent to the technology: (1) it is only applicable to ionizable molecules, (2) weak acids (pKa > 11) and weak bases (pKa < 3) cannot be measured because the ionization is then shifted to a range outside the dynamic range of the pH electrode, and (3) it is not a high-throughput method (1–2 compounds/day)

Saturation Shake-Flask

This method is used as a complementary tool for neutral compounds or for weak acids or bases. Its main advantage is that ionizable as well as neutral compounds can be analyzed. For the former, however, one has to assume that the components of the buffer do not influence the solubility, particularly when one performs a measurement at a pH value where the molecule is strongly ionized.

Prior to each analysis one needs to establish an analytical method and to build a calibration curve. In most cases it can be achieved by reverse-phase HPLC with a C18 or C8 column. The main difficulty of the method lies in the separation of the phases where decantation by centrifugation or filtration is used. In each case, there is some risk that the solubility is affected by this phase-separation step. When decantation is used the solubility is potentially overestimated, while the opposite is true when filtration is used. In today's drug discovery environment one usually prefers to separate the phases by decantation through centrifugation because the large majority of the research compounds are rather hydrophobic ($CLOGP > 3$). Although this method gives good results for good to medium soluble compounds, there are limitations for low or very low soluble compounds due to the analytical challenge to accurately quantify the low amounts of solute (sorption to vials, limit of detection). It is often difficult to accurately measure solubility values lower than 0.005 g/L, and generally impossible when it drops below <0.001 g/L unless one uses a co-solvent extrapolation approach.

Nephelometric Titration

Nephelometric titration has a number of advantages over the classical approaches:

- This is obviously a fast method, relatively easy to integrate into a robotic workstation and allowing for high throughput.
- It requires only a small amount of samples (0.2–0.5 mg for two pHs for assay range of 0.5–200 mg/mL), making possible for solubility determination for early research compounds with limited sample amount.
- The assay simply monitors the sample precipitation in aqueous solutions and is thus widely applicable to almost all compounds, as no chromophore is needed as compared to the LC-UV method. In addition the quantification of kinetic solubility assay is much simpler than LC-UV or MS (no analytical bottleneck).
- Kinetic solubility can also be measured in pre-formulation and customized media (not limited to aqueous).
- The assay dynamic range (0.5–100 μg/mL) covers the vast majority of compounds involved in the early discovery research.

On the other hand, the kinetic solubility assay also suffers from some limitations, in comparison to other existing approaches:

- Nephelometric titration involves 8 or 9 discrete concentration points within a range of 0.5 to 200 mg/mL. Thus the kinetic solubility between two concentration points will be rounded to the closest concentration value, thereby possibly causing deviations.
- The dynamic range of the assay is relatively narrow (2–3 log units) but fortunately falls in the solubility range of interest in drug discovery.
- Kinetic solubility does not characterize the solid. The loss of memory of such information is a consequence of the dissolution of solids in DMSO (the stock solutions for the assay).
- Precautions should be taken for low soluble (<0.01 mg/mL) and highly soluble compounds (>0.2 mg/mL). For the former, kinetic solubility

data tend to be overestimated, possibly due to presence of a small amount of DMSO in the medium or a slow nucleation process of compounds. For the latter, underestimated solubility may be derived presumably because of the presence of a low soluble impurity.

- As a rule, one should keep DMSO within 1% to 2% to avoid interferences due to the presence of the solvent. Highly soluble compounds can be underestimated when a short incubation time was used between mixing and reading. On the other hand, longer incubation time favors adsorption, thus leading to overestimation of solubility.

In summary, although conventional thermodynamic equilibrium solubility remains the method of choice for the determination of aqueous solubility, in particular, for a drug substance in the development stage, kinetic solubility using nephelometer offers a higher throughput and faster screening of water solubility for early drug discovery research. Thus this method is not designated to substitute conventional thermodynamic solubility. Instead, it allows for a prioritization of early drug candidates using their water solubility along with other activity and physicochemical parameters.

96-Well Plate UV Assay

This is a relatively straightforward approach that allows collecting solubility pH profiles of large collections of compounds. The main advantages of the system are:

- No analytical bottleneck (UV detection).
- Full solubility pH profile easy to collect.
- Corrections to take into account the effect of DMSO possible for ionizable substances.
- Possibility to start with solid material or DMSO stock solutions.

There are, however, some drawbacks inherent to the technology used:

- Risk of overestimation of solubility in the presence of a high soluble impurity (no chromatographic step involved). Shape analysis can potentially flag this, but it only works when the impurity has a different UV spectrum than the substance. In practice, this is not necessarily the case with real drug discovery samples.
- Low soluble substances might be underestimated due to loss of material to the filter.

- As in nephelometric titration, the dynamic range of the assay depends on the DMSO stock concentration and the % of organic solvent one tolerates after dilution of that stock.
- The sensitivity of the assay varies among substances as it depends on the strength of the chromophore.
- Compounds lacking UV chromophore cannot be analyzed.

15.3.7. Other Common Sources of Uncertainty

Another common source of uncertainty in solubility determinations is when the compound is highly ionized (>99%) at the pH considered. This usually happens when the pH is more than 2 units above the pKa for an acid or than 2 log units lower than the *pKa* for a base. In this case the solubility pH profile is very steep around that pH, which results in a higher uncertainty on the result. This is the case of Diclofenac at pH 6.8, where we observed some differences within our measurements and between our data and literature values. This limitation is, however, valid for both dissolution template titration method and the saturation shake-flask.

Polymorphism is also known to affect solubility significantly. However, as at the discovery stage the crystalinity of the compounds is usually not investigated, this aspect is generally covered in the early development phase. This means that the solubility of the final drug substance is eventually lower than what is reported from initial drug discovery studies.

15.4. IONIZATION CONSTANTS DETERMINATIONS

Knowledge of the aqueous ionization constant (pKa) of a drug substance is very important in the pharmaceutical industry as it strongly impacts on permeability and solubility, the two components of the biopharmaceutic classification scheme that classify drugs with respect to their potential for oral absorption. Knowledge of the pKa also helps to understand binding mechanisms if the drug binds to complementary charged site in its receptor. Potentiometric titration with a pH electrode remains the method of choice for pKa determination, as it is the most direct method to count protons taken up or liberated upon change of pH. Since many new drug candidates or drug substances are poorly soluble in water and/or have overlapping pKa values, the determination of their aqueous pKa is often challenging.

15.4.1. Potentiometric Titration

Potentiometric titration with an Ag/AgCl double junction pH electrode remains the reference system for pKa determination. It remains, however, a relatively slow procedure as one needs to wait until the electrode's signal stabilizes between consecutive additions of titrant. So far glass electrodes are the most commonly used because of the quality of their response in the pH range of interest. The major drawback of glass electrodes is their relatively slow response time and their size, which prevents significant miniaturization of the titration device. The latter point is likely to be addressed in the midterm with the progresses made in manufacturing high-resolution solid electrodes.

Electrode Standardization

The most critical part in potentiometric titration lies in the proper standardization of the pH electrode. An aqueous HCl solution of known concentration, containing 0.15 M KCl is titrated with standardized KOH in the pH interval 1.8 to 12.2 (blank titration). Data are recorded every 0.2 pH units. After the stirrer is turned on one wait until a stable pH reading is obtained (drift ≤ 0.01 pH/min). In buffered solutions pH readings stabilize typically in 5 to 10 seconds with combined glass electrodes. In unbuffered solutions around pH 7 in a strong acid–strong base titration, readings often do not satisfy the stability criterion. In this case the pH reading is taken anyway after a pre-set time, usually 60 to 90 seconds. Such data are not of good quality; the weighting scheme in the fitting algorithm takes it into account. To establish the operational pH scale,[15] the measuring circuit is first calibrated with a single aqueous pH 7.00 phosphate buffer and the Nernst slope is assumed. The aim of this step is to convert millivolt readings to operational pH. Since all equilibrium constants which will be reported are on a concentration scale with reference to a particular value of fixed ionic strength (0.15), it is necessary to convert the operational pH to pcH (i.e., the pH scale based on concentration $-\log[H^+]$). Since pcH is known at each point in the HCl–water–KCl versus KOH titration, the operational pH reading can be related to the pcH values for example using the multiparametric equation[15–18]

$$pH = \alpha + S pcH + j_H[H^+] + j_{OH}\frac{K_w}{[H^+]}. \tag{15.2}$$

The standardization parameters are determined by a nonlinear least squares procedure. The α term mainly corresponds to the negative log of the activity coefficient of the $[H^+]$ at the working temperature and ionic strength. S refers to the Nernst slope. The j_H term corrects pH readings for the nonlinear pH response due to liquid junction and asymmetry potentials in the acidic region (1.5–2.5), while j_{OH} corrects for high-pH (11–12.2) nonlinear effects. $K_w = [H^+][OH^-]$ is ionic product of water. K_w varies as a function of temperature and ionic strength; reference values are taken from Sweeton et al.[19] Typical values for α, S, j_H, and j_{OH} are 0.14, 1.0000, 0.4, and −0.6, respectively, for a good performing pH electrode.

Sample Preparation

The sample concentration typically ranges between 0.25 and 1 mM. Higher concentrations might be necessary to determine extreme pKa values (<2.5 or >11.5). The most common problem encountered pertains to sample solubility. One can distinguish three cases: (1) the substance dissolves quickly after water–KCl addition, (2) the substance dissolves but extremely slowly, and (3) the amount of substance given exceeds its aqueous solubility. Case 1 represents the straightforward experiment. In case 2 the low dissolution rate can be increased as follows: dissolve the sample in MeOH, and then evaporate the solvent: this leads to the formation of microcrystals or to an amorphous substance that will considerably increase the sample dissolution upon water–KCl addition. In case 3 direct measurement in water–KCl cannot be done and a co-solvent approach is required.

Defining the Assay Strategy

Usually one titrates from acidic to alkaline in order to minimize the interference by carbon dioxide. It is important that the compound stays in solution in the course of the titration. Compound precipitation can lead to significant errors in the pKa determination. This problem is particularly acute when one works with a compound concentration close to its solubility limit. In this case partial precipitation of the substance induces a distortion in the titration curve, while a good fit to the fitting of the experimental can still be achieved. For bases, partial precipitation leads to an underestimation of the pKa. For acids, partial precipitation leads to an overestimation of the pKa.[20] When the compound concentration is way above its solubility limit, a stronger distortion of the titration curve is observed and the experimental data no longer fit with the proposed model. Argon is used to minimize carbon dioxide uptake in the course of the titration.

Water-Insoluble Compounds: The Co-Solvent Approach

When the compound is not soluble in water, a mixed-solvent approach is the way to go. Mixed-solvent solutions of various co-solvent–water ratios are prepared and the psKa (apparent pKa) is determined in each mixture. The aqueous pKa is determined by extrapolation of the psKa values to zero co-solvent. Numerous examples of pKa's determined by extrapolation from mixtures from methanol,[21,22,23] ethanol,[24,25,26] propanol,[27] dimethyl sulfoxide,[28,29] dimethylformamide,[30] acetone,[31] and dioxane[32] are reported in the literature. The mixed-solvent method to determine aqueous pKa's of water-insoluble substances has been described in the literature.[33] The organic solvents of interest are the ones of lower dielectric strength than water. The PCA 101 and the GlpKa titrators support the following co-solvents: MeOH, DMSO, acetonitrile, and dioxane. The instrument is able to re-calculate the four electrode standardization parameters (α, S, j_{H}, j_{OH}) obtained from the aqueous blank titration for each co-solvent–water ratio. K_{w} values are also changed accordingly. Usually three different co-solvent–water ratios are used to extrapolate the aqueous pKa. The GlpKa instrument is able to perform up to three titrations with the same sample, while three samples are necessary with the PCA 101. The extrapolation to zero co-solvent is done by linear regression using the Yasuda-Shedlovsky plot[34,35] as described:[33]

$$psKa + \log[\mathrm{H_2O}] \quad \text{versus} \quad \frac{1}{\varepsilon}. \qquad (15.3)$$

Data Analysis

The key element in the data analysis is the construction of the difference or Bjerrum plot. This plot shows the average number of bound protons versus pcH. The difference plot is obtained from the difference between two titration curves: one is the titration of an ionizable substance and the other is a blank titration. Graphically, the pKa corresponds to the pH where the average number of bound proton equals 0.5 (or a multiple of 0.5 if multiple ionization takes place). Nowadays nonlinear regression calculations are used to derive pKa values form Bjerrum plots.

15.4.2. Spectrophotometric Titration

This method measures the UV change induced by ionization instead of the volume of titrant necessary to change pH. Spectrophotometric titration is a powerful technique that works with relatively low concentrations of analyte (factor 10—lower compared to potentiometric titration), thus preventing sample precipitation for a larger proportion of the samples. The interpretation of experimental data is relatively easy when one considers a single ionization. It becomes significantly more complex if the molecule has several ionizable centers, and even more if they overlap. In this case sophisticated numerical procedures like target factor analysis have to be used.[36]

The drawback of the spectrophotometric technique is that it is bound to ionization induced changes in UV spectra. It is difficult to guess whether a particular ionization will induce an change in UV spectra a priory. Practically, one combines this approach with an *in silico* tool like ACD pKa database, which predicts the number of ionizations and gives estimates of the pKa values. If the number of theoretical ionizations is lower than the number of ionizations obtained from the experimental data, one must re-analyze the sample using the potentiometric titration approach.

15.4.3. pH-Gradient Technology

Traditional methods to measure ionization constants are based on potentiometric or spectrophotometric titration. These methods are inherently slow due to the equilibration time needed after each addition of titrant: typically one titration takes 20 to 40 minutes, and as most samples are poorly soluble in water, one needs to perform three titrations in water–co-solvent mixtures and extrapolate back to aqueous.

The profiler SGA (Sirius Analytical Instruments) measures pKa by a spectrophotometric procedure on a fully automated sample-handling platform, using a continuous pH-gradient instead of an electrode to control the pH. In this setup the molecule is exposed to a continuously changing pH environment and the variation in its UV spectrum within the pH range 2 to 12 can be measured in about 2 minutes.

Principle of the Method

The instrument is composed of:

- Three computer controlled dispensers; acidic and basic buffer reagents, water/sample.
- A degasser, to prevent bubbles in the UV flow cell.
- A diode array detector.
- A back pressure valve to ensure sample to pass at a constant speed.

Spectrophotometry is a powerful method for pKa measurement,[36] as long as the compound contains a chromophore. It is based on the change of absorbance of the sample as a function of pH. The instrument is composed of a three-flow system in which a linear pH gradient is made by appropriate mixing of two buffers, called acidic and basic buffer reagents, that are controlled by the two first dispensers while the third injects the sample at a constant speed. Prior to use, the instrument has to be calibrated to establish the link between the gradient status (or time) and the pH scale. Buffer reagents are commercially available and supplied by Sirius analytical Instruments.

Data Analysis

Two methods can be used:

- AFD : Automated first derivative
- TFA : Target factor analysis

Automated First Derivative. The AFD method calculates the first derivative of the absorbance against the pH curve and assigns the pKa to the pH where a maximum peak is found. AFD is a simple and quick method to calculate pKa, but it is limited to relatively simple molecules. Typically AFD is well suited for monoprotic molecules (or well-separated *pKa's*) that generate a good UV signal. From a practical viewpoint it is not recommended to use AFD when OPH is lower than 1. OPH (overall peak height) is a parameter, that describes the amplitude of the absorbance change induced by the ionization:

$$\text{OPH} = 1000 \times \text{Difference between minimum}$$
$$\text{and maximum values of } dA/dpH.$$

Target Factor Analysis. The TFA method has already been described elsewhere[36] and is briefly described below:

The absorbance A is linked to the concentration of the species present in the solution. Their absorbance and at each wavelength the Beer-Lambert law applies such that:

$$A = C\,E,$$

where C and E are, respectively, the concentration of each independent light-absorbing species as a function of pH and the molar absorptivity of each species as a function of pH and wavelength. At the start of the calculation, C and E are guessed values, whereas A is a measured value. During the calculation the A matrix is whereby deconvoluted using target factor analysis whereby values of the concentration and absorptivity are proposed iteratively until $A - C\,E$ tends to a minimum. At that point, the pKa values required to calculate the distribution of species (C matrix) are assumed to be to correct pKa values.

TFA analysis of the pH/spectra is more time-consuming and requires additional expertise to run the software. It is, however, a more powerful tool that allows one to cope with overlapping pKa's and is able to separate noise from signal by principal component analysis, thus yielding less favorable signal to noise ratios. To some extent TFA is also able to remove signal due to precipitation of the sample, particularly if the number of ionization's is fixed by an independent method (i.e., using an *in silico* method).

Practical Considerations for Data Refinement, AFD or TFA?

Use of AFD. ALL conditions below have to be fulfilled:

- Number of peaks equals number of expected pKa's
- Peaks separated by at least 2 pKa units
- Symmetrical peaks
- OPH <1.0

Use of TFA. TFA is used if ANY of the conditions below applies:

- Overlapping *pKa*'s
- Number of peaks in AFD not equal to number of expected pKa's
- OPH <1

With the currently commercially available technology, a success rate of 60% to 70% can be envisaged with "real-life" compounds from lead optimization programs. Further improvement could be envisaged if one could circumvent precipitation of samples, for example, by addition of a co-solvent in the buffer reagents.

The strength of the technique lies in its speed and degree of automation (samples can be loaded in 96-well plates). However, our experience has shown that successful automated data analysis requires relatively clean data. For this reason we expect improvement in efficiency once a set of co-solvent buffer reagents becomes available.

15.5. PERMEABILITY DETERMINATIONS

15.5.1. Permeability pH Profiles: Ionization, Membrane Retention, and Unstirred Layers

There are a variety of terms used in the literature to express permeability. As experimental permeability can be corrected in a number of ways to account for membrane retention, ionization, and unstirred layer effect, the definitions used in this chapter is given below together with their physico-chemical meaning:

Definitions

$\text{Log}\, P_a$: *Apparent permeability.* It is directly calculated from the concentration in the acceptor concentration

$\text{Log}\, P_e$: *Effective permeability.* This is $\log P_a$ corrected by membrane retention (if any). In absence of membrane retention, $\log P_a = \log P_e$

$\text{Log}\, P_m$: *Membrane permeability.* That is $\log P_e$ corrected by the unstirred layer effect.

$\text{Log}\, P_o$: *Intrinsic permeability.* That is $\log P_m$ corrected by ionization. For neutral compounds $\log P_m = \log P_o$

Membrane permeability depends on the ability of the compound to diffuse through the unstirred layer and the membrane. Intrinsic permeability is linked to membrane partitioning as described below:

$$\log P_o = \log P(\text{mem}) + \log\left(\frac{D}{h}\right),\qquad (15.4)$$

where $\log P(\text{mem})$ is the partition coefficient between the membrane considered and the incubation medium, D the diffusion coefficient of the compound within the membrane and h the membrane thickness.

Effective membrane permeability is related to membrane permeability and unstirred layer permeability as:

$$\frac{1}{P_e} = \frac{1}{P_m} + \frac{1}{P_{u1}} \qquad (15.5)$$

or

$$\frac{1}{P_e} = \frac{C_t/C_n}{P_o} + \frac{1}{P_{u1}}, \qquad (15.6)$$

where C_t/C_n represent the fraction of neutral species. Equation (15.6) can be used to derive $\log P_m$ in terms of pH (and $\log P_o$) for ionizable compounds with known *pKa* values. This is illustrated graphically with Diclofenac permeability pH profile in Figure 15.3. The dynamic range of all permeability assays is limited by the diffusion through the unstirred layers (see Table 15.4). As the thickness of the unstirred layer increases, the dif-

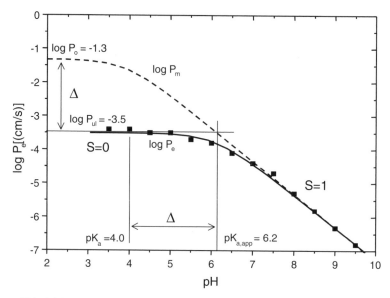

Figure 15.3. Diclofenac effective permeability pH profile corrected for the unstirred layer effect. The solid line represents the effective membrane permeability, while the dotted line represents the membrane permeabiltiy, corrected by the unstirred layer effect.

Table 15.4. Unstirred Layers Thickness and Upper Effective Permeability

UWL thickness (µm)	$\text{Log} P_{e,\text{max}}$ (cm/s)
50	-2.9^a
300	-3.5^b
500	-3.9
800	-4.1
1500	-4.4^c

[a] In vivo (GI).
[b] HDM.
[c] Caco-2 (unstirred).

fusion becomes rate limiting. The same phenomenon occurs in human GIT, although there the layer thickness is relatively small due to the motions of the intestine and the presence of surfactants. If not corrected for the unstirred water layer compounds with vastly different intrinsic permeability might appear similar as shown in Table 15.5. Data in Table 15.5 show that Caco-2 values[37–39] in literature for these highly permeable compounds

Table 15.5. Effect of Unstirred Layers on Assay Dynamic Range

Compound	Caco-2 $\log P_e$	HDM $\log P_{e,max}$	HDM $\log P_o$
Chlorpromazine	-4.7^{42}	-3.6	1.3
Imipramine	-5.3^{43}	-3.7	0.8
Desipramine	-4.6^{42}	-3.7	0.1
Ibuprofen	-4.7^{44}	-3.7	-0.9
Propranolol	$-4.5^{44,45}$	-3.5	-1.6

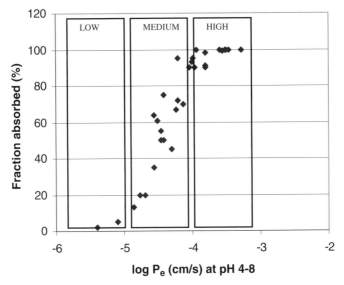

Figure 15.4. Fraction absorbed in humans versus effective permeability for a set of generic drugs measured with artificial membranes.[49,50] The pH gradient in the intestine is accounted for by the highest permeability within the pH range 4 to 8.

do usually not exceed the value calculated for an unstirred layer of 1500 nm, which was reported for unstirred Caco-2 monolayers.[40] Similarly hexadecane membrane (HDM) permeability values reach a plateau at -3.6 for all of these highly permeable compounds, in agreement with the calculations presented in Table 15.4. When solubility is not the rate-limiting step for absorption, there is a sigmoidal relationship between the fraction absorbed and permeability (see Figure 15.4). With respect to the prediction of the fraction absorbed, the highest uncertainty is in the middle part of the calibration curve where the fraction absorbed changes steeply with only modest variations of permeability.

15.5.2. Artificial Membrane Permeability Models

Immobilized Artificial Membranes

Immobilized artificial membrane (IAM) packings are prepared by covalently immobilizing phospholipid monolayers to a solid silica matrix to mimic the surface of cell membranes.[41] In practice, the capacity factors ($\log k$) are determined from capacity factors obtained with different water co-solvent ratios ($\log k'$). Capacity factors ($\log k$) have been shown to correlate with liposome partition coefficient[41] and with Caco-2 monolayers or intestinal perfusion permeability. The advantage of IAM chromatography is that the result is not affected by the presence of impurities. Also there is no technical difficulty associated with detection of a low concentration of sample in the acceptor compartment, which becomes an issue with low permeable compounds with the two methods described next. The major drawback is that the throughput is significantly lower than with multiparallel techniques designed to run in the 96-well plate format.

Filter-Coated Phospholipids in Dodecane

The parallel artificial membrane assay (PAMPA) was first described by Kansy et al.[42,43] The assay consists in measuring the rate of transfer of compounds from a donor to an acceptor compartment that is separated by a porous filter coated with a mixture of phospholipids in dodecane. The exact nature of the membrane is not known and is unlikely to be a well-defined phospholipid bilayer. Interesting correlations between kinetics of transfer and fraction absorbed in humans were, however, observed. The technology was further improved by taking into account membrane retention and the derivation of differential equations to convert flux ratios into permeability values.[14,44] The system is now available in the form of a commercial instrument (PSR evolution series, pION, Inc; *www.pion-inc.com*). Improved correlations with GIT permeability were observed using different lipid compositions designed to closer mimic the nature of native membranes.[45–48]

Filter-Coated Hexadecane Liquid Membrane

In our laboratory we developed a similar model based on a hexadecane liquid membrane.[49,50] The selection of an alkane membrane to model transcellular passive was guided by previous work by several groups who have demonstrated through computer simulation[51,52] or experimentally that the bilayer interior is in many ways well represented by long-chain alkanes.[53,54] Moreover a number of studies have reported a good correlation between alkane/water distribution coefficient and permeation across phospholipid bilayers.[52] The advantage of this approach over the coated phospholipid is (1) its simplicity and robustness and (2) permeability through which the hexadecane layer can be converted into hexadecane/water partition coefficients so that the model describes a well-defined physicochemical molecular property.

The apparent permeability value P_a is determined from the concentration of compound in the acceptor compartment after a given incubation time. Equation (15.8) has been derived from the differential equation (15.2), which describes diffusion kinetics under non-sink conditions. Experiments performed under non-sink conditions allow one to work with substantially higher concentrations in the acceptor compartment. This makes optical (UV) detection possible and greatly simplifies analytics when LC-MS is used. In (15.8) below r is the ratio of the absorbance in the acceptor chamber divided by the theoretical equilibrium absorbance (determined independently), V_R is the volume of the acceptor compartment, V_D is the donor volume, A is the accessible filter area (total filter area multiplied by porosity), and t is the incubation time. Equation 1 is obtained from the differential equation

$$\frac{dc_R}{dt} = \frac{P_a \cdot A}{V_R} \cdot (c_D - c_R),\qquad(15.7)$$

with $c_D(t)$ being the compound concentration in the donor compartment, $c_R(t)$ being the concentration in the acceptor compartment, and

$$P_a = -\frac{V_D}{V_D + V_R} \cdot \frac{V_R}{A \cdot t} \cdot \ln(1 - r).\qquad(15.8)$$

In absence of membrane retention P_a is identical to P_e, the effective membrane permeability.

15.5.3. Caco-2 Monolayers Permeability Model

Transport processes in epithelia have been studied in a variety of organ and tissue preparations from different species. However, native epithelial tissues are complex and contain many different cell types. Monolayers obtained from established epithelial cell lines, on the other hand, have the advantage of being structurally simple, homogeneous, and relatively easy to manipulate. Transport studies in epithelial monolayers are therefore often easier to perform and interpret. The major disadvantage of epithelial cell cultures is that in most situations they are unlikely to entirely mimic a more complex normal epithelium. Also epithelial cells in cell culture may express carriers and enzymes in a different or more variable manner than in vivo. Despite these disadvantages, cell cultures have shown promise as models for intestinal drug absorption.

The most popular commercially available human intestinal cell line for drug absorption studies is Caco-2. Caco-2 cells, which are derived from a human colonic adenocarcinoma, exhibit morphological as well functional similarities to intestinal (absorptive) enterocytes.[55] The cells form tight junctions and express many brush border enzymes (mainly carboxylesterases), some CYP isoenzymes, and phase II enzymes such as glutathione-S-transferases, sulfotransferase, and glucoronidase, namely enzymes relevant in studies of presystemic (drug) metabolism.[56] Many active transport systems that are normally found in small intestinal enterocytes have been characterized. These include carriers for amino acids, dipeptides, and vitamins as well as widely unspecific efflux systems like Pgp and MRP.[57] The unusually high degree of differentiation, together with the fact that these cells differentiate spontaneously in normal serum-containing cell culture medium, has resulted in Caco-2 becoming one of the most popular cell lines in studies of epithelial integrity and transport.

Transport experiments are usually carried out using Caco-2 cells grown on artificial filters (polycarbonate or polyethylene terephtalate) in wells. The chamber system is adapted to allow electrophysiological parameters such as potential difference to be measured in parallel so that cellular integrity can be monitored.

The standard method for the determination of apparent permeability, Pa [cm/min] through Caco-2 cells is the use of Artursson's equation[58]:

$$P_a = \frac{\Delta Q}{\Delta t \cdot A \cdot C_o}, \qquad (15.9)$$

where $\Delta Q/\Delta t$ is the permeability rate [mg/min], C_0 is the initial concentration in the donor chamber [mg/mL], and A is the surface area of the membrane. It must be stated that (15.9) should be used with caution when membrane (or cell) retention takes place. Retention causes a kinetic shift and a change in shape in the concentration/time profile as shown in Figure 15.5 for an artificial membrane system (moves from hyperbolic to sigmoid). In that case the initial slope cannot be used directly to calculate effective permeability, as P_a becomes significantly lower than P_e. The consequence is that effective permeability of lipophilic compounds can be underestimated (and rank ordering of compounds be wrong) if this phenomenon is not properly taken into account.[59,60]

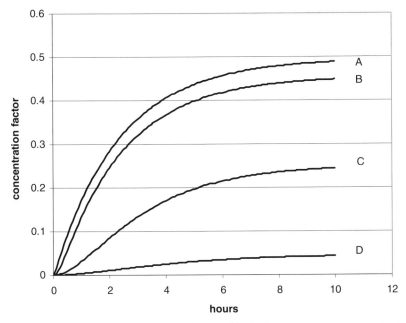

Figure 15.5. Theoretical concentration vs. time profiles in the acceptor compartment calculated for different membrane water distribution coefficients (A) $\log D = 1$; (B) $\log D = 2$; (C) $\log D = 3$; (D) $\log D = 4$. Calculations were performed using a numerical integration approach based on diffusion coefficients in water and membrane. A water/membrane volume ratio of 1000 was used. The absolute concentration in the acceptor compartment is the product of the initial concentration (in donor compartment) times the concentration factor (y axis).

15.5.4. Artificial Membranes versus Caco-2

A commonly asked question is which of the two approaches should one use. In fact a careful analysis of the potential and limitations of the two techniques shows that they are not mutually exclusive but rather complementary.

Artificial membranes are a relatively low-cost assay for fast screening of transcellular passive diffusion. The system allows easy access to permeability pH profiles and unstirred water layer correction. Due to the simple composition of the buffer reagents, the analytics is much easier than with cell cultures. Straight UV detection can be used in most cases and LC-MS (low soluble samples or compounds lacking UV chromophore) is considerably simplified in comparison to Caco-2, as the system matrix is relatively simple. A reasonable correlation has been obtained between artificial membrane and reported human GIT permeability[61–63] in our laboratory (Figure 15.6A).

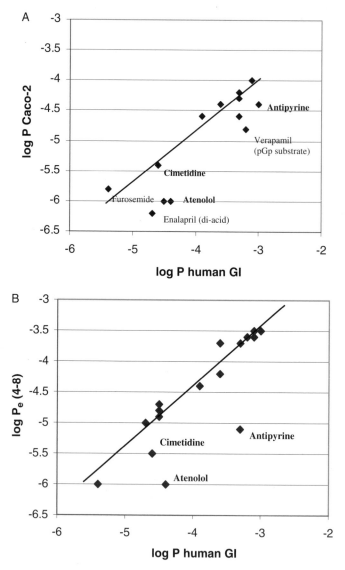

Figure 15.6. (A) Correlation between human GIT and Caco-2 permeability. GIT permeability values are from Refs. [61–63]. Compounds included in the study are amiloride, antipyrine, amoxicillin, atenolol, carbamazepine, cimetidine, desipramine, furosemide, hydrochlorothiazide, ketoprofen, metoprolol, naproxen, piroxicam, propranolol, and ranitidine. (B) Correlation between human GI permeability and artificial membrane permeability (hexadecane artificial membrane). Paracellular transported compounds are given in bold.

As expected, paracellular transported compounds are underestimated as this pathway is not represented in our artificial membrane system. Similar correlations have been obtained by Avdeef using a phospholipid/dodecane mixture.[64]

On the other hand, artificial membranes represent a further simplification of the human GIT than Caco-2 monolayers as a number of features are not present, like the paracellular pathway, active transporters, or efflux systems. Caco-2 is a useful model to study actively transported compounds as well as substrates and inhibitors of efflux systems (PgP, MRP). Caco-2 monolayers also have some potential to model paracellular transport, although the junctions appear tighter than in the upper GIT. The consequence is that some paracellular transported compounds are properly transported while others are underestimated (Figure 15.6*B*).

15.6. PARTITION (LOG*P*) AND DISTRIBUTION COEFFICIENTS (LOG*D*)

The 1-octanol/water partition coefficient (log *P*) has proved to be a useful parameter in quantitative structure-activity relationship.[65] Different approaches like the shake-flask,[66] chromatographic methods,[67] or the filter probe assay[68] are being used to measure log *P*. In the last decade there was a growing interest in an alternative method based on potentiometric titration, originally described by Dryssen[69] in the mid-1950s, that was being developed by Sirius Analytical Instruments (Forest Row, UK). This method is particularly valuable when a substance has no useful analytical chromophore or is partially ionized at physiologically relevant pH's, (usually between 1 and 8). Quite recently another technique based on octanol-coated columns has been launched to measure distribution coefficients at pH 7.4. In our laboratory we have developed a technique based on octanol artificial liquid membranes to measure log *P* (partition coefficient).

15.6.1. Saturated Shake-Flask

Shake-flask is the traditional method used to measure octanol/water partition and distribution coefficients. Although the method is relatively straightforward in its principle, it is often difficult for samples with log *P* values greater than 3 and usually impossible for log *P* values above 4.5, because the quantification of the concentration in the aqueous phase becomes very challenging. HPLC-UV is the most widely used analytical technique for this assay. For lipophilic compounds (log *D* > 2) it is necessary to do corrections to take account of octanol's solubility in water; otherwise, the log *D* (or *P*) value will be underestimated. Each sample, in principle, requires the development of an analytical method with appropriate lower limit of quantification. Generic methods can be used for compound clusters, but this approach is not applicable for high-throughput screening or hit to lead profiling because (1) the physicochemical properties can vary significantly between compounds and (2) one is often confronted to high $\log P$ values in today's drug discovery programs.

15.6.2. Dual-Phase Potentiometric Titration

This method is particularly well suited for ionizable compounds for which lipophilicity varies with pH (see Figure 15.7). The pH-metric technique basically consists of two linked titrations, one without 1-octanol and one in the presence of a given volume of the partition solvent. When the substance partitions into octanol, the aqueous pKa is shifted, and $\log P$ values can be derived from this difference.[70–73] The method covers an exceptional dynamic range of about 8 log units, from −1 to +7. In addition the lipophilicity pH

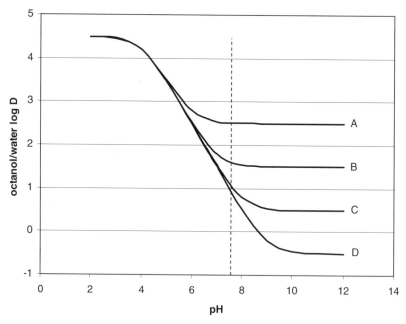

Figure 15.7. Diclofenac octanol/water lipophilicity pH profile ($\log D$) calculated for different ion-pair extraction coefficients. Parameters used for calculation are pKa = 4.0 (acid), $\log P$ (neutral) = 4.5, $\log P_{ion}$ = 2.5 (A); $\log P_{ion}$ = 1.5 (B); $\log P_{ion}$ = 0.5 (C); and $\log P_{ion}$ = −0.5 (D). The dotted line indicates $\log D$ values at pH 7.4.

profile can be derived from the titration data, and it is possible to distinguish between partitioning of the neutral species and ion-pair extraction. Before the experiment, it is necessary to have a relatively accurate value of the ionizazion constant either from a potentiometric or spectrophotometric titration measurement. Before measurement, a calculated $\log P$ value should be obtained to select of the appropriate octanol/water ratio. With lipophilic compounds ($CLOGP > 2$) two titrations must be performed to take into account ion-pair formation; otherwise, $\log P$ could be underestimated. A first titration is performed with a low octanol volume and is followed by a second titration with a higher octanol/water ratio. The key element in the data analysis is, for potentiometric pKa measurements, the construction of the difference or Bjerrum plots. The applicability of this method is, however, limited to ionizable compounds (neutral or weakly ionizable compounds cannot be measured). Another limitation lies in the relatively high sample concentration required by potentiometry (ca. 10^{-4} M), so low soluble compounds are difficult to handle. Most problematic compounds in this respect are compounds that are moderately lipophilic and low soluble. As with all multiparametric fitting procedures a wrong result can be obtained with a good fit. This can happen if the model entered is not right, in particular, in multiprotic substances. Another source of uncertainty with this method comes from the aqueous pKa, which, if not determined with adequate accuracy, leads to uncertainty on the $\log P$. Weak bases and weak acids are also often challenging to measure because their apparent pKa in the presence of the organic solvent tend to be shifted outside the dynamic range of the pH electrode.

15.6.3. Partitioning of Amphoteric Drugs

Lipophilicity profiles of amphoteric drugs are more complicated than partitioning of simple acids and bases. A comprehensive review of the issues encountered in the determination of $\log P/D$ of amphoteric compounds has been published by Pagliara et al.[74]

Ordinary Ampholytes

Ordinary ampholytes have an acidic and a basic ionizable group with pKa (acidic) > pKa (basic). Proteolysis and partitioning of ordinary ampholytes are straightforward when the difference between pKa (acidic) and pKa (basic) is greater than 3. In this case the substance is fully neutral in the pH region between the two pKa values. If, however, ΔpKa is lower than 3, an overlap between the two protonation equilibria takes place, allowing for the existence of a small proportion of the zwitterionic species as discussed next.

Zwitterionic Ampholytes

Zwitterionic ampholytes have an acidic and a basic ionizable group with pKa (acidic) < pKa (basic). A zwitterion can exist in solution on four different electrical states (cation, zwitterion, neutral, and anion). In this case the acid–base equilibria are defined in terms of macroscopic constants— macro pKa (acidic) and macro pKa (basic)—that refer to the stoichiometric ionization and of microscopic constants that refer to the ionization of individual species. The pH-metric method acts as a proton-counting technique and cannot distinguish between zwitterion and neutral species. Therefore the ionization constants delivered refer to macro *pKa*'s and the resulting $\log P$ values must be considered as $\log D_{max}$. These two values can be very close or significantly different, depending on the tautomeric equilibrium between the neutral and the zwitterionic species and the difference between $\log P_n$ and $\log P_z$.

However, it is generally accepted that when it comes to biological correlations the intrinsic lipophilicity of zwitterionic ampholytes is best characterized by their $\log D_{max}$ values (provided by the present technique). In the reports and the database, partition coefficients of zwitterionic ampholytes are reported as $\log P$ values, although they must be considered as $\log D_{max}$. In this case measured $\log P$ are usually lower than calculated $\log P$ as the former relates to a macroconstant (tautomeric equilibria) and the latter to a microconstant (neutral species).

15.6.3. Chromatographic Method

RP-HPLC capacity factors have been shown for a long time to correlate with lipophilicity but have also been criticized as not being a true replacement for octanol/water shake-flask methods.[75,76] In particular, capacity factors respond to solute hydrogen bond acidity, while $\log P$ octanol do not. Often times good correlations were obtained with related analogs or relatively simple molecules, but when the test set was extended to druglike molecules covering a wider physicochemical property space, the correlation with shake-flask $\log P$ were much less convincing. A improved RP-HPLC method that addresses the points above was proposed by Lombardo et al.[77] and seems to cover a wider range of molecules over a broad range of lipophilicity. The approach combines the stationary phase used by Pagliara[76] with the mobile phase used by Minick.[78] In a fist step the method was limited to neutral compounds. More recently the method was extended to neutral and basic compounds.[79]

The principle of the approach by Lombardo is based on a Supelcosil LC-ABZ column (Bonded phase ligand: $Si-(CH2)_3NHCO(CH_2)_{14}CH_3$), which solid phase is end-capped by a small zwitterionic fragment so silanophilic interactions are suppressed. 1-Octanol is added to the mobile phase containing MOPS buffer and MeOH to mimic intermolecular interactions encountered by a molecule in a shake-flask experiment. This also presumably helps to correct for the residual hydrogen bond acidity of the stationary phase. In a typical experiment, capacity factors ($\log k'$) are measured at three MeOH concentrations in the mobile phase (depending on $CLOGP$) and extrapolated to aqueous ($\log k'_w$). $\log P$ values are obtained from a calibration curve where one plots $\log k'_w$ in terms of $\log P$ octanol/water. The method covers $\log P$ within the range 0 to 5.

The main advantages of this approach is that (1) it is relatively fast (ca. 30 min/compound), (2) it covers a wide range of $\log P$ values (allows to access high $\log P$ values that are difficult to obtain using the traditional shake-flask method), (3) requires relatively little material, and (4) is not very sensitive to impurities. There are, however, a number of limitations inherent to the method: (1) Only neutral and monofunctional basic compounds are covered. (2) For bases only $\log D$ values at pH 7.4 are obtained (and not the neutral species lipophilicity). (3) $\log D$ values can be affected by ion-pair extraction (see Figure 15.7). (4) Compounds need to have a UV chromophore for detection. (5) The throughput, although higher than with most techniques described above, is still relatively limited to about 20 compounds a day. (6) The method is restricted to the octanol/water system.

15.6.4. Artificial Membrane Log P

In our laboratory we have been developing an alternative approach for $\log P$ determination based on artificial liquid membrane permeability in 96-well plates. Two aqueous compartments are separated by a thin octanol liquid layer coated on a polycarbonate filter. One follows the diffusion from the donor to the acceptor compartment. As shown in Figure 15.5, the time-dependent accumulation of the sample changes when membrane retention takes place and apparent permeability derived from the concentration at any time t using equation (15.8) becomes lower than effective permeability. We have shown that the higher the lipophilicity, the higher is the difference between $\log P_e$ and $\log P_a$. We have established experimental conditions where $\log P_a$ varies linearly with $\log P$ membrane (= $\log P_{octanol}$ in the case of an octanol membrane) within the 2 to 6 range. The upper limit is imposed by the lower limit of quantification in the acceptor compartment, while the lower limit is imposed by the unstirred water layer permeability of the test samples and the thickness of the unstirred water layers on each side of the membrane. The method works with neutral, acids and bases. From a practical viewpoint, each compound is tested at pH 2.0, 6.0, and 11.0.

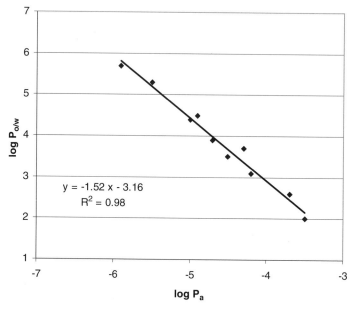

Figure 15.8. Correlation between apparent membrane permeability ($\log P_a$, P_a in cm/s) and octanol/water $\log P$. This calibration curve has been obtained using $V_a = V_d = 0.3\,\text{mL}$, a membrane volume of $3\,\mu\text{L}$, and an incubation time of 4 hours. Compounds included are benzoic acid, furosemide, ketoprofen, propranolol, warfarin, valsartan, nortryptiline, diclofenac, chlorpromazine, and terfenadine.

Calculated pKa values are used to determine at which pH the compound is neutral. Log P_a (neutral) is then used to calculate back the membrane log P partition coefficient. This way one calibration curve covers neutral, acids, and bases (Figure 15.8). One difference with the RP-HPLC based method is that the approach described here measures $\log P$, while the former measures $\log D$ values. Potentially the method can be relatively easily extended to other water/solvent systems. The limitation of the approach is with some ampholytes and with zwitterions, which are not >95% neutral at any pH value througout the 2 to 11 range.

ACKNOWLEDGMENTS

I wish to thank to Dr. Gian Camenisch (Pre-clinical Compound Profiling, Novartis, Switzerland) for stimulating discussions around the Caco-2 monolayer assay and its interpretation. Many thanks also to S. Arnold, X. Briand,

F. Loeuillet, and W. Salamin (Physicochemical Characteristics group, Novartis, Switzerland) for their excellent technical assistance, to Dr. J. Wang (Physicochemical Characteristics group, Novartis, USA) for reviewing this manuscript and to Dr. H.-P. Grimm (GiMS AG, Switzerland) for his help in setting up the numerical simulations of permeability processes.

REFERENCES

1. G. L. Amidon, H. Lennernäs, V. P. Shah, J. R . Crison, *Pharm. Res. 12*, 413–420 (1995).

2. C. A. Lipinski, F. Lombardo, B. W. Dominy, P. Feeney, *Adv. Drug. Delivery Rev. 23*, 3–25 (1997).

3. C. A. Lipinski. *Curr. Drug Disc.*, 17–19 (April 2001).

4. J. B. Dressman, G. L. Amidon, C. Reppas, V. P. Shah, *Pharm. Res. 15* (1), 11–22 (1998).

5. S. H. Yalkovsky, S. Banerjee, *Aqueous Solubility: Methods of Estimation for Organic Compounds*, Marcel Dekker, New York (1992).

6. A. Avdeef, *Pharm. Pharmacol. Commun. 4*, 165–178 (1998).

7. A. Avdeef, C. M. Berger, C. Brownell, *Pharm. Res. 17*, 1, 85–89 (2000).

8. A. Avdeef, J. J. Bucher, *Anal. Chem. 50*, 2137–2142 (1978).

9. P. E. Brooker, M. Ellison, *J. Am. Chem. Soc. 64*, 785–787 (1942).

10. C. D. Bevan, R. S. Lloyd, *Anal. Chem., 72*, 1781–1787 (2000).

11. A. Noyes, W. R. Whitney, *J. Am. Chem. Soc. 19*, 930–934 (1897).

12. A. M. Serajuddin, C. I. Jarowsky, *J. Pharm. Sci. 74*, 2, 148–154 (1985).

13. A. Avdeef, *Curr. Topics. Med. Chem. 1*, 277–351 (2001).

14. A. Avdeef, in B. Testa, H. van de Waterbeemd, G. Folkers, E. Guy, eds., *Pharmacokinetic Optimization in Drug Research*, Verlag Helvetica Chimica Acta, Wiley-VCH, Weinheim, pp. 305–326 (2001).

15. R. G. Bates, *CRC Crit. Rev. Anal. Chem. 10*, 247–27 (1981).

16. A. Avdeef, J. Bucher, *J. Anal. Chem. 50*, 2137–2142 (1978).

17. A. Avdeef, R. S. Sofen, T. L. Bregante, K. N. Raymond, *J. Am. Chem. Soc. 100*, 5362–5370 (1978).

18. A. Avdeef, *Anal. Chim. Acta. 148*, 237–244 (1983).

19. F. H. Sweeton, R. E. Mesmer, C. F. Baes, *J. Solution. Chem. 3*, 191–214 (1974).

20. A. Avdeef, *Pharm. Pharmacol. Commun. 4*, 165–178 (1998).

21 A. L. Bacarella, E. Grunwald, H. P. Marschall, E. L. Purlee, *J. Org. Chem. 20*, 747–762 (1955).

22. T. Shedlovsky, R. L. Kay, *J. Am. Chem. Soc. 60*, 151–155 (1956).

23. L. G. Chatten, L. E. Harris, *Anal. Chem. 34*, 1495–1501 (1962).

24. E. Grunwald, B. Berkowitz, *J. Am. Chem. Soc.* **73**, 4939–4944 (1951).

25. B. Gutbezahl, E. Grunwald, *J. Am. Chem. Soc.* **75**, 565–574 (1953).

26. P. B. Marshall, *Br. J. Pharmacol.* **10**, 270–278 (1955).

27. N. Papadopoulos, A. Avranas, *J. Solution. Chem.* **20**, 293–300 (1991).

28. J.-C. Halle, R. Garboriaud, R. Schaal, *Bull. Soc. Chim. Fr.*, 2047–2053 (1970).

29. K.-S. Siow, K. P. Ang, *J. Solution. Chem.* **18**, 937–947 (1989).

30. E. R. Garrett, *J. Pharm. Sci.* **52**, 797–799 (1963).

31. G. W. K. Cavill, N. A. Gibson, R. S. Nyholm, *J. Chem. Soc.*, 2466–2470 (1949).

32. L. G. Van Uitert, C. G. Haas, *J. Am. Chem. Soc.* **75**, 451–455 (1953).

33. A .Avdeef, J. E. A. Comer, S. Thomson, *Anal. Chem.* **65**, 42–49 (1993).

34. M. Yashuda, *Bull. Chem. Soc. Jpn.* **32**, 429–432 (1959).

35. T. Shedlovsky, in B. Pesce, ed., *Electrolytes*, Pergamon Press, New York, pp. 146–151 (1962).

36. R. I. Allen, K. J. Box, J. Comer, C. Peake, K. Y. Tam, *J. Pharm. Biomed. Anal.* **17**, 699–712 (1998).

37. M. Yazdanian, S. L. Glynn, J. L. Wright, A. Hawi, *Pharm. Res.* **15**, 1490–1494 (1998).

38. S. Chong, S. A. Dand, R. A. Morrison, *Pharm. Res.* **14**, 1835–1837 (1997).

39. M.-C. Grès, B. Julian, M. Bourrie, V. Meunier, C. Roques, M. Berger, X. Boulenc, Y. Berger, G. Fabre, *Pharm. Res.* **15**, 726–733 (1998).

40. I. J. Hidalgo, K. M. Hillgren, G. M. Grass, R. T. Borchardt, *Pharm. Res.* **8**, 222–227 (1991).

41. S. Ong, H. Liu, C. Pidgeon, *J. Chrom.* **A728**, 113–128 (1996).

42. M. Kansy, F. Senner, K. Gubernator, *J. Med. Chem.* **41**, 1007–1010 (1998).

43. M. Kansy, H. Fischer, K. Kratzat, F. Senner, B. Wagner, I. Parrilla, in B. Testa, H. van de Waterbeemd, G. Folkers, R. Guy, eds., *Pharmacokinetic Optimization in Drug Research*, Verlag Helvetica Chimica Acta, Wiley-VCH, Weinheim, pp. 447–464 (2001).

44. A. Avdeef, M. Strafford, E. Block, M. P. Balogh, W. Chambliss, I. Khan, *Eur. J. Pharm. Sci.* **14**, 271–280 (2001).

45. K. Sugano, H. Hamada, M. Machid, H. Ushio, *J. Biomolec. Screen.* **6**, 189–196 (2001).

46. K. Sugano, H. Hamada, M. Machida, H. Ushio, K. Saitoh, K. Terada, *Int. J. Pharm.* **228**, 181–188 (2001).

47. K. Sugano, N. Takata, M. Machida, K. Saitoh, *J. Pharm. Sci. Tech. Jpn.* **62**, 240 (2002).

48. K. Obata, K. Sugano, R. Sat, M. Machida, K. Saitoh, *J. Pharm. Sci. Tech. Jpn.* **62**, 241 (2002).

49. B. Faller, F. Wohnsland, in B. Testa, H. van de Waterbeemd, G. Folkers, R. Guy, eds., *Pharmacokinetic Optimization in Drug Research*, Verlag Helvetica Chimica Acta, Wiley-VCH, Weinheim, pp. 257–274 (2001).

50. F. Wohnsland, B. Faller, *J. Med. Chem.* **44**, 923–930 (2001).

51. J. S. Marrink, H. J. C. Berendsen, *J. Phys. Chem.* **100**, 16729–16738 (1996).

52. R. M. Venable, Y. Zhang, B. J. Hardy, R. W. Pastor, *Science.* **262**, 223–226 (1993).

53. A. Walter, J. Gutknecht, *J. Membrane Biol.* **77**, 255–264 (1984).

54. A Walter, J. Gutknecht, *J. Membrane Biol.* **90**, 207–217 (1986).

55. P. Artursson, K. Palm, K. Luthman, *Adv. Drug Del. Rev.* **22**, 67–84 (1996).

56. T. Prueksaritanont, L. Gorham, J. Hochman, L. Tran, K. Vyas, *Drug Met. Dis.* **24**, 634–642 (1996).

57. A. Tsuji, I. Tamai, *Pharm. Res.* **13**, 963–977 (1996).

58. P. Artursson, J. Karlsson, *Biochem. Biophys. Res. Comm.* **175**, 880–885 (1991).

59. G. A. Sawada, C. L. Barsuhn, B. S. Lutzke, M. E. Houghton, G. E. Padbury, M. F. H. Ho, T. J. Raub, *J. Pharm. Exp. Ther.* **288** (3), 1317–1326 (1999).

60. P. Wils, A. Warnery, A. Phung-Ba, V. Legrain, D. Schermann, *J. Pharm. Exp. Ther.* **269** (2), 654–658 (1994).

61. H. Lennernaes, *Int. J. Pharm.* **127**, 103–107 (1996).

62. H. Lennernaes, *Br. J. Pharmacol.* **37**, 589–596 (1994).

63. U. Fagerholm, M. Johansson, H. Lennernaes, *Pharm. Res.* **13** (9), 1336–1342 (1996).

64. A. Avdeef, *Curr. Topics. Med. Chem.* **1**, 277–351 (2001).

65. C. Hansch, A. Leo, *Substituant Constants for Correlation Analysis in Chemistry and Biology*, Wiley Interscience, New York, pp. 13–43 (1979).

66. A. Leo, C. Hansch, D. Elkins, *Chem. Rev.* **71**, 525–616 (1971).

67. M. S. Mirrlees, S. J. Moulton, C. T. Murphy, P. J. Taylor, *J. Med. Chem.* **19**, 615–619 (1976).

68. E. Tomlinson, *J. Pharm. Sci.* **71**, 602–604 (1982).

69. D. Dryssen, *Sv. Kem. Tidsks.* **64**, 213–224 (1952).

70. F. H. Clarke, N. Cahoon, *J. Pharm. Sci.* **76**, 611–620 (1987).

71. A. Avdeef, *Quant. Struct. Act. Relat.* **11**, 510–517 (1992).

72. A. Avdeef, *J. Pharm. Sci.* **82**, 183–190 (1993).

73. B. Slater, A. McCormack, A. Avdeef, J. E. A. Comer, *J. Pharm. Sci.* **83**, 1280–1283 (1994).

74. A. Pagliara, P.-A. Carrupt, G. Caron, P. Gaillard, B. Testa, *Chem. Rev.* **97**, 3385–3400 (1997).

75. H. Van de Waterbeemd, M. Kansy, B. Wagner, H. Fischer, in V. Pliska, B. Testa, H. van de Waterbeemd, eds., *Lipophilicity in Drug Action and Toxicology*, VCH, Weinheim, pp. 73–87 (1996).

76. A. Pagliara, E. Khamis, A. Trinh, P.-A. Carrupt, R.-S. Tsai, B. Testa, *J. Liq. Chromatogr.* **18**, 1721–1745 (1995).

77. F. Lombardo, M. Shalaeva, K. Tupper, F. Gao, F. M. Abraham, *J. Med. Chem.* **43**, 2922–2928 (2000).

78. D. J. Minink, J. H. Frenz, M. A. Patrick, D. A. Brent, *J. Med. Chem.* **31**, 1923–1933 (1988).

79. F. Lombardo, M. Shalaeva, K. Tupper, F. Gao, *J. Med. Chem.* **44**, 2490–2497 (2001).

CHAPTER

16

SOLUBILITY IN THE DESIGN OF COMBINATORIAL LIBRARIES

CHRISTOPHER LIPINSKI

16.1. INTRODUCTION

The design of combinatorial libraries with drug-like properties is a trade-off between efficient chemistry and therefore better numerical compound output on one side and increasingly difficult chemistry but better physico-chemical properties on the other. This trade-off is not solely a technical issue but is strongly influenced by people attitudes and organizational timing issues. The author believes that there is a hierarchy of properties that need to be controlled in a quality library. Simply put, those properties that are under poorest chemistry control and that are the most difficult to fix in chemistry optimization should be given highest priority. In this context poor chemistry control equates to chemistry SAR that is difficult to control.

Physicochemical profiling of solubility and permeability has recently been reviewed.[1] These properties are difficult but not impossible to control by medicinal chemistry. In a combinatorial chemistry setting, poor aqueous solubility is almost universally a problem[2] and must be explicitly addressed by both computational and experimental intervention strategies. This chapter is not simply a treatise on the highest capacity methods for measuring aqueous solubility in a discovery setting that most closely approximate a lower capacity thermodynamic solubility measurement. Rather it is the author's viewpoint that dealing with the problem of poor drug solubility in an early discovery setting requires an appreciation of recent changes in three related but distinct areas. The areas to be considered are (1) changes in the chemistry method of synthesis and therefore in resultant physical form and aqueous solubility, (2) changes in the method of compound distribution for biological assays and therefore changes in apparent

Analysis and Purification Methods in Combinatorial Chemistry, Edited by Bing Yan.
ISBN 0-471-26929-8   Copyright © 2004 by John Wiley & Sons, Inc.

aqueous solubility and compound concentration, and (3) changes in screening technologies and changes in biological targets and their resulting impact on aqueous solubility. These three areas will be discussed as they relate to aqueous solubility. This will be followed by a general discussion of experimental approaches to measuring aqueous solubility in a discovery setting. Detailed descriptions of varied discovery solubility assays can be found in the excellent review by Kerns.[3]

16.2. CHANGES IN COMPOUND SYNTHESIS

Combinatorial chemistry and parallel synthesis are now the dominant methods of compound synthesis at the lead discovery stage.[4] The method of chemistry synthesis is important because it dictates compound physical form and therefore compound aqueous solubility. As the volume of chemistry synthetic output increases due to combinatorial chemistry and parallel synthesis, there is greater probability that resultant chemistry physical form will be amorphous or a "neat" material of indeterminate solid appearance. There are two major styles of combinatorial chemistry: solid-phase and solution-phase synthesis. There is some uncertainty as to the true relative contribution of each method to chemistry output in the pharmaceutical/biotechnology industry. Published reviews of combinatorial library synthesis suggest that solid phase synthesis is currently the dominant style contributing to about eighty percent of combinatorial libraries.[5] In solid-phase synthesis the mode of synthesis dictates that relatively small quantities of compounds are made.

16.3. CHANGES IN COMPOUND PHYSICAL FORM

The most commonly encountered physical form of compounds in current early drug discovery is noncrystalline. This is important because noncrystalline materials are almost always more soluble in all solvents including aqueous media. In traditional chemistry the melting point was used as a criterion for compound identity and purity. For example, if the compound being made had previously been reported in the scientific literature, one would look for an identical or perhaps higher melting point for the newly synthesized material. A lower melting point was viewed as problematical because of the observation that a contaminant usually causes a melting point depression. A sharp melting point was an indication of compound purity. A broad melting point was an indicator of an impure compound.

In the modern era there is an emphasis on the use of solution spectra such as proton or carbon NMR as an indicator of compound identity or purity. Mass spectral analysis serves to confirm identity. So, in terms of compound identity or purity, the melting point is less important than in previous eras.

When chemists face pressure to increase chemistry synthetic output, they do so by initially deleting noncritical steps in the synthesis/isolation process. Repeated compound crystallization to a sharply melting crystalline solid is time-consuming, is noncritical to compound identity and purity, and is therefore eliminated. This has occurred even if the chemistry is the traditional single compound at a time synthesis. For example, at the Pfizer Global R&D Groton laboratories the percentage of early discovery stage traditionally made compounds with melting point measured was close to 0 in 2000 as opposed to a decade earlier where some 80% of compounds intended for biological screening were registered with melting point information.

16.4. CHANGES IN COMPOUND DISTRIBUTION

Compounds intended for biological screening are increasingly being distributed as solutions in DMSO rather than as "neat" materials. There are two effects of this trend on the apparent compound solubility in DMSO each with different implications. Over relatively short time scales (e.g., up to a day) compounds may appear more soluble in DMSO than thermodynamic solubility would suggest. Over longer time scales compounds may precipitate leading to uncertainty as to the actual concentration of the drug in DMSO stock solution. As for aqueous solubility, compounds appear to be more soluble than a thermodynamic assay would suggest if the initial dissolution in aqueous medium occurs from a DMSO stock solution. This is especially so if the time scale for the solubility assay is short (on the order of tens of minutes). Why are compounds distributed as DMSO solutions? The reason is that it is far easier to automate the distribution of solutions of compound solubilized in DMSO than it is to distribute compounds as "neat" materials.

Robots efficiently pipette measured volumes of liquids. One cannot nearly as efficiently quantitate and transfer a neat (solid) compound. When a compound in DMSO solution is added to an aqueous medium, it is being delivered in a very high energy state relative to the thermodynamically most stable polymorph. The compound is in DMSO solution, so there is no compound crystal lattice to disrupt as part of the aqueous solubilization process. The effect is that the compounds initial apparent aqueous solubility is almost invariably higher than in a thermodynamic solubility experiment. This phenomenon occurs irrespective of any solubilization advantage attendant to the amount (often quite small) of DMSO that may be present in the aqueous medium.

The apparent increase in solubility tends to be time dependent, meaning that the solubility difference tends to decline with time. Exceptions to the phenomenon of higher apparent solubility when compound is added in DMSO solution to aqueous media have been observed when solubility is quantitated by very sensitive particle-scattering methodology as in the use of cell-sorting technology to detect particulate (undissolved solid) by light scattering. The causes of this counterintuitive behavior are unknown, but it is reasonable to conjecture that the phenomenon of lower solubility is due to light scattering caused by aggregates with sizes in the micron to submicron size range.

16.4.1. Compound Physical Form: Ostwald's Rule of Stages

When a newly synthesized compound is first isolated it frequently may exist in amorphous form. As experience is gained in compound isolation, the amorphous form is gradually successively replaced by polymorphs (crystalline forms) of increasing thermodynamic stability. This process occurs so frequently that it has been given a name "Ostwald's rule of stages." The process occurs because in the isolation/crystallization process there is a tension between the importance of kinetic and thermodynamic factors in the crystallization/isolation process. When the knowledge base in the chemistry crystallization process is low (i.e., early in the history of the compound isolation process), it is unlikely that the crystallization conditions are the optimum to produce the thermodynamically most stable form. Rather kinetic processes determine the crystalline form and thermodynamically less stable polymorphs are likely to be encountered. With increasing isolation experience, control of crystallization conditions such as compound concentration, solvent choice, and cooling rate becomes better known, and the probability of isolating the thermodynamically most stable polymorph is increased.[6]

16.4.2. Polymorph Form and Aqueous Solubility

Process changes in chemistry that increase the probability that new compounds will be isolated in amorphous or thermodynamically unstable polymorphic form have the effect of increasing the apparent aqueous solubility of newly synthesized compounds. Ostwald's rule of stages explains the common phenomenon that the physical form that is first isolated for a newly synthesized compound has a low melting point. This low melting point material corresponds to amorphous material or a thermodynamically unstable polymorph. The initially isolated material is in a high-energy state relative to the thermodynamically most stable (highest melting point) polymorph.

The consequence of the high-energy state is that when dissolution in a solvent is attempted less energy is required to disrupt the crystalline/solid state and the compound is therefore more soluble. In general, there are only a few exceptions to the rule that the highest melting point polymorph is the most insoluble. The types of polymorphs that might have a lower melting point but higher solubility (enantiotropic polymorphs) are seldom encountered in a pharmaceutical setting. These exceptions typically occur when the solubility is measured at a temperature moderately close to the compound melting point.

Most pharmaceutical organizations select against compounds with a melting point of less than 100°C or 110°C because the low melting point is associated with problems in accelerated stability testing and with problems in formulation development. As a result the rule that the highest melting point polymorph is the least soluble generally holds for solubility measured at room temperature or body temperature.

Implications for In Vitro Assays

Process changes in chemistry that increase the probability that new compounds will be isolated in amorphous or thermodynamically unstable polymorphic form have profound implications for in vitro assays and, in particular, for high-throughput screening (HTS). In the case of libraries of hundreds or thousands of compounds, the reader can approximate the thermodynamic aqueous solubility that might be expected by performing a calculation using any of the commercially available solubility calculation programs. A discussion of solubility calculation is outside the scope of this chapter.

An aqueous solubility calculator that operates in batch mode using molecular design limited (MDL) format structure data (sd) files as input can be obtained as a free download from a US Environmental Protection Agency (EPA) Web site.[7] Generally, a molar solubility of 10 micromolar or poorer is incompatible with appreciable oral activity for the average permeability, average potency (typically one mg/Kg) drug.[8] It is not unusual to find that half of the compounds in a suboptimal combinatorial library are predicted to have a thermodynamic solubility at or below 10 micromolar when examining sd files of commercially available combinatorial compounds. When noncrystalline combinatorial compounds are screened in

HTS assays, the results are far better than might be predicted from solely thermodynamic solubility considerations, and it is possible to detect fairly insoluble active compounds in the HTS screen.

Success in detecting a poorly soluble compound in an in vitro screen depends on the assay conditions. For example, does the assay contain components likely to solubilize the compound? Success in the HTS assay in detecting activity in an insoluble compound is also dependent on the initial physical form of the compound. An amorphous compound (the predominant physical form in combinatorial chemistry compounds) is far more likely to dissolve to give a concentrated DMSO stock solution than is a crystalline compound to give a DMSO stock solution. For many amorphous compounds the solubility in DMSO will be supersaturated relative to the thermodynamic DMSO solubility of a more stable crystalline form.

Some compounds will stay in solution indefinitely even though the solutions are supersaturated. These are compounds whose concentration lies in a thermodynamically unstable but kinetically stable so-called metastable zone. This phenomenon is encountered only when a supersaturated solution is created in the absence of crystalline solid. It does not occur if dissolution starts from the crystalline solid.

Compounds will precipitate from a DMSO stock solution if the kinetic conditions are appropriate, for example, if enough time passes or enough freeze–thaw cycles are encountered as part of the compound in DMSO storage process. Operationally this writer has encountered the phenomenon of a "working-day" solubility window. Amorphous compounds generally dissolve fairly easily at 30 or even at 60 mM in DMSO but noticeable precipitation can be detected after about a day. This occurs even at room temperature and even in the absence of any freeze–thaw cycle. The one working-day window phenomenon for solubility in DMSO has implications for the detection of HTS actives.

For example, suppose that an HTS assay results in a number of actives. A similarity search is then conducted on the actives and based on the search compounds are submitted to an HTS re-screen. Frequently compounds were found to be active in the re-screen that were inactive in the original HTS assay. The likely reason is the kinetic phenomenon of compound precipitation from medium time length storage of DMSO master plates. Active compounds were not detected in the primary screen because these compounds were not available to the primary screen. They precipitated from the stored master DMSO plates. The re-screen is performed with freshly DMSO solubilized samples (within the one working-day window) allowing a better detection of a truly active compound.

Process changes in chemistry that increase the probability that new compounds will be isolated in amorphous or thermodynamically unstable poly-morphic form have profound implications for in vitro assays in traditional biology laboratories. This author anticipates an increase in the noise level of traditional biology in vitro assays, in particular, as compounds are increasingly distributed as DMSO stock solutions, and anticipates a resultant very negative reaction from traditional biology personnel. The reason for the increased noise level is the increased degree of uncertainty as to the actual concentration of a compound being screened when compounds are distributed as DMSO stock solutions.

The purely technical considerations relating to compound physical form and resultant precipitation from DMSO stocks are similar in HTS assays and traditional biology in vitro assays, but the people considerations are very different. Personnel involved in HTS assays have long been tolerant of the problems attendant to HTS screens. They do not get upset if active compounds are missed in an HTS screen because of poor aqueous or DMSO solubility. To people familiar with HTS assays, it is just part of the cost of doing business, an inevitable trade-off between the greatly enhanced screening capacity of HTS versus an inevitable loss of accuracy. This type of mindset is very different from that likely to be encountered among traditional biologists. Traditional biologists place a premium on in vitro screen accuracy and take pride in assay reproducibility. Uncertainty as to compound concentration is minimized when stock solutions are freshly pre-pared from a "neat" amorphous material or a freshly prepared DMSO stock. In essence with respect to compound distribution, a traditional biology lab works within the one working-day scenario. However, all this changes if the method of compound distribution changes toward an automated dispensing of drug in DMSO stocks.

Compound concentration in an in vitro assay becomes an important noise factor if the compound has precipitated from DMSO as part of the compound storage process. The biologist may interpret assay results as reflective of compound aqueous insolubility. However, the problem may have occurred at an even earlier stage in that precipitation actually occurred from the DMSO stock solution. In addition compound loss from DMSO may be interpreted as a chemical instability (degradation) problem. Considerable semantic confusion can result from the term "instability." To the biologist it may mean compound loss from solution. The chemist may interpret "instability" as a change in chemical structure, namely as a chemical degradation problem. These differences are meaningful because different strategies may be chosen to deal with compound loss from a DMSO stock depending on whether the problem is precipitation or chemical degradation.

Implications for In Vivo Assays

Process changes in chemistry that increase the probability that new compounds will be isolated in amorphous or thermodynamically unstable polymorphic form enhance the probability that in vivo activity will be detected if the compound aqueous solubilization is performed fairly rapidly. Likely the biologist or drug metabolism person will want to ensure oral exposure of a solution to an animal fairly quickly so that dosing occurs before the compound precipitates. Alternately, dosing might be performed with "neat" amorphous solid materials as a suspension. In either of these scenarios the oral dosing with amorphous solid will appear more efficacious than if the oral dosing started with a suspension of the less soluble, thermodynamically more stable crystalline product form.

The oral activity improvement attendant to the amorphous form of the compound may not be detected until a serious attempt is made to obtain crystalline material. This could come as late as the clinical candidate nomination stage. What should be avoided at all costs is the choice of the "best" orally active compound by a comparison of the oral bioavailability of poorly characterized solid forms. The oral absorption component of oral bioavailability (but not the metabolism factor) is extremely dependent on crystalline form. Making a choice using poorly characterized solid forms risks choosing a nonoptimum compound. In fact this error may not even be noticed if only the clinical candidate has been crystallized. Critical to these arguments is the presumption that most in vivo assays will not be conducted with stored compound in DMSO stock solutions.

16.5. SOLUBILITY, POTENCY, AND PERMEABILITY INTERRELATIONSHIPS

Aqueous solubility, potency, and permeability are three factors under medicinal chemistry control that must be optimized to achieve a compound with acceptable oral absorption. Typically a lead (chemistry starting point) is deficient in all three parameters. The interrelationships of these three parameters has been described in a series of publications from Pfizer researchers.[9,10] Figure 16.1 provides a bar graph that depicts minimum acceptable solubility as a function of projected clinical potency and intestinal permeability. A minimum thermodynamic aqueous solubility of 52 µg/mL is required in the most commonly encountered situation where the projected clinical potency is 1 mg/kg and the permeability is in the average range. If the permeability is in the lower tenth percentile as might be found for a peptidomimetic the minimum acceptable solubility is

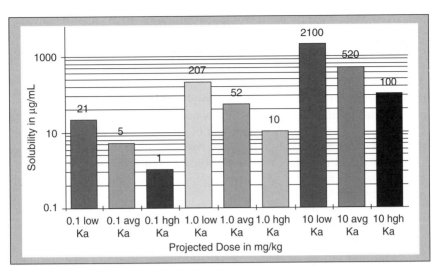

Figure 16.1. Minimum acceptable solubility in μg/mL. Bars show the mimimum solubility for low, medium, and high permeability (Ka) at a clinical dose. The middle three bars are for a 1-mg/kg dose. With medium permeability you need 52-μg/mL solubility.

207 μg/mL. The minimum accepatable solubility is 2100 μg/mL if the potency is poor (and the dose high), for example, as occurs among some HIV protease inhibitors.

16.5.1. Acceptable Aqueous Solubility for Oral Activity: Traditional Definition

A classic pharmaceutical science textbook might have defined poor solubility as anything below a solubility of 1 gram/mL (2 mol/L solution for a molecular weight of 500) at pH 6.5 (or pH 7). This classic view is reflected in the Chemical Abstracts SciFinder 2001® solubility range definitions for solubility calculated using Advanced Chemistry Development (ACD) Software Solaris V4.67. These semiquantitative ranges for molar solubility are very soluble, 1 mol/L < solubility; soluble, 0.1 mol/L < solubility < 1 mol/L; slightly soluble, 0.01 mol/L < solubility < 0.1 mol/L; and sparingly soluble, solubility < 0.01 mol/L. The traditional definition was reasonable for the time frame where many drugs had clinical potency in the 10 to 20 mg/kg per day range. This type of very poor potency (high doses in the gram range) is seldom encountered today except for therapy with some infectious disease and antiviral drugs. The likely reason is the use of mechanism as opposed to phenomenological screening (in vivo screening) and the improved potency attendant to leads resulting from high-throughput screening.

16.5.2. Current Era Definition

In the current era with widespread problems of poor solubility,[4] a compound (drug) with average permeability and a projected clinical potency of 1 mg/kg, needs a minimum aqueous solubility of 50 to 100 μg/ml to avoid the use of nonstandard solubility fixing formulation technology. The guidelines published by Pfizer's Curatolo on maximum absorbable dose are an excellent source for the combination of permeability, solubility, and potency required in an orally active drug.[8]

Development Thermodynamic Solubility as a Benchmark

The quantitation of solubility in the current discovery setting differs markedly from that of a traditional aqueous solubility assay. In the traditional solubility assay the crystalline well-characterized solid is equilibrated in aqueous solution for sufficient time to reach equilibrium (generally 24–48 hours). The solution composition is important in the comparison with a discovery solubility assay. The traditional thermodynamic solubility experiment is performed in the absence of any organic co-solvent. By comparison the discovery solubility assay is frequently performed in the presence of a co-solvent, usually DMSO. The physical state of the starting solid in the development thermodynamic assay is relevant to the comparison with discovery solubility assays. As was previously mentioned, the solids in discovery today are frequently noncrystalline and therefore invariably more aqueous soluble.

The thermodynamic assay time duration is important to the comparison of the time scale of discovery solubility assays where the time scale is frequently much shorter. The development thermodynamic solution experiment is filtered through a filter typically in the micron size range. For example, filters of 1.2 μ are recommended for lipid containing total parenteral nutrition (TPN) solutions and filters of 0.22 μ for nonlipid TPN. The particle size that would be considered as constituting a "solution" in a thermodynamic solubility measurement is important in the context of discovery solubility measured by particle-based light scattering. After separating the solid from the liquid phase by filtration (or possibly by centrifugation), the quantity of compound in the aqueous phase is measured by a quantitative technique, for example, by HPLC with UV detection. Positive standards are used to bracket the aqueous concentration of soluble drug.

Care is taken that the positive standards reflect the ionization state of the drug in solution, since substantial changes in UV extinction coefficient frequently occur as a function of solution pH for an ionizable drug. This point is important in the context of discovery solubility measured by ultraviolet (UV) spectroscopy. In the developmental thermodynamic solubility assay the compound purity is typically quite high. In comparison in the discovery solubility assay the compound purity is typically lower. At the time of writing, typical compound purity criteria for combinatorial chemistry compounds from the better chemistry sources were 90% pure by evaporative light-scattering detection (ELSD) or 80% pure by UV detection (typically ELSD is a less sensitive detector than UV of compound impurities). The compound purity issue is particularly important to the issue of discovery solubility assays using a UV quantitation end point.

16.6. TURBIDIMETRIC/PARTICLE-BASED SOLUBILITY

The presence of a precipitate of insoluble material can be used as an indicator that the solubility limit of a compound has been exceeded. There are two general methods to instrumentally detect the presence of a precipitate. One can detect precipitation using UV absorbance, or one can detect precipitation by directly measuring light scattering with a type of nephelometric light-scattering detector. One can also visually detect light scattering by taking advantage of the Tyndall effect.

The UV absorbance method takes advantage of the fact that particles in the path of the light beam scatter light away from the light detector or actually block the light from reaching the detector. The light-scattering "absorption" curve has the properties of an inverse power curve with the absorption predicted by a curve of the form (constant divided by wavelength raised to the power n). In any particular solubility experiment the constant term and power n tend to be constant until a significant change in the particles occurs (e.g., from agglomeration). In our experiments we encountered values of n ranging from 2.5 to 4.5. Because of the inverse power curve form of the absorption curve the absorption due to light scattering is greatest at short wavelengths. The presence of a confounding UV absorbance from the soluble drug is the major disadvantage of measuring particle based light scattering by UV absorbance. For compounds whose aqueous solution is not visibly colored, this is generally not a problem, since one can measure absorbance due to light scattering at about 500 nm because it is outside of the typical soluble chromophore absorption range. However, colored compounds or compounds with colored impurities can absorb due to soluble chromophore in this range.

The problem can be circumvented by using a UV diode array detector and performing a Fourier transform analysis on an absorbance data array, for example, a range of 64 to 128 absorbance readings at evenly spaced wavelength increments. The Fourier transform analysis relies on the fact that the apparent band width of the absorption light-scattering curve is very much broader than the band width of any soluble absorbance chromophore. The successive terms of the absorbance data Fourier transform can be interpreted in terms of contributions of gradually decreasing bandwidth.

The first term of the Fourier transform essentially defines a baseline shift of very broad bandwidth and is a very sensitive measure of the absorbance due to light scattering. A plot of the first Fourier transform term versus amount of compound added works much better as an index of precipitation than a simple plot of absorbance versus amount of compound added. The UV absorbance method is also somewhat dependent on the instrumental design and works best when there is a large distance between the sample cell and the UV detector.

The presence of a precipitate of insoluble material can be directly detected by an instrument specifically designed to detect a light-scattering signal. Instruments are available that operate in single cell or plate reader format. Single cell instruments can be used as is or adapted to automated measurement. For example, our laboratory was able to develop an automated assay using a modified HACH AN2100 nephelometric turbidity detector. Instrument costs are quite reasonable because simple, robust light-scattering detectors are widely sold for use by water quality engineers to test for clarity in water treatment plants.

It turns out that water clarity is a very good indication of the absence of a variety of pathogenic intestinal parasites that might be found in contaminated water. These types of organisms have effective sizes of about 1 μ and are reliably detected by visible light scattering at 90 degrees to the visible light path. Most mammals (including humans) have through evolution developed a keen ability to detect light scattering (cloudiness) in water. In our benchmarking this author discovered his ability to detect light scattering using the Tyndall effect to be only about one order of magnitude less sensitive than that using a HACH instrument. A nephelometric turbidity detector can be calibrated against a set of commercially available turbidity standards. The detector response is nonlinear to the quantity of compound precipitated. So what is measured is essentially binary data; that is, the solution is clear (no precipitation) or turbidity (precipitation) is detected outside of the baseline (clear) nephelometric turbidity signal.

16.6.1. Types of Turbidity Signal

Two extremes of turbidity signal are detected as compound in DMSO solution is added stepwise to an aqueous medium. In one pattern a gentle steady increase in signal is observed as the solubility limit is exceeded. We find that this pattern most often corresponds to a compound that precipitates because it is excessively lipophilic. In the second pattern a very marked steep increase in turbidity signal is encountered as the solubility limit is exceeded. We find that this pattern most often corresponds to a compound that precipitates because of crystalline packing considerations. We also find that compounds generating the second type of pattern tend to give far more reproducible results in replicate assays. We interpret this as reflecting a very quick precipitation when the solubility limit is surpassed for a compound precipitating because of crystal packing reasons. Large signals are generated with these types of compounds because particle sizes are small and in the optimum size range (about 1 μ) for detection with a visible light source. Lipophilic compounds tend to precipitate as larger particles, which tend to generate a weaker light-scattering signal. Reproducibility is poorer for the detection of precipitates of lipophilic compounds, presumably because of kinetic considerations; that is, there is more variability in the time it takes for precipitation of the lipophilic compound to occur.

The precipitation behavior has similarity to the crystallization behavior of compounds in a chemistry synthesis lab. The more lipophilic compounds are harder to crystallize than the high melting point compounds. A fluorescent compound can generate a false positive signal in a nephelometric turbidity assay. A soluble drug with good fluorescent properties absorbs unpolarized light and emits unpolarized light with a good Stokes shift (i.e., at higher wavelength in all directions). This behavior could then be confused with a light-scattering signal. The potential problem is easily solved by introducing a low wavelength cutoff filter in front of the off axis detector. In our assays at Pfizer we introduced a cutoff filter that prevented detection of soluble fluorescein up to the upper concentration limit in our assay. The filter had no discernable effect on reducing sensitivity to the detection of scattered visible light. Fortunately for pharmaceutical drug solubility screening there are very few drugs that fluoresce above the fluorescein emission range since these soluble compounds would generate a false positive signal.

16.6.2. UV Detector-Based Solubility

The solubility of a compound can be quantitated by measuring the compound solution concentration by UV spectroscopy. This method requires the use of some type of positive standard because there are no reliable computational techniques to predict UV absorbance and extinction coefficient from structure alone. The factors that need to be considered in a reliable UV-based method include the following. A method needs to be in place to correct for any UV spectral absorbance changes between reference and test samples that occur as a result of (1) solvent medium changes between reference and test sample, (2) pH changes between reference and test sample, and (3) confounding UV absorbances between reference and test samples that may result from UV absorbent impurities in the compound whose solubility is being tested. The order of importance of these possible problems and the need for their correction is listed from least to most important. Very detailed discussions of these issues in solubility profiling can be found in the excellent review by Avdeef.[11]

The least problematic issue are UV spectral changes as a function of different solvents between reference and test sample. Solvent effects on UV spectra in solvents of decreased dielectric constant compared to water parallel solvent effects on apparent pKa. The changes are most marked for acids, for example, leading to a numerical increase of up to two pKa units— an apparent decrease in the acidity of the carboxylic acid. Effects on bases are considerably less. The apparent pKa of a base in a reduced dielectric constant solvent might be up to about half a pKa unit numerically lower (less basic). The UV spectra of neutral compounds are generally not markedly affected by changes in solvent dielectric properties. As a result solvent corrections are generally not critical for UV quantitation of libraries of neutral or weakly basic or acidic compounds (frequently combinatorial). The most problematical functional group, namely a carboxylic acid, does not pose a major concern. Among known phase II drugs only about 14% are carboxylic acids and carboxylic acids are frequently under represented relative to this 14% figure in combinatorial libraries.

The issue of UV spectral absorbance changes resulting from pH changes between reference and test sample is more important. The UV spectra of compounds containing ionizable moieties located on or near chromophoric groups can be markedly changed (both in absorption maxima and extinction coefficient) by differences in medium pH. This phenomenon is well known and is the basis of methods for determining compound pKa by plotting spectral absorbance changes versus pH. Correction requires experimental data, namely UV spectra captured at several pH values. Alternatively, batch mode pKa calculations can be used to bin compounds into the neutral and weakly basic and acidic compounds for which there is unlikely to be a potential problem as opposed to the strong acid and base class for which the ionization state can markedly affect the UV quantitation.

16.6.3. Titration-Based Solubility

Aqueous solubility can be measured by potentiometric titration. The theory behind this method and its practical implementation has been pioneered by

Avdeef. A solution of the compound in the pH range where the compound is soluble is titrated toward the pH region where the compound is insoluble. In the soluble region the normal type of data from a potentiometric titration is collected, namely millivolt versus volume of titrant added. This data range serves to define the compound pKa. At and past the compound precipitation point, data corresponding to millivolt versus volume of titrant added continue to be collected. However, the data obtained past the precipitation point defines not the pKa but the compound solubility product. In favorable cases where the precipitation point occurs about halfway through the titration, it is possible to obtain both compound pKa and compound solubility product from a single titration experiment. The theory is very well worked out for this method for almost every reasonable combination of ionizable groups likely to be present in a drug molecule. The analysis of the change in titration data once precipitation occurs requires very high quality potentiometric titration equipment, requires excellent protection against carbon dioxide uptake and requires very sophisticated curve analysis software. All these are commercially available.

The method requires a compound in the low milligrams and works best for very insoluble compounds, which also are the most difficult to quantitate by other methods. The method is relatively slow, since for best results the titration rate is greatly slowed as the precipitation point is reached. The compound is solubilized at the beginning of the assay, either by pH dissolution of solid or by dissolution of a concentrated DMSO stock solution in aqueous media of appropriate pH. As a result there is no control as to which polymorphic form of the compound corresponds to the solubility product. The method is likely to be particularly useful in the solubility classification of compounds according to the FDA bioavailability waiver system because of the extensive validation of the method, its reproducibility, and its particular applicability to compounds with the poorest solubility.

16.7. SOLIDS NOT CHARACTERIZED IN EARLY DISCOVERY

As previously discussed, the compound's form differs markedly from early discovery to the late discovery and development interface. The early discovery compound is poorly characterized as to crystalline form. It may be non-solid, amorphous, or even crystalline but uncharacterized as to polymorphic form. The late discovery/development interface compound is crystalline as defined by phase contract microscopy or powder X-ray diffraction and its polymorphic and salt form is frequently characterized. This difference has profound implications for the design of a discovery solubility assay.

The key question is this: Is it better to design an early discovery solubility assay as a separate type of experiment, or is it better to try to automate a traditional thermodynamic solubility assay to handle the very large number of compounds likely to be encountered in early discovery? Another way to state this question is: Does it make sense to run a thermodynamic solubility assay on poorly crystalline early discovery compounds? This is the type of question about which reasonable people could disagree. However, this author does have a distinct opinion. It is much better to set up a distinctively different solubility assay in early discovery and to maintain a clear distinction between the assay type appropriate in early discovery and the assay type appropriate at the late discovery / development interface. Two issues are relevant to this opinion. One relates to the need for a solubility assay to reflect/predict early discovery stage oral absorption and the other relates to people/chemistry issues.

An early discovery solubility assay is most useful to chemists at the stage where oral absorption SAR is not present or poorly defined. This typically might occur after reasonable in vitro activity has been discovered and attempts are underway in drug metabolism to establish oral absorption or in biology to demonstrate oral activity. The early discovery solubility assay is most likely to predict the early drug metabolism or biology studies when the solubility protocol mimics the manner of dosing in drug metabolism and biology. Typically this means that starting compound is dissolved in DMSO solution, and short read times in the low tens of minutes are used in the solubility assay. These solubility assay conditions mimic the dosing conditions in early drug metabolism and biology. This type of reasoning is a very hard sell for a scientist with a strong development or analytical background because it means setting up a solubility assay that breaks every pharmaceutical sciences textbook rule about the qualities of a "proper" solubility assay.

16.7.1. Role of Chemistry in Solubility Improvement

Chemists have the primary responsibility of altering the compound chemical structure so as to improve on a solubility problem. Anything that undermines this primary responsibility of the chemist is harmful to overall research productivity. In my experience, chemists (perhaps unconsciously) are always eager to obtain data that makes their compounds look better. For example, chemists might seek solubility data in simulated intestinal contents in order to obtain better (higher) solubility values. This exercise is of no value if it does not somehow result in compounds with better properties being made. Unfortunately, the search for better appearing solubility data can actually be counterproductive. This happens if the new solubility data does not solve a solubility problem and takes the pressure of the chemist to change the chemical structure toward a better profile. The chemist "people" factor is a major reason for avoiding a "thermodynamic" solubility assay on early discovery compounds. The term "thermodynamic assay" has a connotation associated with developmental rigor and accuracy. There is a grave danger that the chemist will uncritically accept the solubility value from a "thermodynamic assay" in early discovery on amorphous material or on material in a DMSO solution.

As was previously discussed, solubility assays on compounds in high-energy forms (DMSO solution or amorphous) almost always overestimate the aqueous solubility. If the chemist uncritically accepts an inflated early stage solubility as truth, there will be little pressure to produce an acceptable solubility compound once the material is characterized in the crystalline form. It is much better to keep the solubility assays distinct: an early assay for prediction of absorption SAR, and an efficient thermodynamic assay for the late discovery/development interface. In our discovery organization in Groton, CT, we force this distinction by refusing to run our automated thermodynamic solubility assay on compounds that are not crystalline. The traditional "thermodynamic" solubility assay, starting from crystalline material, can be considerably improved by applying more modern dissolution technology with particular care to reducing the aqueous volume and hence the sample size requirement.[12]

16.8.   DMSO SOLUBILIZATION IN EARLY DISCOVERY

Adding compounds solubilized in DMSO to aqueous medium as part of a discovery solubility assay can lead to two types of solubility assays with different uses. At one extreme, the quantity of DMSO is kept very low (below 1%). At this low level of DMSO the solubility is only slightly affected by the DMSO content. For example, data from a poster by Ricerca[13] suggest that a DMSO content of 1% should not elevate apparent solubility by more than about 65%. At 5% DMSO this group reported an average solubility increase of 145% due to the DMSO content.

Solubility in an early discovery assay containing 1% DMSO can however exceed "thermodynamic" solubility by a lot more than 65%. However, this is very likely due to the time scale. Studies by the Avdeef (pIon Inc) group show a close approximation of early discovery solubility (quantitated by UV) to literature thermodynamic solubility if the early discovery assay is allowed to approach equilibrium, for example by sitting overnight. The availability of DMSO stock solutions places limitations on the concentration range that can be achieved without exceeding the one percent DMSO limit.

For example, if the DMSO stock is 30 mM, a 1 to 100 dilution yields a final concentration of 300 μM. This corresponds to a concentration of 150 μg/mL (for MWT = 500), which is above the solubility limit required for oral activity in an average permeability, average 1 mg/kg potency heterocycle. However, the upper concentration range will not be high enough at 1% DMSO if something like an HTS DMSO stock is used with a concentration of 4 mM. The resulting aqueous concentration at 1 to 100 dilution is only 40 μM (which corresponds to 20 μg/mL solubility (for MWT = 500). This is not high enough a solubility to ensure oral activity for the average permeability, average potency heterocycle. A minimum solubility of 50 μg/mL is required.

At the other extreme of adding compounds dissolved in DMSO in an early discovery assay, the goal is to maximize assay throughput and to generate a relative solubility ranking and the DMSO content is allowed to exceed one percent. For example, workers at Glaxo have employed an HTS solubility assay keeping the DMSO content at a fixed 5%.[14] If the addition is of aqueous medium to DMSO stock, a series of increasing dilutions can be performed with a plate reader nephelometric assay being run at each dilution stage. Very high throughputs can be obtained. The assay is completely compound sparing if it is implemented as part of a normal process of compound preparation for assay testing. The disadvantages are fairly clear. It may be difficult to compare compound solubility across chemical series. At high DMSO percentages the link to thermodynamic solubility is lost, and it may be difficult to conclude whether a compound has sufficient solubility to have the potential for oral activity.

16.9. LIBRARY DESIGN, IN SILICO–EXPERIMENTAL DIVERGENCE

Library design viewed as a computational process and the experimental implementation of a library design can result in quite different outcomes. In a library design one might consider a core template that can be modified by a variety of substituent groups. The physicochemical profile of the library is determined by the chemical structure of the core and the range of physicochemical properties of the substituent groups. For example, there might be only one core in a design but many substituent groups. As a result the physicochemical envelope of the library is very much influenced both by the availability of the chemical pieces that will become the substituents and by the physicochemical properties of those substituents. If the experimental implementation (synthesis) of a design is 100% successful, then the actual experimental physicochemical envelope profile is the same as in the library design. In actuality the chemical synthesis is seldom 100% successful. So what matters is whether there is any bias in the experimental synthesis success rate such that the experimental library differs significantly from the design library. A priori, the chemical synthesis success rate is usually not known, so in general, libraries are overdesigned. More compounds are designed than will actually ever be synthesized.

The experimental synthesis success rate almost always biases the experimental library, so the physicochemical profile relative to aqueous solubility is significantly worse than that in the design library. The fewest chemistry problems are found in lipophilic substituent moieties lacking polar functionality. In almost all cases polar functionality is electron withdrawing, so reactions of a substituent moiety like reactive amination, acylation, and nucleophilic substitution proceed poorly. Blocking and deblocking of a polar group adds to the complexity and length of a synthesis. As a result polar reagents that require blocking and deblocking are experimentally selected against. Robotic pipettors perform poorly or not at all on slurries of precipitates, so any factor that increases the insolubility of a reagent in an organic solvent will bias the library outcome.

16.9.1.   Two Causes of Poor Solubility

Poor solubility in an organic solvent arises from two quite different factors: solvation energy and disruption of intermolecular crystal packing forces in the crystalline reagent. Solvation of a lipophilic reagent in an organic solvent is typically not a problem. But disruption of intermolecular crystal packing forces is very much a problem in an organic solvent, especially if the reagent has a high melting point. This type of problem is most likely to be present in a reagent with polar hydrogen bond acceptor/donor functionality. Thus the reagent insolubility problem tends to bias a library toward a more lipophilic and hence more aqueous insoluble profile.

To accommodate diversity considerations, a range of substituent moieties is selected. A large structural range translates into a broad molecular weight distribution. The combination of reagent solubility and diversity considerations results in an experimental library that is biased toward higher lipophilicity and higher molecular weight relative to the design library. The bias occurs because high lipophilicity and high molecular weight are the worst combination of "rule of 5" parameters in terms of leading to poor aqueous solubility.

16.9.2. Library Design—Importance of the Rate-Determining Step

Effectiveness of designing for adequate aqueous solubility depends on whether chemistry protocol development or chemistry production is rate determining. If chemistry production is rate determining, there will be excess validated protocols relative to library production. This means that protocols can be prioritized as to the attractiveness of the compound solubility profile and the least attractive protocols from a solubility perspective may never be translated into actual library production. However, often protocol development and not library production is the rate-determining step. This eventuality is unfortunate because there is an understandable reluctance to discontinue chemistry synthetic efforts due to poor experimental solubility profile if considerable chemistry effort has already been expended. Consider the following situation. The effort toward library production is 70% complete. The experimental solubility profile is poor. Would you discontinue completion of library synthesis because of poor solubility if 70% of the chemistry effort had already been completed? So a key issue becomes how much chemistry experimental effort takes place before exemplars are experimentally profiled in solubility screens?

If protocol development is rate determining, the effectiveness of experimental solubility assays depends on how the early exemplars are synthesized. In theory, the most effective method would be to obtain a well-spaced subset of the library in an experimental design sense. A traditional noncombinatorial synthesis would accomplish this but would not fit in well with a combinatorial optimization process. A possible way around this problem is to institute some type of early automated cleanup of combinatorial exemplars from partially optimized reaction schemes. This is not a tidy solution because the most efficient process would be an automated cleanup on the entire library after the optimization process was complete.

The least effective method of providing samples for experimental solubility profiling is a late-stage selection from the optimized combinatorial libraries. It is least effective, and not because of chemistry efficiency considerations. A late-stage selection from the optimized combinatorial libraries is actually chemistry efficient. However, the inefficiency comes from the people aspect. The data comes too late to prevent poor solubility compounds from being made. The timing problem in obtaining combinatorial exemplars is one of the driving forces that makes computational solubility profiling so attractive.

Poor Solubility—Business Aspects

There is a business aspect here for a company selling combinatorial libraries if the combinatorial libraries contain compounds experimentally verified to have poor solubility properties. What do you do with the poor solubility compounds? Do you sell them to an unsuspecting customer, or do you swallow the cost by not selling these compounds? At the present time there is still a way out. Not all customers want all compounds to have good ADME properties. There is still a sizable market (about 30% but dwindling) for compounds perceived to be possibly active in an in vitro assay regardless of aqueous solubility properties. The argument is that one cannot proceed anywhere without that first "lead." So try to get the "lead" first and then worry about solubility later. I am not a proponent of this view, but I can see how people who are involved in screening against very difficult targets might take this position.

If library production is rate determining, then the effectiveness of experimental solubility assays becomes much simpler. Operationally it is fairly straightforward to de-prioritize a library that profiles poorly in solubility assays if there are more protocols than can be translated into production. One simply executes the best protocols in a solubility sense and drops from production those that profile the worst. There is a clear message here if one really cares about solubility properties. The manning and effort should be greater on the protocol validation stage as opposed to the library production stage. This solubility derived message is exactly contrary to that which one receives at the vendor trade shows. The vendor trade shows tend to emphasize the production side. This is hardly surprising if there is much more money to be made in selling production hardware and technology as opposed to selling tools for protocol design and validation.

16.9.3. Prioritizing Assays

I believe that assays are not all equal in terms of contributing to drug quality. Design is most important for those properties under the poorest chemistry control. Good chemistry control equates with the chemists' ability to control SAR. Good control of SAR means that a chemist can make a small molecular change with resultant large change in the measured property. Poor control of SAR means a loss of relationship between molecular structure and the measured property. I believe it is inadvisable to totally filter out compounds with poor properties if they can easily be fixed by chemistry. The goal in drug research is to discover inherently active compounds with appropriate properties. There is little value to being so restrictive that few inherently active compounds will be found.

Chemistry SAR for Solubility is Poor

Chemistry control of aqueous solubility is poor. The good news is that if a compound has poor aqueous solubility, methods do exist to fix the problem in terms of pharmaceutical formulation technology. However, these are always expensive in time and manning, and depending on the degree of the solubility problem, they may have limited or no precedent in terms of existing approved products. By far the preferred solution to poor solubility is to fix it in chemistry. Formulation fixes are a last resort. The bad news is that in general, poor aqueous solubility is by far the most common physicochemical problem in combinatorial libraries. It takes no effort at all to introduce poor aqueous solubility to a combinatorial compound library. The combination of high molecular weight and high lipophilicity outside the "rule of five" limits[15] is an almost certain guarantee of poor aqueous solubility. Based on our experimental screening, lipophilicity above the "rule of five" log P limit of 5 by itself carries with it a 75% chance of poor aqueous solubility.

Industry Solubility Changes

The definition of "poor solubility" has come downward in the combichem/HTS era. A classic pharmaceutical science textbook might have defined poor solubility as anything below a solubility of 1 g/mL at pH 6.5 (or pH 7). In today's era most drug researchers would be very excited by a solubility as high as 1 g/mL. In general, with average permeability and a projected clinical potency of 1 mg/kg, a drug needs a minimum aqueous solubility of 50 to 100 μg/mL to avoid the use of nonstandard solubility-fixing formulation technology. We find the guidelines published by Pfizer's Curatolo[16] on maximum absorbable dose to be an excellent guide for the combination of permeability, solubility, and potency required in an orally active drug.

Chemistry control of aqueous solubility is poor because, except for a few very specific cases, the chemistry SAR is blunt. In this respect control of solubility like that of permeability is poor. Solubility due to excessive lipophilicity improves only gradually as the lipophilicity is moved in the desired downward direction. Trying to decrease lipophilicity by incorporating polar functionality may or may not work. The potential solubility improvement attendant to introducing polar hydrophilic functionality can easily be more than counterbalanced by a decrease in solubility, because of increased intermolecular crystal packing forces arising from the new polar functionality. In our experimental solubility screening, about 60% of poor solubility is unrelated to excessive lipophilicity. This 60% of poor solubility arises from high crystal packing forces due to intramolecular interactions in the solid crystal state, which make disruption of the crystal energetically costly. So the blunt SAR feature in solubility comes from the phenomenon that solubility improves only gradually (or not at all) as a physicochemical property (lipophilicity) is moved in a desired downward direction.

16.9.4. Solubility and Compound pKa

Changing the pKa of an acidic or basic group in a molecule so that more of the compound exists in the ionized form at physiological pH lowers log D (at about pH 7) and, in general, should improve aqueous solubility. The improvement in solubility is limited, however, if the solubility of the neutral form of the compound (the inherent solubility) is very low. The situation is worsened if the starting pKa is far from 7. We find this to be a particular problem with weak bases. Weakly basic pyridines, quinolines, quinazolines and thiazoles seem to be frequent members of combinatorial libraries.

The extent of poor aqueous solubility may be experimentally under-estimated in a combinatorial library. No combinatorial library is purified by traditional crystallization. The vast majority of compounds purified by an automated process will be isolated in amorphous form. Compounds in an amorphous solid form exist in a much higher energy state than a true crystalline solid, and aqueous solubilities of amorphous solids are always higher than that of crystalline solids. This phenomenon may only be recognized if there is a high degree of interest in a combinatorial compound. The combinatorial compound is scaled up and purified by crystallization. The newly crystallized compound can easily be an order of magnitude more insoluble and hence more poorly absorbed than the original sample.

Aqueous solubility can be improved by medicinal chemistry despite the blunt SAR feature. However, to improve solubility takes commitment to a combination of computational and experimental interventions and a real effort on chemists' part to incorporate solubility information into synthesis design. The importance of rapid experimental feedback is particularly important given the current inability to computationally predict poor solubility arising from crystal packing interactions. It is critical not to miss a serendipitous improvement in solubility attendant to a molecular change. Because of the blunt SAR feature the easiest way to improve solubility with respect to library design is to try to design the better solubility profile as best as possible right at the beginning.

16.9.5. Phony HTS Hits and Solubility Screening

In vitro potency has always been under excellent chemistry control, the hallmark of good control being tight chemistry SAR. With respect to combinatorial chemistry several exceptions should be noted. Compounds are often encountered as leads in HTS screening and can be characterized as "phony HTS leads." These types of compounds should be avoided at all cost as templates in combinatorial chemistry. A very common attribute of these leads is that the chemistry SAR is flat and fuzzy if compounds are subjected to lead optimization. Large chemistry changes can be made with only very small changes in activity. Often these types of problems can be avoided by similarity searches on the initial apparent lead. A loss or gain of activity related to a small structural change over the initial lead is a good sign. A change in activity of a factor of 10 between two compounds differing by only a single methyl group is a classic example of good SAR. A flat SAR among analogs is not a good sign. Often these "leads" are not very active, such as in the low micro molar range. Sometimes the exact same compound shows up active in multiple HTS screens. Alternatively, as we have observed, members of a well-defined structural series appear as apparent HTS actives but not necessarily the exact same compound across different HTS screens. Some phony HTS "leads" are removable by compound quality filters. The mechanisms for the "phony HTS lead" phenomenon are largely unknown. Some of our experience suggests that nonspecific lipid membrane perturbation of intramembrane protein receptor targets could be a factor. Experienced medicinal chemistry "intuition" works quite well in avoiding "phony HTS leads," but it would clearly be very advantageous to have computational filters for this problem. Software, even if it worked no better than the chemists' intuition, would be very advantageous from the people viewpoint. Biologists generally do not understand nor appreciate the chemists' exquisitely tuned sense of what constitutes a "good" chemical structure. So a lot of hard feelings and miscommunication between chemists and biologists could be avoided by a computation that merely mimics the chemists' structural intuition.

The phenomenon of "aggregation" as a cause of phony HTS leads has been described by Shoichet.[17] Compounds screened at relatively high doses (e.g., 30 μM) appear reproducibly as hits or leads across unrelated biological screens. These weakly active compounds cannot be optimized in medicinal chemistry and are a great time waster if pursued in chemistry. The formation of aggregates in the 0.1 to 1.0 μ range was the common feature found in these compounds. Aggregates of this size might pass through a filter used to separate solid phase from solution phase in a traditional thermodynamic solubility assay, and hence an aqueous phase containing these aggregates would in a traditional thermodynamic solubility sense be classified as "soluble." However, it would clearly be advantageous to avoid these kinds of compounds. Aggregates in this size range are easily detected by light scattering at 90 degrees from the incident light. Even relatively unsophisticated equipment is adequate as would, for example, be used by a water quality engineer to measure the clarity of water at a treatment plant. The ability to detect these aggregates is an advantage of quantitating solubility by light scattering. Prior to the Shoichet paper the detection of small aggregates by light scattering would have been viewed as a "false positive," that is, as an incorrect deviation between a proper thermodynamic measurement and the result of a light-scattering measurement. After the Shoichet publication the light-scattering detection of aggregates can be viewed as an advantageous. The advantage being the ability to detect the time wasting "phony HTS" leads or hits.

16.9.6. Solubility Is of Low Dimensionality

ADME and therefore solubility space is of low dimensionality. For example, if one computes the types of physicochemical properties likely to be important to oral absorption, one seldom finds more than six or perhaps seven significant independent properties (e.g., in a principal components analysis). Typically these are properties related to size, lipophilicity, polarity, hydrogen bonding properties, and charge status. The low dimensionality of ADME space explains the effectiveness of simple filtering algorithms like the "rule of five." The low dimensionality of ADME space contrasts with the very high dimensionality of chemistry space. For example, description of a large diverse chemical library by electrotopological parameters does not result in low dimensionality. A principal components analysis on a large diverse chemical library might find that eight independent properties (components) derived from electrotopological properties described less than 50% of the variance in chemistry space.

16.9.7. Solubility and Permeability Interrelationships

Poor aqueous solubility, a compound related factor rather than an assay related factor, has a major effect by introducing "noise" into permeability screening and hence has an effect on making computational model building very difficult. It must be stressed that the compound solubility factor virtually never appears as an explicit consideration in the published permeability literature. Compound sets are published that are used to validate in vitro cell based absorption assays. Validation usually means obtaining an acceptable correlation between human fraction absorbed data and in vitro permeability data. The absorption data always include the experimentally very well controlled but compound number limited human fraction absorbed data that are used to define absorption ranges in the FDA bioavailability waiver guidelines. This limited compound set is then supplemented with additional compounds chosen from published human absorption literature. In our own work we have been able to accumulate literature human fraction absorbed data on a total of about 330 compounds. Larger data sets of up to about 1000 compounds exist, which are based on published reference texts[18] or intensive literature searches supplemented by detective work, to differentiate the absorption and metabolism components in oral bioavailability.[19] The hallmark of compounds with human absorption data is that they are very well behaved compounds from a "drug-like" viewpoint. The fraction absorbed is very heavily biased to the high percentage absorbed range, and the compounds are almost universally quite soluble in aqueous media. This simply reflects the compound quality filtering process that must be passed for a compound to enter the types of studies likely to generate human fraction absorbed data. In short, literature compound permeability validation sets are completely appropriate and say a lot about assay issues in a permeability screen, but they have almost no relevance to assay reproducibility issues related to poor compound solubility.

Figure 16.2 sets the stage for the types of solubility among currently synthesized compounds that are likely to be submitted to a permeability screen like a Caco-2 assay. In this type of assay a variety of biological transporters are present that mediate both absorption and efflux. The movement of a compound through the Caco-2 polarized cell layer through the action of

μg/mL	μm (MWT_300)	μm (MWT_400)	μm (MWT_500)	μm (MWT_600)	
1	3.33	2.50	2.00	1.67	30% of Groton cpds are
3	10.00	7.50	6.00	5.00	in this solubility range
5	16.67	12.50	10.00	8.33	▼
10	33.33	25.00	20.00	16.67	10% of Groton cpds
20	66.67	50.00	40.00	33.33	are in this solubility
30	100.00	75.00	60.00	50.00	range
40	133.33	100.00	80.00	66.67	▼
50	166.67	125.00	100.00	83.33	Solubility acceptable
60	200.00	150.00	120.00	100.00	for 1 mg/kg potency
70	233.33	175.00	140.00	116.67	
80	266.67	200.00	160.00	133.33	60% of Groton cpds are
90	300.00	225.00	180.00	150.00	in this solubility range
100	333.33	250.00	200.00	166.67	
200	666.67	500.00	400.00	333.33	▼
300	1000.00	750.00	600.00	500.00	FDA 1 mg/kg solubility
500	1666.67	1250.00	1000.00	833.33	in 250 mL glass of water
1000	3333.33	2500.00	2000.00	1666.67	▼

Figure 16.2. Solubility of synthesized compounds.

these transporters can be saturated if the drug concentration is high enough. So it is important to screen at a physiologically relevant concentration. If the dose is too low, the permeability estimate will be too low because the importance of efflux transporters will be overestimated.

Figure 16.2 maps the acceptable solubility ranges as defined by pharmaceutical science to the molar concentration range of biological screening. For an average potency compound of about 1 mg/kg, the screening dose in a Caco-2 screen should be somewhere in the range of 100 μM. This concentration is the minimum required for adequate absorption. However, pharmaceutical industry caco-2 screening doses are typically 10 to 25 μM. This dose range is chosen for very practical reasons. If the assays were run at 100 μM, a very high incidence insoluble or erratically soluble compounds would be encountered. In caco-2 screening in our Groton, USA laboratories, one-third of compounds screened at 10 μM are insoluble in an aqueous medium. When one-third of compounds screened in an assay are insoluble in aqueous media, assay reproducibility becomes a major issue. I think it is entirely reasonable to question the value of permeability screening of combinatorial libraries given their general tendency toward poor solubility.

The reader can get a quick idea of whether poor solubility might be a confounding factor for permeability screening of a combinatorial library. In my experience the existing batch mode solubility calculation programs generate very similar and quite reasonable solubility histogram profiles when run on thousands of compounds (although I would not trust numerical prediction results for small numbers of compounds). Experimental permeability screening (especially if it is manning intensive) might not be worthwhile because of the solubility noise factor if a very significant fraction of the library is predicted to be insoluble at 10 μM (the low end of the typical screening concentration range).

16.10. CONCLUSION

In summary, poor aqueous solubility is the single physicochemical property that is most likely to be problematical in a combinatorial library. It can be avoided in part by intelligent use of batch mode solubility calculations. The solubility problem is not simply a technical issue in library design. It is exacerbated by chemistry synthesis considerations and by the timing of the availability of combinatorial exemplars. Formulation fixes are available unless the solubility is extremely poor, but these should be avoided as much as possible.

REFERENCES

1. A. Avdeef, *Curr. Top. Med. Chem.* **1** (4), 277–351 (2001).
2. C. A. Lipinski, *Curr. Drug Disc.* (April), 17–19 (2001).
3. E. H. Kerns, *J. Pharm. Sci.* **90**, 1838–1858 (2001).
4. R. A. Wards, K. Zhang, L. Firth, *Drug Disc. World* (Summer), 67–71 (2002).
5. R. E. Dolle, *J. Comb. Chem.* **3**, 477–517 (2001).
6. J. Bernstein, R. J. Davey, J. Henck, *Angew. Chem. Int. Ed.* **38**, 3440–3461 (1999).
7. Syracuse Research Corporation's EPI suite of software. Includes the KowWin aqueous solubility calculator is available at *http://www.epa.gov/oppt/exposure/docs/episuitedl.htm.*
8. C. A. Lipinski, *J. Pharm. Tox. Meth.* **44**, 235–249 (2000).
9. K. C. Johnson, A. C. Swindell, *Pharm. Res.* **13**, 1795–1798 (1996).
10. W. Curatolo, *Pharm. Sci. Tech. Today* **1**, 387–393 (1998).
11. A. Avdeef, *Curr. Topics Med. Chem.* **1**, 277–351 (2001).
12. C. S. Bergstrom, U. Norinder, K. Luthman, P. Artursson, *Pharm. Res.* **19**, 182–188 (2002).
13. Estimated from poster data. Presented in "Rapid screening of aqueous solubility by a nephelometric assay," J. R. Hill, C. M. Curran, N. S. Dow, K. Halloran, V. Ramachandran, K. G. Rose, D. J. Shelby, D. A. Hartman, Ricerca Ltd., AAPS Denver, CO (2001).
14. C. D. Bevan, R. S. Lloyd, *Anal. Chem.* **72**, 1781–1787 (2000).
15. C. A. Lipinski, F. Lombardo, B. W. Dominy, P. J. Feeney, *Adv. Drug Del. Rev.* **23** (1–3), 3–25 (1997).
16. W. Curatolo, *Pharm. Sci. Tech. Today* **1** (9), 387–393 (1998).
17. S. L. McGovern, E. Caselli, N. Grigorieff, B. K. Shoichet, *J. Med. Chem.* **45** (8), 1712–1722 (2002).
18. Advanced Algorithm Builder, *http://www.pion-inc.com/products.htm.*
19. Oraspotter, *http://www.zyxbio.com.*

CHAPTER

17

HIGH-THROUGHPUT DETERMINATION OF LOG D VALUES BY LC/MS METHOD

JENNY D. VILLENA, KEN WLASICHUK,
DONALD E. SCHMIDT JR., and JAMES J. BAO

17.1. INTRODUCTION

Lipophilicity is an important determinant of the suitability of a compound as a drug candidate, as it is directly linked to absorption, permeation, disposition and bioavailability.[1–2] Lipophilicity is related to the thermodynamic partition coefficient, P, which is the ratio of the equilibrium activity (or concentration in dilute solution, i.e., $<10^{-4}$ M) of the neutral species in a water-immiscible organic solvent to the equilibrium activity of the neutral species in water. Usually the partition coefficient is expressed as the form of logarithm, $\log P$. The desirable $\log P$ values of drug candidates are less than 5.[3]

The partition coefficient P can be estimated using commercially available software programs that break the molecules into their substructures and then summarize the fragment values and add correction terms for intramolecular interactions.[4–6] However, the most reliable results still come from experimental determinations. A common way of determining the log P value is to add a water-immiscible organic solvent, such as octanol, into water or buffer solution at various pHs.[7] Under this condition the apparent partition coefficient (or distribution coefficient D) is a better representation of the lipophilicity of an ionizable compound at this specific pH.[8] Thus lipophilicity is often expressed as $\log D$ rather than $\log P$. With partition coefficient and pKa values (calculated or experimentally derived) available, $\log D$ values can also be calculated. In our experience, $\log D$ values that depend on calculated pKa values are rarely in agreement with the experimental values, except perhaps in highly analogous chemical series. Therefore it is still common to measure $\log D$ values experimentally.

Analysis and Purification Methods in Combinatorial Chemistry, Edited by Bing Yan.
ISBN 0-471-26929-8   Copyright © 2004 by John Wiley & Sons, Inc.

17.1.1. Determination of Log D Values

There are many different methods for measuring log D values experimentally. The most classic method is the shake-flask method, which involves measuring the concentrations of the species in each of the octanol and buffer phases separately after they have reached equilibrium.[9] This procedure is labor intensive and requires a large amount of material, which limits the overall throughput and makes the method unsuitable for current drug discovery programs.

Several new methods, direct or indirect, have emerged as alternatives to the traditional flask method.[10] Direct methods, such as the shake-flask method, are based on measuring the concentrations of a species in the two phases at thermodynamic equilibrium.[11] Indirect methods either involve a third medium, such as a generator column,[12] or use a totally different equilibration system, such as the reversed stationary phase in an HPLC column,[13] or the micelles in micellar electrokinetic capillary chromatography (MECC).[14] Thus indirect methods rely on empirical correlations between two-phase partitioning and other partitioning phenomena and need to be calibrated against known compounds under the same conditions that will be used to test compounds. This correlation may yield erroneous and misleading results, particularly when applied to structurally diverse compounds. In addition it is hard to know if the two phases in indirect methods are at thermodynamic equilibrium. Therefore direct measurement of the concentrations of analytes at equilibrium is the preferred method.

The simplest way of measuring the concentrations in the two phases is to read the absorbance by a spectrometry method. Unfortunately, spectrometric methods are suited only for compounds that either are very pure or have unique chromophores. For most pharmaceutical compounds at discovery stage, it is difficult to obtain pure compounds. A separation technique, such as HPLC, can avoid interferences from impurities, eliminating the purity and/or unique chromophore requirements,[15] and allow for UV/Vis detection. Even with HPLC separation, UV/Vis detection still has its limitations. First, even with photodiode array (PDA), UV/Vis may be not specific enough to identify co-eluting impurities, especially when the concentration of the sample is relatively low or the compound of interest has low absorbance. Second, it is very hard to identify the peak of interest when there are multiple peaks. An example is shown in Figure 17.1 where the biggest peak in a UV trace is actually not from the compound of interest. Third, the sensitivity of UV/Vis detection is limited, thus limiting the range of log D values that can be determined by this method. The measurable range of the log D values is limited by the lowest concentration of the test substance that can be reliably measured in either phase. Therefore a more

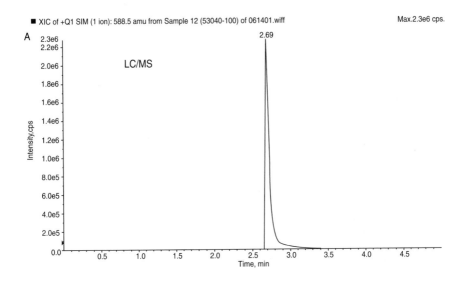

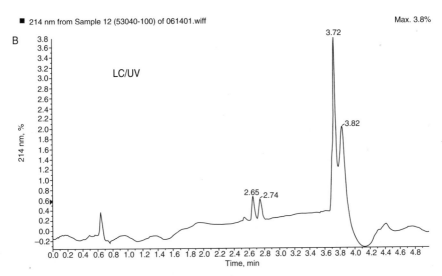

Figure 17.1. Chromatograms obtained from on-line (*A*) MS and (*B*) UV detection with the UV detector placed ahead of the MS detector. The UV scale is expressed as percentage with 100% at 1 V. The LC/MS data confirmed a relatively small peak in the UV trace at 2.65 minutes was actually the peak for the compound of interest.

sensitive and more specific detection technique, such as mass spectrometry (MS), is more appropriate.[16–18]

17.1.2. LC/MS Method Development

Burkhard and Kuehl first reported the use of liquid chromatography with MS detection for the determination of *n*-octanol/water partition coefficients of eight tetrachlorinated planar molecules in 1986.[16] They used electron impact (EI) or negative chemical ionization (CI) mode with multiple-ion-selection (MIS) mass analyses. Later Wilson et al. used an electrospray ionization (ESI) method and analyzed several hundred pharmaceutical compounds.[17] Wlasichuk et al. recently reported a new method for determining a broad range of log *D* values (−2 to +5.5) for several hundred pharmaceutical compounds using an atmospheric pressure chemical ionization (APCI) technique.[18] This method took advantage of APCI for better quantitation[7–21] and obtained a reasonably wide range of linearity. In addition to the use of APCI, they also developed a dilution strategy to minimize the number of experiments needed for obtaining log *D* values. Therefore this method can be routinely used for log *D* determination with high accuracy and improved efficiency.

17.1.3. High-Throughput Log *D* Determination

While the LC/APCI-MS log *D* method improved throughput compared with previous methods, improvements were required to meet the demand for faster and better results in the pharmaceutical industry. To achieve this, it was mandatory to develop a totally new working process. For example, even with the simplified procedure as described previously,[18] the method is still relatively labor intensive. Most of the current libraries come in the format of 96-well or 384-well plates making it impossible to manually analyze all of these samples, as was done previously with individual compounds. Automated liquid handling devices have to be used to accommodate these plates. In addition not all fractions obtained from the combinatorial library contain meaningful compounds for log *D* measurements. Thus it is necessary to be able to selectively choose different compounds from a single plate. This requirement imposes challenges to the automated multiple-needle liquid handler.

In collaboration with the high-throughput purification group, a new working process involving sample preparation, processing, and injection was developed (Figure 17.2). Specifically, during the sample preparation stage, samples are prepared in 96-well plates, which are called the parent plates and have individually movable wells. When samples are loaded into

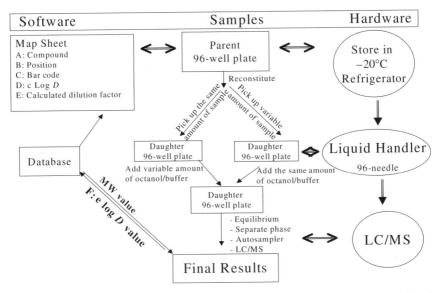

Figure 17.2. Flowchart showing the possible working process related to the 96-well plate based log D determination.

these parent plates, plate maps are generated simultaneously. The plate map lists the compound identification number, the location (row and column) on the plate, and the barcode for the specific well. The compounds of interest are rearranged into new 96-well plates, which are daughter plates.

There are two different ways of rearranging the daughter plate. One is to use the same amount of sample for each well but add different amounts of octanol and/or buffer to account for differences in the log D values and to make the final concentrations of the compounds within the linear range. The other approach is to vary the amount of sample, depending on the calculated log D values, but to keep the amount of octanol and/or buffer the same. Either way, the calculated log D values for these compounds are required prior to rearrangement of the movable wells. In theory, the first approach is possible if the wells are large enough to hold the solution. In practice, the wells have very limited volume, which limits the amount of octanol and buffer that can be added into the wells. Therefore a further dilution step may be necessary. In the second approach it is not always easy to obtain the proper amount of compounds to start the analysis. Therefore some compromises may have to be made to make the process work smoothly. For example, a fixed amount of sample can be used for all of the samples. This will avoid the problem of the second approach. Then, the samples can be diluted differently to match the calculated log D ($clogD$) values. This will solve the volume difference problem in the first approach.

Compounds with similar $clogD$ values can be plated on the same plate for the convenience of adding similar amounts of liquid (octanol or buffer) simultaneously in 96-wells. After equilibrium, compounds in the octanol and the aqueous buffer can be transferred to new 96-well injection plates, which are placed in the autosampler for direct injection into the LC/MS. The results are analyzed, and the experimentally determined log D values ($elogD$) are stored in the database. If necessary, samples taken from one or two phases may be diluted before being placed inside the autosampler vials. The specific dilution factors can be selected following strategies described previously[18] and the 96-needle Apricot pipettor can be used to add octanol and buffer instead of a manual pipette.

The experimental procedures and the results obtained with this new working process are summarized here.

17.2. EXPERIMENTAL

17.2.1. Reagents and Chemicals

Boric acid, citric acid, sodium acetate, Rhodamine 6G, and sodium dihydrogenphosphate were purchased from Sigma (St. Louis, MO). Dimethyl sulfoxide (DMSO), trifluoroacetic acid (TFA), sodium hydrogenphosphate, and octanol were obtained from Aldrich (Milwaukee, WI), and HPLC grade acetonitrile and water were ordered from VWR Scientific (Brisbane, CA).

All other commercial compounds (alprenolol, amitriptyline, atenolol, atropine, bupivacaine, caffeine, clomipramine, diazepam, haloperidol, lidocaine, mexiletine, nortriptyline, propranolol, scopolamine, and verapamil) were purchased from Sigma (St. Louis, MO). Compounds were obtained from Theravance, Inc. (previously Advanced Medicine, Inc.).

17.2.2. Buffer and Buffer-Saturated Octanol Preparation

The vast majority of assays used phosphate buffer at pH 7.0. Phosphate at pH 7.4 was used for the standard drugs from commercial sources to enable comparison with published values. All buffers were 10mM concentration and were saturated with octanol prior to use. Approximately 20mL of octanol was added to 500mL of each buffer, and the solutions stirred overnight. The phases were allowed to separate, unstirred, for 24 hours.

Similarly, 500 mL samples of octanol were saturated with 20 mL of the corresponding buffer at pH 7.0 for assays at pH 7.0.

17.2.3. Sample Preparation

Manual Sample Preparation

Test compounds were prepared as 20 mM stock solutions in DMSO. For partitioning, either 1 or 10 μL of compound stock was added into 1.7 mL polypropylene micro-centrifuge tubes. Then 500 μL of octanol and 499 or 490 μL of the appropriate buffer was added to these tubes to total 1 mL. The tubes were capped, vigorously vortexed for 30 seconds, then mixed end-over-end for 1 hour at room temperature on a Laboratory Rotator (Glas-Col Model RD4512, Terre Haute, IN) at a speed setting of 70%. The tubes were centrifuged at 15,000 g for 5 minutes to separate the phases, and transfer pipettes were used to remove the upper (octanol) and lower (buffer) phases to autosampler vials. When needed, the phases could be diluted with octanol or buffer before injection and analysis by HPLC/MS.

96-Well Plate Sample Preparation

Samples were prepared in a 96-well plates (Matrix Technologies, Hudson, NH) containing 2 micromoles of test compound in each well. The 20 mM stock solutions of the test compounds in DMSO were prepared by adding 100 μL of DMSO to each of the wells.

For partitioning, either 1 or 10 μL of compound stock was added to 2.0 mL 96-well plates (Nunc, Naperville, IL). For wells with 1 μL of compound stock, another 9 μL of DMSO was added. Then 500 μL of octanol and 490 μmL of 10 mM phosphate buffer at pH 7.0 were added, to a total volume of 1 mL, using the 96-needle Apricot pipettor (Model PP-550MS-XH, Apricot Designs, Inc., Monrovia, CA). The 96-well plate was sealed with a pre-slit well cap (Nalge Nunc, Rochester, NY), and then mixed end-over-end for 1 hour at room temperature for the manual sample preparation. The plate was then centrifuged at 4300 rpm for 10 minutes using a Sorvall centrifuge to separate the phases. The 96-needle Apricot pipettor was used to transfer the upper (octanol) and lower (buffer) phases to new 96-well plates. Alternatively, a Rainin EDP3 12-channel pipette (Rainin Instrument, Oakland, CA) was used to transfer samples from the equilibrium plate to the autosampler plates. In some instances the phases were diluted with octanol or buffer using the 96-needle Apricot pipettor before injection and analysis by LC/MS.

The selection of either 1 or 10 µL of compound stock was related to the $c \log D$ value of the compound. For compounds with an absolute $c \log D$ value equal or smaller than 2, 1 µL of compound stock was used. For compounds with $-2 \leq c \log D \leq 2$, 10 µL was used. For wells with 1 µL of initial stock solution, 9 µL of DMSO was added to bring all samples to 10 µL solution before the addition of octanol and buffer.

When the 96-needle Apricot pipettor was used to transfer the upper (octanol) and lower (buffer) phases to new 96-well 2 mL Nunc plates (Nalge Nunc, Rochester, NY), it was often necessary to dilute the octanol and/or the buffer phase 10× and 100× before injection and analysis by LC/MS. For compounds with $-2 \leq c \log D \leq 2$, both buffer and octanol phases were directly injected without dilution. For compounds with $c \log D > 2$, the buffer was undiluted but the octanol phases were diluted 100× before injection. For compounds with $c \log D < -2$, the buffer phases were diluted 100× while the octanol phase remained undiluted. Since there were very few compounds with $c \log D < -2$, dilution of the aqueous phase was rarely needed. Therefore one assay plate, in a typical experiment, would generate four plates, that is, one undiluted buffer, one undiluted octanol, one 10× octanol, and one 100× octanol after the phases were transferred and diluted. Then the required buffer and octanol phases were transferred to at least two new 96-well 2 mL Nunc plates using a Rainin EDP3 12-channel pipette.

17.2.4.   LC/MS

All of the LC/MS experiments were performed on a PE150 Sciex LC/MS system (Applied Biosystems Inc., Foster City, CA). For chromatography, the HPLC consists of a PE 200 Autosampler (Norwalk, CT), and LC-10ADVP high-pressure binary pumps and an SPD-10AVP UV-VIS Detector from Shimadzu (Columbia, MD). A Rheodyne Synergi 2-position, 6-port fluid processor (Phenomenex, Torrance, CA) was used for diverting the early effluent in the chromatographic process. All of the chromatographic peripherals were controlled using the same software controlling the MS, Analyst (Version 1.1), from Applied Biosystems, Inc.

The HPLC columns used for this work were 50 × 4.6 mm packed with 3 µm particles with Luna C18(2) stationary phase from Phenomenex. The gradient LC/MS analyses used two mobile phases: (A) 5% acetonitrile in water and (B) 5% water in acetonitrile. For ESI, both mobile phases contained 0.25% formic acid, while for APCI both contained 0.01% trifluoroacetic acid. All samples were analyzed using a flow rate of 1.0 mL/min having a linear LC gradient between 5% and 80% B over 2.5 minutes in a total run time of 5 minutes. All injections were 10 µL.

The mass spectrometer used for this study, manufactured by Applied

Biosystems, was a PE Sciex API 150EX single quadrupole mass spectrometer fitted with a Heated Nebulizer® interface run at 450°C for APCI or a Turbo Ionspray® source at 250°C for ESI. All analyses used SIM (Selected Ion Monitoring), operating in positive ion mode, with unit resolution up to m/z 1000. For ESI-MS, curtain gas was set at 9–12 L/min, and the nebulizer gas was set at 6 to 10 L/min. The voltage on probe was 5500. For APCI, curtain gas was set at 12, and the setting for the auxiliary gas was 10. The pressure of the nebulizer gas was kept constant at 80 psi.

17.2.5. Apricot Personal Pipettor Evaluation

The accuracy and precision of the 96-needle Apricot pipettor was evaluated by assessing the linearity of the diluted solutions. Since both the octanol and aqueous buffer have no absorbance, a dye, Rhodamine 6G, was added to the sample (octanol or buffer) for absorbance detection at 525 nm. A SpectraMax 250 (Molecular Devices, Sunnyvale, CA) 96-well plate reader was used to measure the absorbance of the dyed samples.

17.3. RESULTS AND DISCUSSION

17.3.1. 96-Well Plate Based Method Development

Since an LC/MS method had already been developed,[18] the focus of this method development was on evaluating the consistency of the 96-well plate based sample preparation and uniformity of results from the various wells in the 96-well plate, namely the edge effect. However we choose to compare the results, it is necessary to evaluate the precision and accuracy of the 96-channel pipettor first.

Precision and Accuracy of Apricot Personal Pipettor

Dilution of Octanol Layer. The Apricot personal pipettor can operate in 96-channels simultaneously. Since octanol is a very viscous solvent, it is necessary to ensure that the right amount can be aspirated and delivered by the pipettor. An experiment was designed to check the linearity of the dilution by adding Rhodamine 6G, a dye with an absorbance at 525 nm, to the octanol. Several dilutions (100×, 200×, 400×, 800×, 1600×, 3200×, and 16,000×) were performed on a 96-well plate. A SpectraMax 96-well plate reader was used to measure the dyed octanol. Figure 17.3A shows a linear dilution curve when using the Apricot personal pipettor.

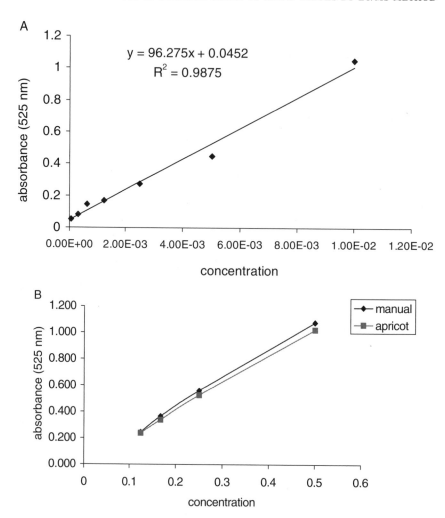

Figure 17.3. Linearity of the serial dilution by 96-needle Apricot pipettor: (*A*) Octanol serial dilution by Apricot pipettor and (*B*) manual dilution compared with Apricot pipettor dilution of aqueous buffer.

In addition the precision of the dilution was evaluated at different concentration levels with another set of samples (data not shown here). For the samples 10-fold diluted from four levels of original samples at 0.125 mM, 0.167 mM, 0.25 mM, and 0.5 mM, the %RSD ($n = 4$) of the dilution with the Apricot pipettor were 1.5%, 2.0%, 6.4%, and 3.5%, respectively. The %RSDs of manual dilution for the same set of samples were 7.5%, 5.1%,

4.6%, and 6.1%, respectively. There is no significant difference between manual dilution and the Apricot pipettor.

Dilution of Buffer. Precision and accuracy of the dilution by the Apricot personal pipettor for aqueous buffer were also evaluated using the same dye. The original samples at four different concentrations, 0.125, 0.167, 0.25, and 0.5 mM, were diluted 10-fold and 100-fold by both 12-channel pipettor and the Apricot personal pipettor. These results show that the Apricot pipettor picked up 90% to 99% of the amount picked up by the manual pipettor. The %RSD ($n = 4$), which is the precision of the dilutions, varied between 2% and 9%. A reasonably good correlation between these two methods at the four levels of concentrations is shown in Figure 17.3B.

It can be concluded from previous data that the Apricot dilution is equivalent to the manual dilution. Therefore it is expected that using the Apricot will not introduce significant errors to the final log$D$ determination.

Different Types of Mixing

One of our main concerns was how well the samples were mixed when a 96-well plate was used instead of micro-centrifuge tubes. Several mixing methods were tested and the log$D$ values were compared. As a control the samples were also prepared in micro-centrifuge tubes and mixed by vortexing for 30 seconds followed by 1 hour end-over-end mixing. The mixing methods used were (1) end-over-end for 1 hour, (2) end-over-end for 4 hours, (3) horizontal shaker for 1 hour, and (4) vertical shaker for 1 hour. A Laboratory Rotator at a speed setting of 70% was used for end-over-end mixing, while a VX-2500 multi-tube vortexer (VWR Scientific, Brisbane, CA) was used for both the horizontal and vertical shaking. Figure 17.4 shows that log$D$ values obtained from the four mixing methods were very close. One mixing method, namely the end-over-end for 1 hour, was used for the sample determinations because it results in less emulsion formation during the mixing process.

Comparison of Different Sample Preparations

In order to evaluate the sample preparation procedures in the 96-well plate based assay, several commercial compounds along with two Theravance compounds were prepared using both the manual (tube) method and the 96-well plate method. Among the compounds tested, the differences are extremely small, with the biggest difference being 0.29, which is within the expected variation range of the method. Therefore the results indicate that the log$D$ values obtained by these two sample preparations are identical.

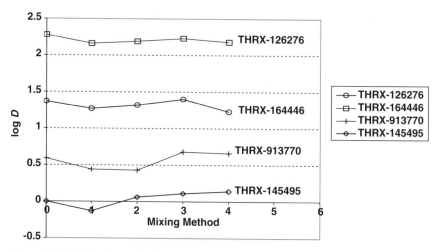

Figure 17.4. Log *D* values obtained from four mixing methods. 0 = tubes; 1 = end-over-end 1 h; 2 = end-over-end 4 h; 3 = horizontal shaker 1 h; 4 = vertical shaker 1 h.

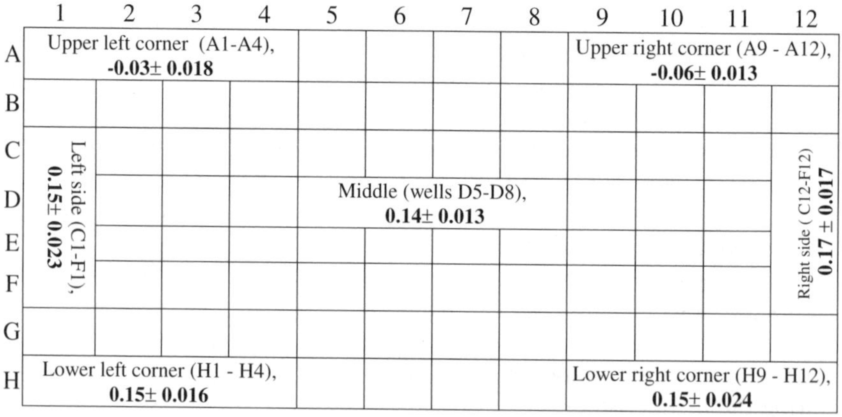

	1	2	3	4	5	6	7	8	9	10	11	12
A	Upper left corner (A1-A4), -0.03± 0.018								Upper right corner (A9 - A12), -0.06± 0.013			
B												
C	Left side (C1-F1), 0.15± 0.023										Right side (C12-F12) 0.17 ± 0.017	
D					Middle (wells D5-D8), 0.14± 0.013							
E												
F												
G												
H	Lower left corner (H1 - H4), 0.15± 0.016								Lower right corner (H9 - H12), 0.15± 0.024			

Figure 17.5. Evaluation of edge effect using Scopolamine as the model compound.

Edge Effect Experiment

Since a 96-well plate is a two-dimensional device, an experiment was designed to ensure the reliability of the results from wells at different positions. This experiment tested the edge-to-edge effect of the 96-well plate when using the Apricot pipettor for solvent addition and dilution. The compound used for this experiment was scopolamine, and the plate was set up as shown in Figure 17.5. Log *D* values of scopolamine obtained from different locations of the plate are summarized in Figure 17.5. The log D values were consistent throughout the plate, except those obtained from the upper left corner and the upper right corner, which were slightly lower than the rest. However, the absolute difference between any two groups was less than 0.23, which is within the acceptable accuracy range of the current method.

17.3.2. LC/MS Method Development

An LC/MS method, in general, has a significantly higher signal-to-noise ratio than that of a LC/UV method. More important, MS offers unique specificity that no other technique can. For example, Figure 17.1 shows the comparison of the LC/MS and LC/UV data. It is very difficult to pinpoint exactly which peak is the peak for the compound of interest. Based on peak height, the peak at 3.72 minutes would be selected in frame B. However, MS data indicate that the peak at 3.72 minutes does not have the right mass. If we take retention time as a guide that eliminates the peaks at 3.72 and 3.82 minutes, three possibilities exists: 2.55, 2.65, and 2.74 minutes. If the right peak were chosen, the small size of the peak would make it difficult to provide accurate quantitative information.

On the other hand, LC/MS can easily identify which peak is correct. We can see from Figure 17.1A that when selective ion monitoring (SIM) is used for detection, only one peak with the correct mass can be detected. The significantly higher signal-to-noise ratio of the MS data makes the quantitation of the LC/MS method very reliable.

However, LC/MS, though specific, may not always be accurate in automatically determining the exact amount of compound present. Other factors have to be considered as well. For example, in these experiments LC/MS combined with electrospray ionization (ESI) generated a nonlinear response curve. As the concentrations in the two phases, octanol and aqueous, are often different, this nonlinearity makes it difficult to determine the log D values by simply comparing the ratios of the peak areas of the compound in two phases.

Fortunately, when atmospheric pressure chemical ionization (APCI) instead of ESI is used as the ionization method, a reasonably large range of concentrations give linear responses for all of the compounds tested. Thus APCI appears to be better than ESI as an ionization method for quantitation purposes. APCI is generally regarded as being more robust and less susceptible to ion suppression than electrospray ionization.[18–19] It was found that at least two orders of magnitude for linearity can be easily obtained by the LC/APCI-MS method. For example, Figure 17.6A shows that the response of lidocaine is linear ($R^2 \geq 0.999$) within the concentration ranges of 0.1 to 12.5 µM (125-fold). Figure 17.6B shows that the response of three

A

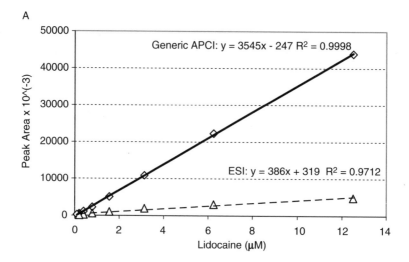

B

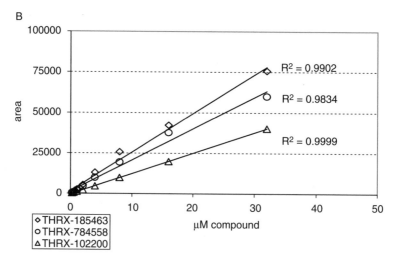

Figure 17.6. Relationship between peak area and the concentrations of compounds by LC/APCI-MS. (*A*) Lidocaine between 0.1 and 12.5 μM; (*B*) Threravance compounds between 0.25 and 32 μM.

randomly picked compounds, THRX-784558, THRX-185463, and THRX-102200 are linear ($R^2 > 0.98$) within the tested range of 0.25 to 32 μM (128-fold). This more than two-orders of magnitude linear range is wide enough for quantitative analysis, and the analysis can proceed without constructing a standard response curve first if concentrations of compounds in the two phases can be adjusted to within the range.

17.3.3. Dilution Strategy for Log D Determination by LC/APCI-MS

The major objective of the current study was to increase throughput of the assay. Ideally a single linear response curve could be constructed to quantitate various analytes over a broad range of concentrations. Unfortunately, MS responses vary from compound to compound, making it impossible to construct a universal response curve for multiple compounds. Fortunately, determination of the absolute concentrations of the compound is not necessary, which would require an absolute calibration curve for each compound. Rather, the goal is to find the ratio of the concentrations of the compound in two phases. As long as the concentrations of the compound in the two phases are within the linear range, the log D value can be determined, as the LC has already taken care of the difference in the sample matrix (i.e., octanol vs. buffer).

A new dilution strategy was developed to bring the sample concentrations in the two phases as close as possible. After dilution, the final concentrations of the analyte usually are less than 20 μM, which is within the linear range for most compounds. Thus errors associated with the nonlinearity of the response curves will be minimal in determining the log D values.

In order improve throughput, c log D values for in-house compounds were used as a guidance in the selection of dilution factors. For compounds with $-2 \leq c$ log $D \leq 2$, only 1 μL of compound stock solution (20 mM) was added into a total of 1 mL octanol and aqueous buffer mixture (1:1), and both phases were analyzed directly without any dilution. If c log $D <$ -2 or > 2, 10 μL of stock solution was used, and the aqueous or octanol phases, respectively, were diluted 100-fold before analysis by LC/MS. Thus the combination of using c log D values for references and performing limited dilutions has enabled generation of experimental log D values with a minimum lab work. In 96-well plate based sample preparations, a similar dilution strategy was used to generate the log D data.

17.3.4. High-Throughput Working Process

Figure 17.2 depicts the proposed working process for the high-throughput log D determination prior to implementing this work in the lab. During the course of the work, modifications were made to make the process work smoothly. The final process is described in Figure 17.7. Two μmoles each of the compounds to be evaluated are submitted in 96-well plates (Matrix Technologies, Hudson, NH). These plates have removable wells that can be

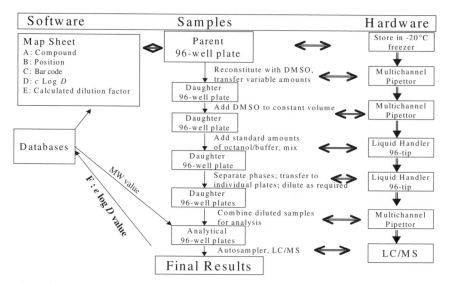

Figure 17.7. Adopted working process flowchart related to the 96-well plate based log*D* determination.

rearranged. This capability is needed to combine samples from different plates into one 96-well plate enabling use of the 96-channel liquid handler. When the samples are submitted, a file containing information on the specific samples, such as the corporate ID, lot number, name of submitter, and plate ID, is also submitted along with the plate (step 1 in Figure 17.8).

Molecular weights and Q1 mass are obtained and the log *D* values calculated (step 3 in Figure 17.8). Depending on *c* log *D* values, compounds were divided into two major groups, i.e., in or out of the −2 to 2 range (step 4 in Figure 17.8). If the *c* log *D* value is within the −2 to 2 range, 1 µL of stock samples will be used. If the *c* log *D* value is out of the range, 10 µL of stock samples will be used.

Following the sequence of the sorted *c* log *D* values, samples are arranged in 96-well plates for actual analysis. The compounds are arranged in a 96-well plate as shown in step 5 of Figure 17.8. This is the plate map for placing samples in the plate. After the samples are arranged, proper amounts of octanol and aqueous buffers are added into the wells and fully mixed to equilibrate. After reaching equilibrium, the samples from both the octanol and aqueous phases are transferred to autosampler plates using EDP3 multi-channel pipettes. If required, the Apricot pipettor would be used to dilute the samples appropriately. General rules for dilution require no dilution for either layer if *c* log *D* is <2. If *c* log *D* is >2, then the

Step 1: **Obtained plate map associate with each plate**

ROW	COLUMN	THRX #	Lot Number	Barcode	Plate B	Plate name
B	1	THRX-228191	1	44557	*	lli_782_78-1 Plate Map
C	1	THRX-967078	1	44560	*	lli_782_78-1 Plate Map

	Well	THRX #	Lot	Submitter	Plate
D	E1	THRX-228191	1	Craig H.	CHu-673-158
E	G1	THRX-287480	1	Craig H.	CHu-673-158

Step 2: **Create a list of all samples submitted**

Step 3 **Obtained a) MW and Q1 mass and** **b) c log D using templates and the THRX #**

THRX #	Lot	Mol Weight	Q1 Mass
THRX-228191	1	596.4304	482.96
THRX-287480	1	671.5434	558.07
THRX-120650	1	594.4579	480.98
THRX-991652	1	652.4945	539.02

a b

THRX #	cpH4	cpH7	cpH9
THRX-228191	3.8	3.8	4.3
THRX-287480	0.3	2.3	3.9
THRX-120650	2.6	2.6	3.5
THRX-991652	2.8	5.4	6.5

Step 4: **Sort sample by plate map, Q1 mass and clog D**

Step 5: **Arrange samples in 96-well plate**

THRX #	Lot	Submitter	Plate	Q1 Mass	clog D
THRX-107414	1	Li Li	lli_782_78-1 Plate Map	496.98	1
THRX-633481	1	Li Li	lli_782_78-1 Plate Map	468.93	1.1
THRX-242551	1	Craig H.	CHu-673-183	549.02	1.7

	1	2	3	4	5	6	7	8	9	10	11
A	THRX-125467	THRX-191052	THRX-690281	THRX-462775	THRX-201405	THRX-138131	THRX-56392	THRX-564857	THRX-354635	THRX-117977	THRX-101335
B	THRX-588179	THRX-114337	THRX-178785	THRX-142189	THRX-310000	THRX-133166	THRX-110600	THRX-555699	THRX-606654	THRX-629445	THRX-196220
C	THRX-152967										
D	THRX-505398										
E	THRX-115587										
F	THRX-211946										
G	THRX-104730										
H	THRX-186256										

	1	2	3	4	5	6	7	8	9	10	11	12
A	aq. THRX-125467	aq. THRX-191052	aq. THRX-690281	aq. THRX-462775	aq. THRX-201405	aq. THRX-138131	aq. THRX-56392	aq. THRX-564857	aq. THRX-354635	aq. THRX-117977	aq. THRX-101335	aq. THRX-178828
B	α THRX-588179	α THRX-114337	α THRX-178785	α THRX-142189	α THRX-310000	α THRX-133166	α THRX-110600	α THRX-555699	α THRX-606654	α THRX-629445	α THRX-196220	α THRX-691154
C	aq. THRX-152967	aq. THRX-194150	aq. THRX-166465	aq. THRX-106364	aq. THRX-789181	aq. THRX-211775	aq. THRX-147573	aq. THRX-741894	aq. THRX-211281	aq. THRX-767717	aq. THRX-127636	aq. THRX-895326
D	α THRX-505398	α THRX-181415	α THRX-463551	α THRX-438181	α THRX-195074	α THRX-890008	α THRX-101566	α THRX-360257	α THRX-155307	α THRX-413995	α THRX-813176	α THRX-352126
E	aq. THRX-115587	aq. THRX-209279	aq. THRX-122179	aq. THRX-799113	aq. THRX-110378	aq. THRX-731519	aq. THRX-698187	aq. THRX-989399	aq. THRX-249558	aq. THRX-202865	aq. THRX-110806	aq. THRX-322556
F	α THRX-211946	α THRX-17...	α THRX-900054	α THRX-413614	α THRX-156622	α THRX-154738	α THRX-936678	α THRX-753006	α THRX-100800	α THRX-776541	α THRX-900054	
G	aq. THRX-104730	aq. THRX-...588...										
H	α THRX-186256	α THRX-...56...										

Step 6:
- **Perform dilution of samples using Apricot Personal Pipettor**
- **Transfer octanol and aqueous phases using a multichannel pipette into a new 96-well plate**

Figure 17.8. Data flow path of general preparative steps for 96-well plate assay.

octanol layer will be diluted 100-fold but the buffer layer will remain undiluted. If $c \log D$ is < -2, then the buffer layer will be diluted 100-fold but the octanol layer will remain undiluted. Samples are then injected into the LC/MS system.

17.3.5. Log D Values of Model Compounds

A total of 15 commercially available compounds, with previously measured log D values ranging from –2.36 to 3.64, were analyzed. The results are compared with the log D values obtained by the manual method (Table 17.1). The manual results for these compounds have been re-assayed recently in direct comparison to the results obtained by the 96-well plate method. The manual results turned out to be very close to the initial results. For a total of 16 compounds, the average difference was only 0.13 (data not shown). This small difference indicates that the log D values are reliable.

The log D results for compounds obtained by these two methods are listed in Table 17.1. Figure 17.9 illustrates the results for Theravance compounds. Note that the log D values obtained by the two methods are almost identical.

Table 17.1. Log D Values of Model Compounds Obtained with Single-Tube and 96-Well-Based Method

Drug Name	Volume of Stock (µL)	Dilutions Factors	Log D Values by the 96-Well Method	Log D Values by the Manual Method	Difference in Log D Values
Scopolamine	1	a;o	0.15	0.16	0.01
Mexiletine	1	a;o	0.68	0.55	0.13
Alprenolol	1	a;o	0.8	0.64	0.16
Propranolol	1	a;o	0.97	0.95	0.02
Atropine	1	a;o	–0.53	–0.54	0.01
Caffeine	1	a;o	–0.07	–0.03	0.04
Atenolol	10	a100;o	–2.06	–2.08	0.02
Lidocaine	10	a;o100	1.58	1.66	0.08
Bupivacaine	10	a;o100	2.40	2.50	0.10
Verapamil	10	a;o100	2.42	2.56	0.14
Haloperidol	10	a;o100	3.1	3.08	0.02
Amitriptyline	10	a;o100	2.78	3.02	0.24
Nortriptyline	10	a;o100	1.69	1.69	0.00
Diazepam	10	a;o100	2.7	2.69	0.01
Clomipramine	10	a;o100	3.67	3.69	0.02
THRX-382460	10	a;o100	2.35	2.39	0.04
THRX-124806	10	a;o100	4.57	4.28	0.29

Note: a = undiluted aqueous; o = undiluted octanol; o100 = 100-fold dilution for the octanol phase.

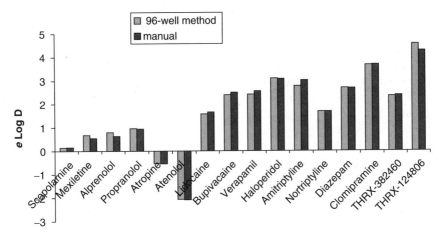

Figure 17.9. Comparison of manual and 96-well plate assays.

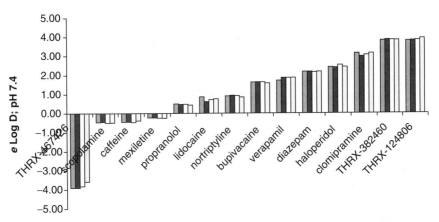

Figure 17.10. Reproducibility of the $\log D$ determination using 96-well plate and 96-needle Apricot pipettor.

17.3.6. Reproducibility of the Assay

Several commercial compounds and Theravance compounds were prepared ($n = 4$) and assayed. Figure 17.10 shows the results of model compounds that include the three Theravance compounds. Similar results were obtained for more Theravance compounds. These figures demonstrate that this method is reproducible for $\log D$ determination.

17.4. CONCLUSIONS

An LC/MS method with APCI is a powerful tool for determining log D values of many different compounds. This method has been used to determine the log D values of more than a thousand compounds. This LC/APCI-MS method is reliable and reproducible. It can be applied to compounds with a wide range of log D values (at least 8 units). More important, LC/MS can significantly improve throughput. When LC/UV was changed to LC/MS, the throughput was increased three- to fourfold. Primarily the improvement was attributed to reduced data processing time and assay repetition. However, sample processing still takes up a lot of the experimental time. If 96-well plates are used instead of the single centrifuge tubes, and a 96-needle Apricot pipettor instead of the single-channel pipette, the throughput can be increased another three- to fourfold.

REFERENCES

1. V. Pliska, B. Testa, H. van de Waterbeemd, *Lipophilicity in Drug Action and Toxicology*, VCH, Weinheim (1996).
2. S. D. Kramer, *Pharm. Sci. Technol. Today* **2**, 373–380 (1999).
3. C. A. Lipinski, F. Lombardo, B. W. Dominy, P. J. Feeney, *Adv. Drug Del. Rev.* **23**, 4–25 (1997).
4. A. Leo, C. Hansch, D. Elkins, *Chem. Rev.* **71**, 525–616 (1971).
5. R. F. Rekker, *Pharmacochemistry Library*, Vol. **1**, Elsevier, New York (1977).
6. C. Hansch, A. J. Leo, *Substituent Constants for Correlation Analysis in Chemistry and Biology*, Wiley, New York (1979).
7. C. Hansch, P. P. Maloney, T. Fujita, R. M. Muir, *Nature* **194**, 178–180 (1962).
8. D. L. Ross, S. K. Elkinton, C. M. Riley, *Int. J. Pharm.* **88**, 379–389 (1992).
9. C. Hansch, A. Leo, D. Hoekman, *Exploring QSAR*, American Chemical Society, Washington, DC, Vol. **1**, pp. 118–122 (1995).
10. L. Danielsson, Y. Zhang, *Trends Anal. Chem.* **15**, 188–196 (1996).
11. A. Avdeef, K. J. Box, J. E. A. Comer, C. Hibbert, K. Y. Tam, *Pharm. Res.* **15**, 209–215 (1998).
12. H. DeVoe, M. M. Miller, S. P. Wasik, *J. Res. Natl. Bur. Stand. (US)* **86**, 361–366 (1981).
13. K. Valko, C. My Du, C. Bevan, D. P. Reynolds, M. H. Abraham, *Curr. Med. Chem.* **8**, 1137–1146 (2001).
14. B. J. Herbert, J. G. Dorsey, *Anal. Chem.* **67**, 744–749 (1995).
15. J. E. Garst, W. C. Wilson, *J. Pharm. Sci.*, **73**, 1616–1623 (1984).
16. L. P. Burkhard, D. W. Kuehl, *Chemosphere*, **15** (2), 163–167 (1986).

17. D. M. Wilson, X. Wang, E. Walsh, R. A. Rourick, *Comb. Chem. High Throughput Screen Sep.*; *4* (6), 511–519 (2001).

18. K. Wlasichuk, D. Schmidt, D. Karr, M. Mamman, J. Bao, *J. Chromatogr.*, submitted.

19. B. K. Matuszewski, M. L. Constanzer, C. M. Chavez-Eng, *Anal. Chem.* *70*, 882–889 (1998).

20 R. King, R. Bonfiglio, C. Fernandez-Metzler, C. Miller-Stein, T. Olah, *J. Am. Soc. Mass Spectrom.* *11*, 942–950 (2000).

21. C. Hansch, A. Leo, D. Hoekman, *Exploring QSAR*, American Chemical Society, Washington, DC, Vol. *2*, pp. 1–216 (1995).

INDEX

CHEMICAL ANALYSIS

A SERIES OF MONOGRAPHS ON ANALYTICAL
CHEMISTRY AND ITS APPLICATIONS

J. D. Winefordner, *Series Editor*